THE ENGINEERING BOOK

Books by Marshall Brain

How God Works

How Stuff Works

More How Stuff Works

What If?

How Much Does the Earth Weigh?

The Teenager's Guide to the Real World

Manna

The Meaning of Life

THE ENGINEERING BOOK

From the Catapult to the Curiosity Rover: 250 Milestones in the History of Engineering

Marshall Brain

STERLING
New York

An Imprint of Sterling Publishing
1166 Avenue of the Americas
New York, NY 10036

ISBN 978-1-4549-0809-8

Distributed in Canada by Sterling Publishing
c/o Canadian Manda Group, 664 Annette Street
Toronto, Ontario, Canada M6S 2C8
Distributed in the United Kingdom by GMC Distribution Services
Castle Place, 166 High Street, Lewes, East Sussex, England BN7 1XU
Distributed in Australia by Capricorn Link (Australia) Pty. Ltd.
P.O. Box 704, Windsor, NSW 2756, Australia

For information about custom editions, special sales, and premium and corporate purchases, please contact Sterling Special Sales at 800-805-5489 or specialsales@sterlingpublishing.com.

Manufactured in China

6 8 10 9 7 5

www.sterlingpublishing.com

This book is dedicated to Dr. Louis Martin-Vega, Dean of the College of Engineering at North Carolina State University. Louis first introduced me to the idea of spreading a positive message about engineering as a powerful force for human advancement, and has encouraged me to do so in many different ways.

And to

The millions of engineers who, step by step over the centuries, have designed and implemented the modern world that we all enjoy today.

Contents

Introduction *10*

30,000 BCE Bow and Arrow *16*
3300 BCE Hunter/Gatherer Tools *18*
2550 BCE The Great Pyramid *20*
2000 BCE Inuit Technology *22*
1400 BCE Concrete *24*
625 BCE Asphalt *26*
438 BCE Parthenon *28*
312 BCE Roman Aqueduct System *30*
100 BCE Waterwheel *32*
79 Pompeii *34*
1040 Compass *36*
1144 Basilica of Saint Denis *38*
1300 Catapult *40*
1372 Leaning Tower of Pisa *42*
1492 Square-Rigged Wooden Sailboats *44*
1600 The Great Wall of China *46*
1620 Gunter's Chain *48*
1670 Mechanical Pendulum Clock *50*
1750 Simple Machines at Yates Mill *52*
1773 Building Implosions *54*
1784 Power Loom *56*
1790 Cotton Mill *58*
1794 Cotton Gin *60*
1800 High-Pressure Steam Engine *62*
1823 Truss Bridge *64*
1824 Rensselaer Polytechnic Institute *66*
1825 Erie Canal *68*
1830 Tom Thumb Steam Locomotive *70*
1835 Combine Harvester *72*
1837 Telegraph System *74*
1845 Mass Production *76*
1845 Tunnel Boring Machine *78*
1846 Sewing Machine *80*
1851 America's Cup *82*
1854 Water Treatment *84*
1855 Bessemer Process *86*
1856 Plastic *88*
1858 Big Ben *90*
1859 Oil Well *92*
1859 Modern Sewer System *94*
1860 Louisville Water Tower *96*
1861 Wamsutta Oil Refinery *98*
1861 Elevator *100*
1869 Transcontinental Railroad *102*
1873 Cable Cars *104*
1876 Telephone *106*
1878 Power Grid *108*
1879 Carbon Fiber *110*
1885 Supercharger and Turbocharger *112*
1885 Washington Monument *114*
1886 Statue of Liberty *116*
1889 Eiffel Tower *118*
1889 Hall-Héroult Process *120*
1890 Steam Turbine *122*
1891 Carnegie Hall *124*
1891 Zuiderzee Works *126*
1893 Two-Stroke Diesel Engine *128*
1893 Ferris Wheel *130*
1897 Diesel Locomotive *132*
1899 Defibrillator *134*

1902 Air Conditioning 136
1903 EKG/ECG 138
1903 The Wright Brothers' Airplane 140
1905 Engineered Lumber 142
1907 Professional Engineer Licensing 144
1908 Internal Combustion Engine 146
1910 Laparoscopic Surgery 148
1912 *Titanic* 150
1913 Woolworth Building 152
1914 Panama Canal 154
1917 Laser 156
1917 Hooker Telescope 158
1919 Women's Engineering Society 160
1919 Under Friction Roller Coaster 162
1920 Kinsol Trestle Bridge 164
1920 Radio Station 166
1921 Robot 168
1926 Heart-Lung Machine 170
1927 Electric Refrigeration 172
1931 Empire State Building 174
1935 Tape Recording 176
1936 Hoover Dam 178
1937 Golden Gate Bridge 180
1937 *Hindenburg* 182
1937 Turbojet Engine 184
1937 Magnetically Levitated Train 186
1938 Formula One Car 188
1939 Norden Bombsight 190
1939 Color Television 192
1940 Tacoma Narrows Bridge 194
1940 Radar 196
1940 Titanium 198
1941 Doped Silicon 200
1942 Spread Spectrum 202
1943 Dialysis Machine 204
1943 SCUBA 206
1944 Helicopter 208
1945 Uranium Enrichment 210
1945 Trinity Nuclear Bomb 212
1946 ENIAC—The First Digital Computer 214
1946 Top-Loading Washing Machine 216
1946 Microwave Oven 218
1946 Light Water Reactor 220
1947 AK-47 222
1947 Transistor 224
1948 Cable TV 226
1949 Tower Crane 228
1949 Atomic Clock 230
1949 Integrated Circuit 232
1950 Chess Computer 234
1951 Jet Engine Testing 236
1952 Center-Pivot Irrigation 238
1952 3D Glasses 240
1952 Ivy Mike Hydrogen Bomb 242
1953 Automobile Airbag 244
1956 Hard Disk 246
1956 TAT-1 Undersea Cable 248
1957 Frozen Pizza 250
1957 Space Satellite 252
1958 Cabin Pressurization 254
1959 Desalination 256
1960 Cleanroom 258
1961 T1 Line 260
1961 Green Revolution 262
1962 SR-71 264
1962 Atomic Clock Radio Station 266
1963 Retractable Stadium Roof 268
1963 Irradiated Food 270

1964 Top Fuel Dragster 272
1964 Drip Irrigation 274
1964 Natural Gas Tanker 276
1964 Bullet Train 278
1965 Gateway Arch in St. Louis 280
1965 Cluster Munition 282
1966 Compound Bow 284
1966 Parafoil 286
1966 Pebble Bed Nuclear Reactor 288
1966 Dynamic RAM 290
1967 Automotive Emission Controls 292
1967 *Apollo 1* 294
1967 Saturn V Rocket 296
1968 C-5 Super Galaxy 298
1968 Boeing 747 Jumbo Jet 300
1969 Lunar Landing 302
1969 ARPANET 304
1969 Space Suit 306
1970 LCD Screen 308
1970 *Apollo 13* 310
1970 Fiber Optic Communication 312
1971 Anti-Lock Brakes 314
1971 Lunar Rover 316
1971 Microprocessor 318
1971 CANDU Reactor 320
1971 CT Scan 322
1971 Power Plant Scrubber 324
1971 Kevlar 326
1972 Genetic Engineering 328
1973 World Trade Center 330
1975 Router 332
1976 The Concorde 334
1976 CN Tower 336
1976 VHS Videotape 338
1977 Human-Powered Airplane 340
1977 Tuned Mass Damper 342
1977 Voyager Spacecraft 344
1977 Trans-Alaska Pipeline 346
1977 MRI 348
1978 Nitrous Oxide Engine 350
1978 Bagger 288 352
1979 *Seawise Giant* Supertanker 354
1980 Flash Memory 356
1980 M1 Tank 358
1980 Stadium TV Screen 360
1981 Bigfoot Monster Truck 362
1981 Space Shuttle Orbiter 364
1981 V-22 Osprey 366
1982 Artificial Heart 368
1982 Neodymium Magnet 370
1983 RFID Tag 372
1983 F-117 Stealth Fighter 374
1983 Mobile Phone 376
1983 Gimli Glider 378
1983 Ethernet 380
1984 3D Printer 382
1984 Domain Name Service (DNS) 384
1984 Surgical Robot 386
1984 Container Shipping 388
1984 Itaipu Dam 390
1985 Bath County Pumped Storage 392
1985 International Thermonuclear Experimental Reactor (ITER) 394
1985 Virtual Reality 396
1986 Chernobyl 398
1986 Apache Helicopter 400
1990 World Wide Web 402
1990 The Hubble Space Telescope 404
1991 Lithium Ion Battery 406
1991 Biosphere 2 408

1992 Low-Flow Toilet *410*
1992 Stormwater Management *412*
1993 Keck Telescope *414*
1993 *Doom* Engine *416*
1994 Channel Tunnel *418*
1994 Digital Camera *420*
1994 Global Positioning System (GPS) *422*
1994 Kansai International Airport *424*
1995 *Toy Story* Animated Movie *426*
1996 Ariel Atom *428*
1996 HDTV *430*
1997 Prius Hybrid Car *432*
1998 International Space Station *434*
1998 Large Hadron Collider *436*
1998 Smart Grid *438*
1998 Iridium Satellite System *440*
1999 Wi-Fi *442*
2001 Segway *444*
2003 *A Century of Innovation* *446*
2004 Millau Viaduct *448*
2005 Bugatti Veyron *450*
2005 Georgia Aquarium *452*
2006 Palm Islands *454*
2006 AMOLED Screen *456*
2007 Smart Phone *458*
2008 Quadrotor *460*
2008 Carbon Sequestration *462*
2008 Engineering Grand Challenges *464*
2008 Martin Jet Pack *466*
2008 Three Gorges Dam *468*
2009 National Ignition Facility *470*
2009 Earthquake-Safe Buildings *472*
2009 US President's Limousine *474*
2010 Solar Impulse Airplane *476*
2010 BP Blowout Preventer Failure *478*
2010 Burj Khalifa *480*
2010 Harry Potter Forbidden Journey Ride *482*
2010 Alta Wind Energy Center *484*
2010 Tablet Computer *486*
2010 The X2 and X3 Helicopters *488*
2011 Fukushima Disaster *490*
2011 Self-Driving Car *492*
2011 Instant Skyscraper *494*
2011 Watson *496*
2012 Curiosity Rover *498*
2012 Human-Powered Helicopter *500*
2013 NCSU BookBot *502*
2014 Ivanpah Solar Electric Generating System *504*
2016 Venice Flood System *506*
c. 2020 Vactrains *508*
c. 2024 Brain Replication *510*
c. 2030 Mars Colony *512*
c. 3000 Things We Have Yet to Engineer *514*

Notes and Further Reading *516*
Index *523*
Photo Credits *528*

Introduction

Simply look around you, wherever you happen to be sitting right now, and chances are that you're surrounded by engineered objects. If you are sitting in a normal office chair, engineers played a big role in its creation. Engineers helped in the design and manufacture of the fabric on the chair, the foam underneath the fabric, the framework that holds the foam and fabric together, the pieces of plastic that make up the armrests, the mechanisms that allow the chair to go up and down and tilt forward and backward, the base of the chair, and the wheels that let it roll around.

You may be sitting in a room where the paint on the wall is engineered, as is the wallboard underneath the paint. The gypsum in the wallboard may come from a power plant, where one engineer designed a scrubber to extract sulfur from the smokestack by turning that sulfur into gypsum, and then another engineer designed the factory that took the gypsum and turned it into wallboard. In your room, chances are that the air you are breathing has been cleaned by an engineered filter attached to an engineered HVAC system, which is controlled by an engineered thermostat so that the temperature is always right. The engineered fan in the HVAC unit gets its power from the engineered power grid, which connects back to the engineered power plant that may have produced the gypsum.

Sitting nearby you may have a number of electronic devices, all created by a wide variety of engineers. For example, you may have a smart phone or tablet on the table, a nice HDTV on the wall, and a digital clock that sets itself using a radio station in Colorado. Electrical engineers, software engineers, industrial engineers, and mechanical engineers make those kinds of things possible.

Stepping outside, you see a road populated with dozens of complex automobiles, perhaps with a remarkably safe passenger jet flying overhead at 550 mph (885 kph). Under the road are sewer pipes, water mains, storm water drains, telephone and cable TV wires, and gas pipelines. All these are engineered, too.

And then there are the radio waves. Completely invisible to you, you are surrounded by thousands of different radio signals, all made possible by engineers. Every AM radio station, FM radio station, and television station in your area is flowing past you right now in the form of radio waves on different frequencies. Every cell phone in your area is in constant communication with its local tower. A smart phone typically has multiple radios for voice calls, along with a separate radio system for Wi-Fi and another for Bluetooth. Every other Wi-Fi hotspot and Bluetooth device in the nearby area is bathing you in radio waves, as is every tablet, laptop, and desktop computer using

Wi-Fi. Then there are hundreds of satellites overhead transmitting GPS signals, satellite TV signals, Iridium phone signals, weather satellite signals, and so on. Plus there are many more radios out there: fire and police radios, water meter radios, temperature and rainwater sensor radios, remote controls for the garage door opener and the key fob that unlocks the car door. It just boggles the mind if you think about it. Not one bit of this would be possible without engineers who bring it all to life, along with a regulatory structure that keeps all of the radio signals from interfering with each other.

Engineers are an amazing group of people who make our modern world possible. In the United States there are about two million engineers practicing their craft, for the most part invisibly and without much fanfare. But if we didn't have these engineers, we would be back in the Stone Age.

Here we are, about to embark on a fascinating journey together into the world of engineering. It might be helpful to answer the question: what, exactly, is engineering? A good starting point would be the *Random House Webster's Unabridged Dictionary*, which defines engineering in this way: "the art or science of making practical application of the knowledge of pure sciences, as physics or chemistry, as in the construction of engines, bridges, buildings, mines, ships, and chemical plants."

Here is another definition, from *Merriam-Webster's Collegiate Dictionary*: "the application of science and mathematics by which the properties of matter and the sources of energy in nature are made useful to people."

Another good way to understand engineering is to think about all the engineering disciplines that you find in industry, or that you find being taught at a large engineering university. For example, if we look at the College of Engineering at a big school like North Carolina State University, we find these departments:

Biological and Agricultural Engineering (BAE)

Biomedical Engineering (BME)

Chemical and Biomolecular Engineering (CBE)

Civil, Construction, and Environmental Engineering (CCEE)

Computer Science (CSC)

Electrical and Computer Engineering (ECE)

Industrial and Systems Engineering (ISE)

Forest Biomaterials (FB)

Integrated Manufacturing Systems Engineering Institute (IMSEI)

Materials Science and Engineering (MSE)

Mechanical and Aerospace Engineering (MAE)

Nuclear Engineering (NE)

Operations Research (OR)

Textile Engineering, Chemistry and Science (TECS)

There are also some specialized areas. For example, petroleum engineers deal with oil drilling and refining. Nanoengineers work in the emerging area of nanotechnology. And so on.

As you read these definitions and the list of disciplines, you start to form a mental image of what engineers do for society. Civil engineers who design bridges, for example, have been trained to understand the mathematics, software tools, best practices, and regulations having to do with the design and construction of safe, reliable, long-lasting bridges. We see their work all around us. The Golden Gate Bridge in San Francisco is considered to be an engineering masterpiece, as is the Millau Viaduct in France. Occasionally engineers get it completely wrong, however, as was the case with the Tacoma Narrows Bridge. From these mistakes, engineers learn valuable lessons that they codify and apply to all future bridge projects. Engineering is a profession where the members constantly communicate with and learn from each other so that technology advances.

Some engineered objects can be extremely simple. For example, the cast aluminum wheel of an automobile is a single piece that has been engineered to handle the load of a car along with all the forces from cornering, hard braking, and potholes. Others can be quite complex, like the car itself, or an airplane, made up of thousands of different parts that all work together reliably. There are also engineered systems, like the antilock braking system, or the airbag system, made up of different pieces that connect together to accomplish a task or solve a problem.

Then there are larger engineered systems that we refer to as architectures. Computer engineers frequently use the term "computer architecture" to denote the many parts and their relationships to each other. Or think about the Apollo moon missions. Millions of people and millions of components came together to create an architecture for the mission, with many related parts: the Saturn V rocket, the Command and Service Module, the Lunar Excursion Module, the space suits, the moon rover, and everything else. If any of those parts did not work properly, the whole mission could fail and the lives of the astronauts would hang in the balance, as was seen in *Apollo 13*. Similarly, the power grid has an architecture that includes power plants, transmission networks, and distribution networks. The cell phone system has its own architecture consisting of towers, handsets, and a complex signaling protocol that ties them together.

Another key role played by engineers is the reduction of costs through mass production, along with the act of engineering out unneeded materials, time, and processes. A great example of this is the clothing we wear. It used to be that people did every bit of the work to produce clothing. People planted cotton, hoed the plants, harvested the bolls, and then picked the seeds out of the fibers, all by hand. Then they cleaned the cotton, dyed it, carded it, spun it, wove it on a hand loom, and sewed the cloth together with a needle and thread. Clothing was expensive because each article represented hundreds of hours of manual labor. Today engineers have mechanized nearly every part of the process, along with making a wide variety of synthetic fibers available, to the point where clothing is very affordable. The same has happened with every type of product, so that we can now all afford to carry around small computers called smart phones, with high-definition (HD) screens and powerful cameras connected to a worldwide network filled with millions of servers that offer the answers to nearly any question a person might have. None of this would have ever happened without engineers.

What is the difference between a scientist and an engineer? Scientists are charged with understanding how the universe works. They do research and answer fundamental questions about nature. For example, a scientist may discover that curved glass bends light, and then come up with mathematical equations to characterize the bending. From that scientific discovery we get a lens. A scientist or inventor might then put lenses together to create an optical device like a microscope or telescope. But eventually it is time to take the next step. If you want to create a million telescopes at an affordable price, or you want to create a giant telescope weighing many tons and you need it to automatically rotate very smoothly to track a star as the earth rotates, this is when the engineers take over.

Scientists discovered that the right amount of carbon in iron creates steel. Engineers use the steel to build bridges, skyscrapers, cars, supertankers. When steel is too heavy or too weak, engineers have many other materials they can use: aluminum, titanium, carbon fiber, Kevlar, plastics, and so on.

You will often hear someone exclaim, "this is beautifully engineered" or "that's a really well-engineered piece of equipment!" Examples are myriad, including the SR-71, the Ariel Atom, or the Parthenon. But it could be something far simpler, like the shifting mechanism in a car's transmission or even a perfectly weighted and silky-smooth control knob. The exclamation happens when engineers achieve a high level of elegance in their designs, or an impressive level of reuse and parsimony, or a kind of refinement that takes the user's breath away on first encounter, or a perfect match between form and function. We see beauty in engineering, just like we do in art and nature.

In all of these different ways, engineers move society forward. They create new technologies that make our lives better. They lower costs so more people can participate. They build things that amaze us, and make us proud to be human beings. They create things on which our modern lives depend.

This is a book of engineering milestones—a celebration of engineering, really. We start at the very beginning, back with the first engineered objects that human beings created. And then we move forward through time, watching engineering advance with achievement after achievement (plus a few blunders along the way—even engineers make mistakes).

In each case, the milestone will have a date. There is a little leeway here: Should it be the date of conception? Of the first patent? Of "first flight"? Of the first widespread popularization? I have tried to pick a "best" date when fitting things into the timeline.

Should these engineering achievements be associated with a specific person? I tend to shy away from this idea in most cases, primarily because engineering is a team sport. Take the steam engine as an example. Is there any one person we can attribute the steam engine to? If so, is it the first Greek who made steam shoot out of a pot? Is it the first person to create any useful steam engine? Or the first person to harness high-pressure steam? What about the person who made the most powerful steam engine, or the most popular one? The fact is that the traditional steam engine played a huge role in the advancement of society for about a century, and thousands of engineers contributed to steam's success. I don't want to pick any single person out of that crowd. Most engineering achievements work that way—many people contribute to the project, so no single person deserves specific credit.

Finally, you may notice that this timeline is very heavy on the twentieth century. Why is that? It's because the twentieth century was an amazing time in terms of engineering. Airplanes did not exist prior to the twentieth century. During those hundred years, they went from rickety wooden contraptions flying at walking speed to sleek aluminum jets flying thousands of miles at supersonic speed. And many other technologies followed the same course: cars, spacecraft, air conditioning, computers, networks, skyscrapers, television, nuclear materials, etc. These technologies all sprang into existence in the twentieth century and then advanced with incredible speed. It all leads to the skew; most of the engineering disciplines listed above did not exist until the twentieth century.

It is my hope that, in this book, you discover 250 examples of engineering that amaze you, and that help you see the full gamut of the art, so that you can appreciate everything that engineers do for us as we live our lives. Let's step into the remarkable world of engineering.

Acknowledgments

Simply stated, this book would not exist without the tireless efforts of Kate Zimmermann. Between the editing, the photographs, the additions and deletions and all of the minor tweaking along the way, Kate is the one who made this into a book. I am very grateful to her for everything she has done.

Thanks also to my wife Leigh and the kids Ian, Johnny, Irena, and David for their good humor and encouragement during the writing process.

And as mentioned in the dedication, I am very thankful that Dr. Louis Martin-Vega started me down this path. Without his influence, this book would have never occurred.

Bow and Arrow

If we were to go back in human history as far as we can, when would we find the first example of engineering? If someone makes a tool, is that engineering? Yes, but there has to be some kind of line. If a person picks up a rock to crack a nut, that is tool use but it is not really what we think of as engineering. When something is engineered, there is more to it. Therefore, the bow and arrow probably qualifies as the first engineered object. And its use is indeed ancient—starting 30,000 years ago or more.

The bow and arrow is a surprisingly clever piece of technology. It is the first device we know of that stores energy for later release. It is the first projectile weapon. And it can be fashioned from objects readily available in nature. A piece of wood combined with a string made of fibers, skin, or sinew handles the energy storage. A piece of wood tipped with bone or stone and stabilized with feathers acts as the projectile.

As humankind's first projectile weapon, think of how useful the bow and arrow is. If a person is hunting a deer or rabbit, a bow and arrow gives the human a fighting chance. Compare the bow and arrow to throwing a rock or a spear. Rocks and spears work for only a short distance, are not particularly accurate, and telegraph the human's position with the windup. With a bow and arrow, the human can fire silently from a hidden position without any windup, with accuracy, and at a decent range. A bow and arrow changes the game for a hunter.

By 1400 CE, the bow and arrow was highly refined. Archers in England were able to use longbows to fire ten arrows per minute. Arrows leave the bow at 100+ mph (160+ kph) and fly 1,000+ feet (300+ meters). At a range of 60 feet (18 meters), metal-tipped arrows can punch through armor.

Guns reconceptualized weaponry (and have since led to high-performance weapons like the **AK-47**), but there is no doubt that the 30,000-year run for the bow and arrow is a record for technological dominance.

SEE ALSO Catapult (1300), AK-47 (1947).

In Egypt's Old Kingdom period (third millennium BCE), the single arched bow was developed.

3300 BCE

Hunter/Gatherer Tools

Otzi (c. 3300 BCE)

One thing that engineers and the engineering mindset create is new technology—useful objects that solve problems. Animals living in the wild do not create novel technology of any complexity—it is a distinctly human trait that comes from our ability to identify problems and then invent solutions for them.

The development of technology is something that happens early in many human cultures. We get a glimpse of the technology available about five thousand years ago because of a man today known as Otzi, who died in 3300 BCE but was preserved almost perfectly in mummy form in a glacier and discovered in 1991. On his person, Otzi was carrying many pieces of technology of his era. He wore, for example, grass-insulated shoes with bearskin soles and deerskin uppers. He also had clothing (hat, coat, pants, belt) made of animal skins. Thread made of sinew held the skins together.

His tools are even more surprising. The most impressive is a copper axe with a yew handle. He was also found with a dagger with a flint blade, wooden handle, and a sheath attached to his belt. He carried a **bow**, although it seems it was not yet finished and did not have a string. To go with the bow he had a quiver with arrows and arrow shafts. The arrowheads are made of flint, and feathers were attached to stabilize the arrows in flight.

Apparently he had a backpack with an internal frame and a bag made of animal hide. Inside the backpack were birch bark containers, and one was probably used to carry embers to start a fire. He also carried a net, some string, a thong device perhaps for carrying dead birds during a hunt, and fungus thought to be used medicinally.

Given the era, and the state of European civilization at the time, the technology is stunning. It shows how deeply seated the engineering mindset is in the human brain. In order to be carrying a copper axe, for example, it implies the ability to mine and refine copper ore and then cast copper objects from molten copper. It is a surprising level of technology to achieve in a primitive culture.

SEE ALSO Bow and Arrow (30,000 BCE), Inuit Technology (2000 BCE), Bessemer Process (1855).

Dutch artists Adrie and Alfons Kennis created this reconstruction of Otzi the mummy based on the latest forensic research.

2550 BCE

The Great Pyramid

When we think about engineers who are working today, they are usually working on something that benefits society. They might be designing a bridge, a consumer device, or a new vehicle. Not so with the Great Pyramid in Egypt. Even today, thousands of years after its construction, the Great Pyramid is one of the biggest, heaviest, tallest things human beings have ever built. Yet it functionally accomplishes nothing.

Engineers did not jump out of bed one day ready to build the Great Pyramid. They built a few test pyramids over the course of a century. The Pyramid of Djoser is a classic step pyramid 200 feet tall. The Maidum pyramid is a classic step pyramid with several of the steps filled in to start making a smooth pyramidal form. The Bent pyramid started with one slope, and then the engineers realized that it would not work, so they changed the slope midway. The Red pyramid gets the shape right, but is 140 feet shorter than the Great Pyramid.

Then engineers were ready to build the Great Pyramid. They cleared off the sand on a 13-acre (5.26 hectare) site to expose bedrock for the pyramid's foundation. They oriented the pyramid almost perfectly north. Then they laid the base layer of stones, measuring 756 feet (230 meters) square. The stones came from quarries along the Nile River.

The engineers had to do something fascinating during construction. They had to visualize the chambers, hallways, and shafts that would exist in three dimensions in the body of the pyramid, and they had to build them layer by layer during the pyramid's construction. This is the same kind of methodology that an engineer with a **3D printer** uses today.

Eventually the pyramid rose to a pinnacle at 481 feet (146 meters). The world's most gigantic monument was complete. The Great Pyramid stands out as an engineering triumph.

SEE ALSO Parthenon (438 BCE), Basilica of Saint Denis (1144), Washington Monument (1885), Eiffel Tower (1889), 3D Printer (1984).

The Great Pyramid of Giza, pictured, is the largest and oldest pyramid in the Giza Necropolis.

2000 BCE

Inuit Technology

Engineering seems to be something wired into the human brain. Many human cultures are quite adept at developing innovative technologies to solve problems they experience. This happened in spades in the Inuit culture in Northern Canada and Greenland.

Although no one is sure exactly when the Inuit peoples arrived in the area, it is thought to have occurred sometime before 2000 BCE. The Inuit lived above the arctic tree line, in an extremely harsh climate, and developed at least a dozen unique technologies to deal with the environment and to help in providing food and shelter.

One of their key requirements is clothing that can protect against winter temperatures that regularly plunge below 0°F (-17°C). Inuit parkas, boots, and gloves provide that protection. Made of animal hides with the fur on the inside to improve insulation and avoid wetting, Inuit garments are works of art and beautifully engineered.

Another area of innovation is the igloo, able to provide shelter in the most extreme arctic conditions. Using a snow saw, Inuit can build igloos in just an hour or two to erect a quick shelter. Given more time, these ice domes can be 13 feet (4 meters) in diameter and 10 feet (3 meters) high.

An Inuit technology widely adopted in the West is the kayak. In its original form, a wood frame bound together with sinew is covered in de-haired sealskins. The Inuit perfected the idea of rolling the kayak back over if it capsized.

Inuit snow goggles carved of wood provide protection against snow blindness on bright days. They consist of an opaque mask with narrow slits to significantly cut down on incoming light.

The Inuit are adept at crafting knives, **arrow**heads, and harpoon heads from materials like bone and stone. The toggling harpoon head is particularly insightful. Once embedded, the head shifts from parallel to perpendicular to make the harpoon's accidental removal nearly impossible.

Together this suite of technologies make it possible for the Inuit to thrive in the harsh arctic climate. Each technology embodies unique engineering discoveries polished to a high art and then handed down orally from generation to generation.

SEE ALSO Bow and Arrow (30,000 BCE), Mars Colony (c. 2030).

In this 1924 photo, an igloo is constructed as children and dogs look on.

formulation, it cracked easily. The city of **Pompeii** was built mainly with Roman concrete.

In today's world, concrete is an incredibly important material. Civil engineers use concrete to build roads, bridges, dams, skyscrapers, runways, canals, and foundations on a massive scale. If you take it by weight, concrete is the number-one building material in the world by far.

It's easy to understand why concrete is so popular: it gives engineers the ability to pour a liquid into a mold and create something similar to solid rock that can last for centuries. By adding steel rebar or pretensioned steel, the strength of concrete improves dramatically and makes it possible to create beams 100 feet (30 meters) long or more. Add to that the fact that most of concrete's weight comes in the form of sand and gravel, and you have a material that is inexpensive compared to alternatives. At today's prices, concrete costs less than three cents per pound (0.45 kg).

Concrete has four ingredients: one part Portland cement, two parts sand, and three parts gravel with enough water to make a paste-like mix. The Portland cement, when mixed with water, acts like a glue that binds the sand and gravel into a dense solid. This is not like a glue that dries, however. It is more like a calcium-silicon-water epoxy that hardens through a chemical reaction. This reaction gives off heat and it is slow. It takes concrete several weeks of curing to reach reasonable strength. This is why you will often see workers pour a concrete foundation and then disappear for a month. They are waiting for the concrete to cure to the point where they can put weight on it.

Although concrete is simple to make, it is important to do it right. When a road or foundation is poured, engineers will often take a cylindrical sample and do a crush test to confirm its compressive strength.

SEE ALSO Pompeii (79), Woolworth Building (1913), Millau Viaduct (2004).

Asphalt

Where would cities be without asphalt? It is a wonder material for building roads and parking lots because it is cheap, easy to work with, smooth, seamless, and durable. Engineers can spec out an asphalt road that can handle millions of cars for a decade or more. Asphalt is so popular that all but 6 percent of America's roads are made of the material. One of the earliest known uses of asphalt goes back to 625 BCE in Babylon.

One reason for asphalt's popularity is its simplicity. Asphalt has three basic components: sand, gravel, and bitumen. Although found occasionally in nature, bitumen today comes from **refineries**. It is separated out from crude oil in the same way as other petroleum distillates. Gasoline and bitumen contain the same atoms (carbon and hydrogen), but the carbon chains are immensely long in bitumen. Therefore bitumen is approximately a solid at room temperature.

It is possible to make asphalt by hand. You could even make a small amount in your kitchen oven. Put some sand and gravel on a cookie sheet. Put it in a 300°F (150°C) oven long enough for it to heat evenly and dry out. Now place a lump of bitumen on it and continue heating until the bitumen melts. Stir thoroughly to mix bitumen and gravel together. You have asphalt. Your kitchen will stink and you will never get that cookie sheet clean, but now you can fix a small pothole.

Engineers have refined the creation of asphalt and the equipment that handles it to push down the cost of roads. For example, the Strategic Highway Research Program created mixing and construction guidelines for SuperPave, the asphalt recipe used in many highways today. On a big road project, engineers will often erect a portable asphalt plant near the construction site to make delivery easier and more consistent.

One little-known fact about asphalt is that it is the most recycled material, by weight, in the US. Construction crews grind it out and re-add bitumen to make new roads. If it weren't for asphalt, engineers would need to use **concrete**, which is much more expensive. This makes asphalt one of the most popular construction materials in the world.

SEE ALSO Bessemer Process (1855), Wamsutta Oil Refinery (1861), Titanium (1940).

The process for the creation of asphalt has been refined since its first use in Babylon in 625 BCE, when Herodotus recorded it being used as mortar.

438 BCE

Parthenon

The Parthenon is a structure that is both beautiful and beautifully engineered. It has stood the test of time, and is amazing to us today because it was built in an era when so little technology was available to provide assistance. It also tells us something about the people who conceived it.

The Parthenon was created in 438 BCE as an immense temple to the goddess Athena. It once housed an enormous statue of her rendered in gold and ivory. There is actually a reproduction of this statue inside a replica of the Parthenon in Nashville, TN.

The basic structural idea the engineers used was fairly simple, while also being magnificently executed. The outer perimeter consists of marble columns—eight across the front and back, seventeen along the sides. Across the tops of the columns are marble lintels, and then a lot of decoration. The original structure had a roof made of wooden **trusses** covered in clay roof tiles. It also had interior walls that created a room for the statue.

One thing that people marvel at today is the fact that engineers designed curves into the Parthenon, apparently as a kind of reverse optical illusion. So the floor of the temple is not flat—it is subtly higher in the middle. The columns do not stand straight; instead they lean in very slightly. The corner columns do not match the others—they are slightly wider and closer to the others. Everything "looks right," but the only reason it looks right is because everything is a bit wrong. The wrongness was built in to create the rightness of appearance.

So today, when we think of the greatest Greek temple, we think of the Parthenon. Not because it was the biggest, or the best preserved, but because of its perfection. The Greeks have gone to a tremendous amount of trouble recently to repair some of the damage that has been inflicted over the centuries and restore the grandeur created by the original engineers and craftsmen.

SEE ALSO The Great Pyramid (2550 BCE), Basilica of Saint Denis (1144), Truss Bridge (1823), Washington Monument (1885).

The Parthenon is widely regarded as the most perfect example of Greek architecture.

312 BCE

Roman Aqueduct System

Appius Claudius Caecus (c. 340 BCE–273 BCE)

Sometimes a group of people have big, pressing needs that can be solved by engineering. Such was the case in ancient Rome, and the problem was the water supply. The year is approximately 300 BCE and Rome is growing. But the water supply stinks. Literally. Water from underground has a bad taste, and water from the Tiber River is loaded with pathogens.

To solve the problem, Roman engineers commissioned by censor Appius Claudius Caecus developed aqueducts. The first one, called Aqua Appia, is a perfect example. The engineers found a large, clean spring about 10 miles (16 km) outside Rome. Located at a higher elevation than Rome, gravity could do the work of moving the water toward the city. Roman engineers cut trenches or dug tunnels (often through solid rock) and then lined them with waterproof mortar. If a valley got in the way, the engineers built a bridge to carry the channel. The channel sloped gently downward all the way to the city.

What to do about mud and sediment in the water? The water flowed slowly through wide, deep pools so particles could settle out. How to maintain the tunnels and clean them out? Vertical shafts connected the tunnels to the surface. What if too much water surged through the system? The tunnels had overflow vents to drain away extra water.

The Aqua Appia aqueduct is thought to have delivered 20 million gallons (76 million liters) of water per day to Rome. Once inside the city, the water from an aqueduct could flow into large, elaborate public fountains, to public baths, into pipe systems to residences, or into the sewer system. The sewers carried waste out of the city and kept Rome remarkably clean.

Even with 20 million gallons of water a day, Rome outgrew the supply. So the engineers built more aqueducts. Over the course of five hundred years, there were eleven aqueducts feeding Rome, the longest one stretching 56 miles (90 km). The entire system brought perhaps 300 million gallons (1.1 billion liters) of water per day to over a million people. It was an amazing achievement and it led to later innovations such as the **modern sewer system**.

SEE ALSO Modern Sewer System (1859), Desalination (1959), Stormwater Management (1992).

This ancient aqueduct is now the Pont du Gard Bridge near Remoulins, France.

100 BCE

Waterwheel

Before the introduction of the steam engine, the diesel engine, and the electric motor, if people wanted to build a factory or use a large tool of any sort that went beyond a hand tool, they needed something to provide the power. Engineers could and did put humans in big hamster-wheel-like affairs (treadwheels) to spin horizontal shafts. They also could have people or horses walk in circles to turn a vertical shaft.

But the innovation that reliably provided a source of continuous power was the waterwheel. And the Romans appear to be the first to have exploited it in about 100 BCE. There are multiple Roman sites that show their engineering prowess, but the most impressive was the multi-wheel flour mill at Barbegal in France.

On a steep hillside, Roman engineers arranged two sets of eight mills with sixteen vertical overshot waterwheels. Because of the hillside arrangement, the water leaving one wheel could feed into the next wheel down.

The horizontal shaft of a waterwheel would connect to a cog wheel so that: 1) the direction of the shaft rotation could switch from horizontal to vertical for the millstone and 2) the rotational speed of the millstone could be two or three times faster than the waterwheel.

It is estimated that the 16 mills at this site could produce perhaps 10,000 pounds (4,500 kilograms) of flour each day. A pound of flour would make a loaf of bread. The 10,000 loaves of bread per day fed the nearby Roman city of Arelate (present-day Arles), which had a population of perhaps 30,000 people.

The Romans also used water to power reciprocating sawmills for wood or stone.

At the start of the Industrial Revolution in America 1,700 years later, water was still the power source. Both vertical and horizontal waterwheels provided the power for the first factories. Therefore factories needed to be located where falling water was available in sufficient quantity. So, for example, the first factory of the Industrial Revolution was located at Pawtucket Falls in Rhode Island. Until **steam engines** became popular, every factory needed falling water to provide the power.

SEE ALSO Roman Aqueduct System (312 BCE), Simple Machines at Yates Mill (1750), Cotton Mill (1790), Mass Production (1845).

This undershot water wheel of a water mill is found in Portogruaro, a town on the river Lemene in the Province Venice in Veneto, Italy.

79

Pompeii

Roman cities were prime examples of early engineering prowess. Especially when the Roman engineers built new cities from scratch; they were highly evolved, orderly, planned metropolises able to support tens of thousands of people comfortably.

Pompeii had been under Roman rule for more than a century when it was buried under volcanic ash in 79 CE. The ash preserved the city like a time capsule and lets us see how Romans lived 2,000 years ago in their engineered cities.

The water and sewer systems were important elements of a Roman city. Water came in via free-flowing **aqueducts**. It was distributed to citizens through pipes and public fountains. Excess water, human waste, and storm water flowed into a belowground sewer system.

Roads were extremely important for letting people, animals, and carts move around the city. The city streets were paved with stones and laid out in a grid pattern much like they are in a modern city. Sidewalks lined the streets and were covered to shade pedestrians.

The public baths were important both for hygiene and socializing. Many incorporated an ingenious heating system called a hypocaust. The floor of the bath was raised up on tile pillars with a 3-foot (1 meter) gap underneath. Smoke and heat from a fire would flow under the floor and through the walls to heat the bath to temperatures as high as 120°F (49°C).

A Roman city also contained shops, workshops, bakeries, markets, a forum with its public temples and government offices, an amphitheater, and a performing theater. The citizens of the city lived in private homes or apartments.

The building materials available to the engineers were stone, concrete, brick, tile, and wood. The Pompeii amphitheater, for example, is the oldest stone amphitheater in the Roman Empire. A typical home's walls were covered with plaster on the inside and often painted, with stucco on the exterior. Roof **trusses** were used, covered in roofing tiles.

Engineers had created the height of urban luxury for the citizens of these Roman cities, delivering everything that large numbers of people needed to live comfortable lives.

SEE ALSO Parthenon (438 BCE), Roman Aqueduct System (312 BCE), Waterwheel (100 BCE), Truss Bridge (1823).

The ruins of Pompeii, seen in this aerial view of what is now Naples, Italy, show the scope of the original city.

1040

Compass

Imagine the problem you had if you were a traveler in uncharted territory five hundred years ago. If you had a **clock** and you could see the sun, you could get a decent sense of direction. Or at night, if you could see the stars and you had some time, you could also get a sense of direction. But wouldn't it be great if you could pull a device out of your pocket and it would immediately tell you your direction of travel in an instant?

A compass seems so simple today. Any kid can make one in five minutes with a sewing needle, a refrigerator magnet, a little piece of foam, and a bowl of water. Rub the needle on the magnet to magnetize the needle and float it on the foam in the bowl. The needle will rotate, and you'll instantly know where magnetic north is.

But rewind a thousand years and it's not so easy. First you need an iron needle. Which means you need iron—no small thing given the technology required to smelt the ore. Assuming you can find the ore and know what it means. Then you need the ability to shape the iron into a needle, which requires tools and some finesse. And then you need a magnet. Where will you get one of those? Electromagnets don't exist yet. There are naturally occurring magnets called lodestones, but they are rare and you need to find one.

Everything came together in China. During the Song Dynasty around 1040, they began to produce iron, and iron needles, which they combined with lodestones to make the first compasses. They hung their needles on strands of silk rather than floating them in a bowl of water.

From an engineering standpoint, compasses were incredibly important to early surveying efforts. If you needed to lay out a railroad, a canal, or the boundary lines on a piece of land, a good compass was essential. There was no other easy, accurate way to get your bearings in the middle of nowhere.

All because the core of the earth is itself a magnet, and instrument makers used that fact to create quality direction-finding instruments.

SEE ALSO Gunter's Chain (1620), Mechanical Pendulum Clock (1670), Bessemer Process (1855), Atomic Clock (1949), Global Positioning System (GPS) (1994).

The magnetic compass was developed during the Song era of China. A modern magnetic compass is shown here.

1144

Basilica of Saint Denis

When we think of the buildings known as cathedrals, the image that usually comes to mind is the Gothic cathedral, made of stone at an incredible scale with massive stained glass windows. These structures are marvels in several respects, but perhaps are most interesting because they represented a significant step forward in terms of architecture and engineering. Although **The Great Pyramid** and other such monumental structures existed, the world had never seen buildings this tall and open, with such gigantic windows and so much light. The St. Denis Cathedral, which opened in France in 1144, is considered to be the first example of this architectural form.

There were two engineering innovations that made the Gothic cathedral possible. The first was the Gothic arch, or pointed arch, which replaced the rounded Roman arch. The pointed arch sends much more of the weight it supports down vertically rather than flattening the arch horizontally. But it does not send all of it downward. This is where the second innovation, the flying buttress, comes in. The buttress pushes in horizontally to counteract the arch's desire to push outward. The flying buttress stabilizes the arch, making the ribbed vault and stone ceiling possible. The buttresses also stabilize the tall walls. Because they are on the outside of the building, buttresses do not get in the way of the windows.

With these two innovations, engineers could build thin stone walls to incredible heights and leave huge holes in the walls for windows. A typical Gothic cathedral is well over 100 feet tall on the inside.

This is not to say that building these cathedrals was an easy task. Workers and craftsmen had to quarry and carve tons of stone. It all had to be hoisted, fitted, and locked in place. A cathedral project could take a hundred years or more.

The typical human being in this timeframe had never seen a building this gigantic, with so much interior volume and such an amazing amount of glass. Engineers had created a completely new way for people to think about structures.

SEE ALSO The Great Pyramid (2550 BCE), Burj Khalifa (2010).

The Basilica of Saint Denis is considered to be one of the first examples of Gothic architecture.

1300

Catapult

The word "engineer" entered the English language in the early 1300s. It denoted a person who built military engines, also known as siege engines. These were various machines used to lay siege to a walled city, castle, or fortress. Siege engines included things like battering rams, ballistas (giant **crossbows**), and catapults.

Although military engines got their start around 400 BCE with the Greeks and Romans, we tend to think of siege weapons being unleashed upon castles in medieval battles. This explains the timing of the introduction of the word "engineer."

At that time, two types of catapults were popular: the mangonel and the trebuchet. The mangonel relied on a torsion device to store energy, while the trebuchet relied on a weighted arm. The trebuchet in particular was quite powerful, able to sling stones weighing 300 pounds (136 kg) or more at castle walls to crush them. The range was hundreds of yards. When not shooting projectiles, catapults could fire incendiaries, animal carcasses, or diseased human bodies.

The basic mechanics of both catapult designs are straightforward. A trebuchet stores energy in the rise of a heavy weight attached to a long arm. In a big trebuchet, the arm might be 60 feet (18 meters) long and the weight could be as heavy as 10 or 12 tons.

The mangonel used hundreds of tightly twisted rope strands. Cocking the catapult involves pulling the arm down 90 degrees, adding even more torsion to the ropes. When released, the ropes would spring back to launch the projectile.

In 1304, engineers built what is believed to be the largest trebuchet ever for a siege that was taking place in Scotland. The trebuchet's name was War Wolf. By repeatedly launching projectiles weighing 300 pounds (136 kg), one of the walls at Stirling Castle crumbled.Engineers won the battle by using mechanical engineering to crush heavily fortified stone walls.

SEE ALSO Bow and Arrow (30,000 BCE), AK-47 (1947), Cluster Munition (1965).

Trebuchet catapults at the Castle of Castelnaud, Dordogne (Perigord), Aquitaine, France.

1372

Leaning Tower of Pisa

Bonanno Pisano (Dates Unavailable)

The **St. Louis Arch**, the **Washington Monument**, the **CN Tower** in Toronto, and the **Burj Khalifa** in Dubai all have massive foundations. The leaning tower of Pisa is an example of why the foundation is so important. Bonanno Pisano is given credit for the original architecture of the tower. Construction began in 1173, but since its completion in 1372, legions of engineers have spent centuries trying to fix the foundation problems that were baked in from its start.

Pisa is located at the confluence of two rivers, on land that is soft and wet. A modern engineer would probably drive piles deep down into the unstable soil, until they anchored in stable soil or rock. This is how the city of **Venice** was built, using piles made of wood. With the Tower of Pisa, it looks like the builders simply dug a trench and used standard masonry footings. With soil so soft, this foundation is inadequate.

So why is the tower standing at all? It is thought that a budgetary fluke saved it. When the tower was three stories tall, construction was interrupted for a century due to a lack of funding. This delay allowed the soil under the tower to consolidate and stabilize. Then, when construction resumed, attempts were made to straighten out the tower by making one side taller than the other. But the tower started leaning more and more.

There have been several attempts to stabilize the lean over the years, none successful and several that made the lean worse. It wasn't until the twenty-first century that engineers found a solution in two parts. The tower leans to the south. So they used a process called soil extraction on the opposite side. They drilled down diagonally with augers and pulled out soil from underneath the north side. Gravity caused the tower to settle toward the north as the cavities filled in. They did not want to straighten the tower—that would kill tourism. They reduced the lean enough to bring the tower back into its safe zone. Then engineers installed a drainage system to extract excess water from the soil around the tower.

Even though early engineers made mistakes initially, and nearly toppled the tower with several bad remediation ideas, their successors eventually found a working solution that saved the tower, lean and all.

SEE ALSO Washington Monument (1885), Gateway Arch in St. Louis (1965), CN Tower (1976), Burj Khalifa (2010), Venice Flood System (2016).

This photo of the Leaning Tower of Pisa dates back to circa 1890.

1492

Square-Rigged Wooden Sailboats

When we think back to the age of discovery, the age of sail, or the age of pirates, we automatically invoke images of square-rigged wooden sailboats. Columbus's Niña, Pinta, and Santa Maria are perhaps the best-known ships of the genre, but there were thousands of ships like these transporting people and goods around the world. These boats were made in shipyards by craftsmen, who learned their trade through apprenticeship and experience.

But in mindset they were engineers, designing and building custom vessels using a common basic architecture. That architecture dominated shipbuilding until steel refined using the **Bessemer Process** completely reconceptualized the design and construction of ships in the late nineteenth century.

The architecture was surprisingly straightforward and consisted of four parts. First there was the main spine of the ship, known as the keel, which ran down the center of the vessel's bottommost point. This was a stout square timber carved from a single log, or joined together from multiple logs, depending on the length of the ship. At the front of the keel, multiple pieces of wood were joined together to make the curving stem extending up and forward from the keel. At the back, a sternpost joined the keel and rose vertically.

Second, attached to the keel was a set of sturdy ribs or frames made from multiple pieces of wood, called futtocks, joined together. The ribs attached to the keel with the help of pieces called floors. Ribs determined the shape of the outer hull.

Third, the top deck and any interior decks were formed with horizontal beams attached to the ribs. Finally, the outer surface of the hull, as well as the deck flooring, was made of one or more layers of planking. The planks attached to the ribs or beams with trunnels—pieces of wood dowel seated with wedges driven into their ends. Rope soaked in tar would be driven between planks to seal the gaps, and then sealed over with more tar.

Once the hull was complete, the masts could be seated and then their sails and rigging added. This basic wooden architecture—an engineered solution to the wooden boat problem—served humankind well for several centuries.

SEE ALSO Bessemer Process (1855), *Seawise Giant* Supertanker (1979), Container Shipping (1984).

A painting of the Santa Maria at sea by Michael Zeno Diemer (1867–1939).

1600

The Great Wall of China

The **Great Pyramid** is certainly impressive by any standard, especially given the technology available at the time.

But from the standpoints of engineering, logistics, project management, sheer grandiosity, and persistence, nothing on earth really compares to the Great Wall. It is hard to believe that human beings could ever be organized long enough and well enough to engineer and build something of this scale.

Imagine that we decided to build a wall of stone and brick 20 feet (6 meters) wide and 20 feet high, along with substantially larger watch towers every 1,000 feet (300 meters) or so, from Washington DC to Los Angeles (a distance of approximately 2,600 miles or 4,200 km) and then back again. That is the scale of the Great Wall. By volume of material, the Great Wall might be 100 times larger than the Great Pyramid. It is difficult to know for sure because large sections of the wall have eroded, collapsed, or been dismantled.

The Great Wall, completed in 1600, was built to solve a problem. Various nomadic groups of non-Chinese people, including the Mongols and the Manchus, were trying to invade China. The wall was meant to provide a line of demarcation and defense. It was built over a long period of time—1,000 years or more. The Ming Dynasty, from 1368 through 1644, stands out as a time when the 5,500 miles (8,860 km) was consolidated, linked, reinforced, and standardized to create the wall we know today.

The basic architecture for a wall section is fairly straightforward. Two thick stone or brick walls were built 20 feet apart, and then the gap filled with dirt and rubble. Paving stones or bricks along the top of the wall covered the dirt fill to create a path approximately 17 feet (5 meters) wide.

With anything we build today, we have assistance from machines—bulldozers, **tunnel boring machines**, **tower cranes**, dump trucks. The Great Wall will stand out as a singular achievement because, despite its scale, it was built instead by millions of hands.

SEE ALSO The Great Pyramid (2550 BCE), Tunnel Boring Machine (1845), Tower Crane (1949), Burj Khalifa (2010).

Several walls within what became the Great Wall were being built as early as the seventh century BCE.

1620

Gunter's Chain

Edmund Gunter (1581–1626)

Surveying is important to the engineering of any large construction project. For example, it is impossible to imagine laying out a railroad or **canal** without professionally trained and careful surveyors helping with the project. When you think about a project as big as the **transcontinental railroad**, which had the added complexity of two railroads being built separately with the plan to meet in the middle of the country, it is an impossible task without surveyors.

It is hard to determine when surveying started, however. The Egyptians certainly knew something about surveying, because the siting, angles, and dimensions of the **Great Pyramid** are nearly perfect. The Romans had surveyors helping to lay out their roads, **aqueducts**, and cities as complex as **Pompeii**. But something like modern surveying started with the invention of the Gunter's chain, which was introduced by mathematician Edmund Gunter in 1620. The Gunter's chain is a tool for accurately measuring distances. It has 100 links in the chain with a total length of 66 feet (20 meters). A piece of land one chain wide by ten chains long is an acre.

Early surveyors needed three pieces of equipment: a Gunter's chain, a plane table with transit or circumferentor, and a surveying **compass**. With these tools they could orient themselves, sight lines, record angles, and accurately measure distances. This was sufficient for laying out boundary lines. To measure grades or elevations, a topographer's rod was used.

Today all of this is handled with a computerized, **GPS**-enabled device called a "total station." It includes the transit/theodolite and combines it with EDM (electronic distance measuring). A helper stands with a reflector and, once sighted, the total station measures and records angles and distances. There are also robotic total stations to eliminate the need for a helper. Compared to the tools that preceded them, total stations make surveying on civil engineering projects incredibly easy. But the principles are all the same: get your bearings, then measure angles, distances, and elevations. With these techniques, civil engineers can lay out the biggest projects people can imagine.

SEE ALSO The Great Pyramid (2550 BCE), Roman Aqueduct System (312 BCE), Compass (1040), Erie Canal (1825), Transcontinental Railroad (1869), Global Positioning System (GPS) (1994).

This antique brass theodolite, pictured, measures horizontal and vertical angles and was an essential tool for surveying until replaced by modern tools like the total station.

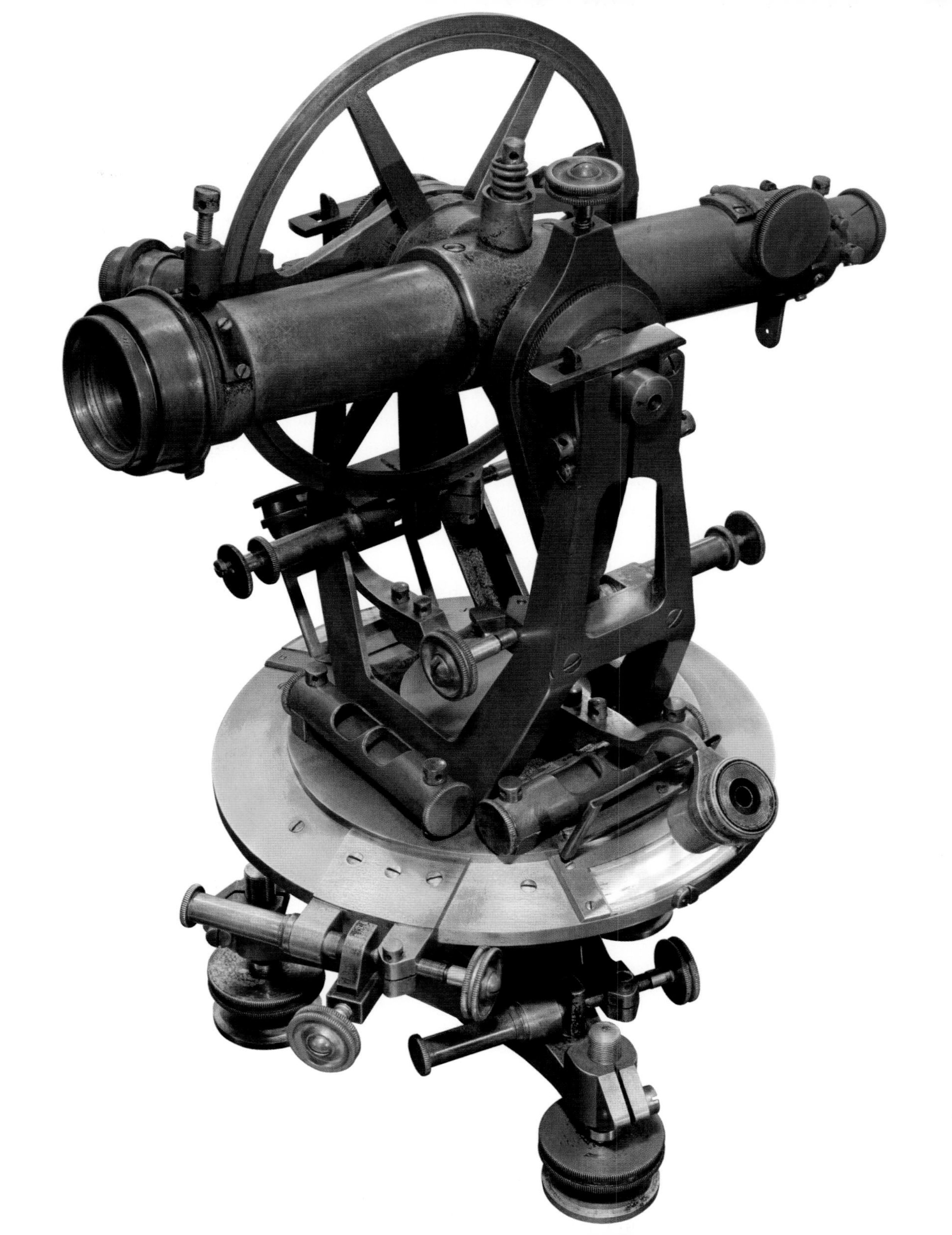

1670

Mechanical Pendulum Clock

Robert Hooke (1635–1703), **Joseph Knibb** (1640–1711)

When we think of the people who design and build clocks, we tend to think of them as clock makers. But if you think about it, a clock maker is a mechanical engineer, working with springs, gears, and mechanical oscillators to create a timekeeping device. The pendulum clock was one of the first timekeeping devices to work with reasonable accuracy.These clocks ranged in size from small boxes to giant towers, the precursors to structures like **Big Ben**.

The key innovation was the anchor escapement mechanism, developed by Robert Hooke around 1657 and perfected by clockmaker Joseph Knibb in 1670. This was the linchpin in a system to convert the motion of a mechanical oscillator into the movements of the clock's hands. The oscillator could be a swinging pendulum (typically seen in a wall clock or grandfather clock) or a wheel that rotates back and forth (typically seen in a watch).

The engineer needs to accomplish two things with the pendulum. First, enough energy—in the form of a little push during each swing—needs to be added to overcome the pendulum's loss of energy to things like air resistance. This keeps the pendulum swinging indefinitely. Second, each pendulum stroke must convert to the correct angular movement of the second hand. The mechanism that accomplishes these two things is called an escapement, and it works in conjunction with a spring or a falling weight to obtain the energy to push the pendulum and move the second hand.

Having arranged a pendulum, a source of energy, an escapement, and a second hand, then the rest of the clock is just appropriate down-gearing to spin the minute and hour hands at their correct rates. A mechanical clock is born. Every mechanical clock you have ever seen is simply an engineer's creative rendering of the four fundamental parts—oscillator, energy source, escapement, and gearing—to get the clock's hands to move properly.

SEE ALSO Big Ben (1858), Atomic Clock (1949), Atomic Clock Radio Station (1962), Global Positioning System (GPS) (1994).

The mechanical pendulum clock, pictured here, was developed based on a number of innovations.

1750

Simple Machines at Yates Mill

In mechanical engineering there is the concept of a simple machine. As classically defined, there are six simple machines: the lever, the ramp, the wedge, the wheel and axle, the pulley, and the screw. The idea is either to change the direction of a force, or to provide a multiplier. With a lever, for example a crowbar, the long end moves a large distance with a smaller amount of force, while the short end moves a short distance with a lot of force.

Mechanical engineers often use other simple devices to change direction or multiply: mechanisms like gears and gear trains, wheels with belts and chains, cranks and cams. Then there are springs and weights to store energy, motors and engines to add energy, etc.

One place where all of these devices combine together in a visible, visceral way is the traditional water-powered gristmill. These mills dotted the American countryside in the 1700s and 1800s. Wake County, North Carolina, once had 70 such mills, one of which (Yates Mill) first opened in 1750 and is still in operation as a museum today.

These mills used rotating millstones for grinding grain into flour. A wagon would pull up to the mill, using wheels and quite likely a ramp, to unload bags of grain. A rope and pulley would unload the bags and hoist them into a hopper. The power source for the mill was falling water, translated into rotational energy by a vertical overshot waterwheel. The waterwheel's rotating shaft would turn a large cogwheel, which would engage a small cogwheel. This wooden gear would shift the rotational force from horizontal to vertical and increase the vertical shaft speed.

A belt system might come off one of the shafts to power things like an Archimedes screw or conveyor belt. A crank system or cam system might help create back and forth motion to clean the grain or sift the flour.

Think about any mechanical system you see today created by engineers: a **car's engine**, a **sewing machine**, a **clock**, etc. They're the same basic elements applied over and over again in myriad different ways. Simple machines form the foundation of mechanical engineering.

SEE ALSO Waterwheel (100 BCE), Mechanical Pendulum Clock (1670), Sewing Machine (1846), Tower Crane (1949), Nitrous Oxide Engine (1978).

Diagram of six different mechanical systems, with arrows showing the direction of the forces involved.

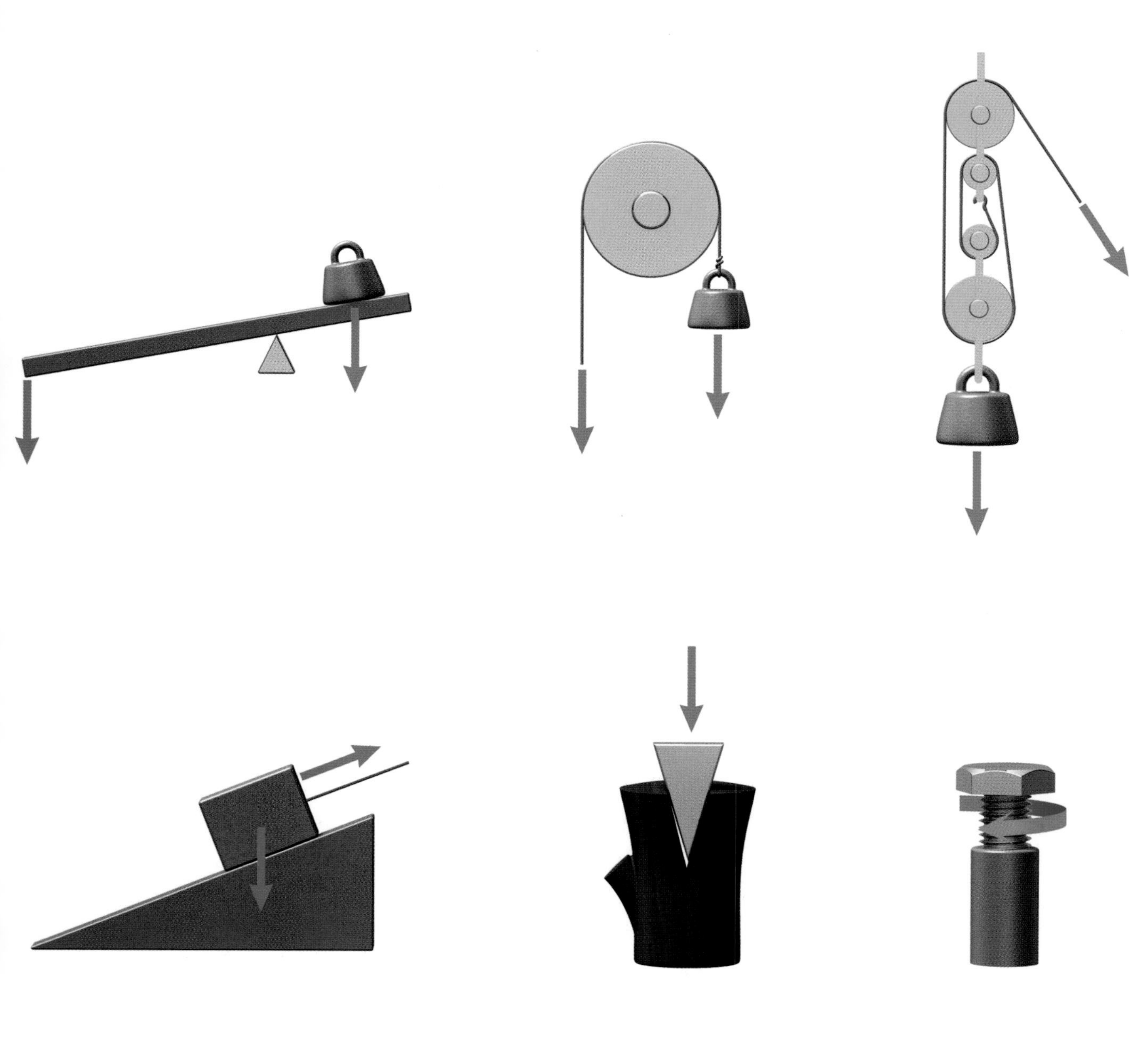

Building Implosions

There is a funny thing about building implosions. On the one hand we have a group of architects and engineers who originally designed a building to stay standing, even in the case of catastrophic events like hurricanes, fires, and earthquakes (hence modern innovations such as **earthquake-safe buildings**). With a building implosion, another set of engineers must defeat all of that hard work and bring the structure down as efficiently and safely as possible. The goal is for the building to fall straight down and land in a pile roughly the size of the building's foundation.

So how do they do it? Initial efforts were crude—massive explosions leveled the building, as in the 1773 destruction of Holy Trinity Cathedral in Waterford, Ireland. Today, this simplistic approach is frowned on because large explosions cause collateral damage to other nearby buildings.

In a modern building implosion, some of the exterior of the original building will be removed manually because it has value. Remaining exterior walls may be perforated or removed entirely. Interior structural beams that keep the building standing will be exposed. And this is where the real engineering comes in.

The easiest thing would be to simply cut all the beams and let the building fall. However, this approach is unsafe and would probably, because of the timing, cause the building to fall over rather than collapse in a pile. Instead, engineers carefully analyze the structure and the loads to understand how the building needs to collapse.

Some of the beams are partially cut to weaken them. Then explosives are attached to the columns at carefully calibrated positions. The explosives are called "shaped charges," which focus the explosive force in specific directions. The effect is to cut the **steel** support beams precisely where required. The shaped charges also reduce the total amount of explosives needed. This is important to demolition engineers because large explosions can damage adjacent buildings and infrastructure, for example by breaking windows.

Once all of the charges are set, they are wired back to a controller. The explosions throughout the building are carefully ordered and timed so that they occur in the correct sequence. Gravity does the rest and the building collapses.

SEE ALSO Basilica of Saint Denis (1144), Woolworth Building (1913), World Trade Center (1973), Earthquake-Safe Buildings (2009).

These photos demonstrate the process of building implosion.

1784

Power Loom

Edmund Cartwright (1743–1823)

In the early 1800s, the Industrial Revolution begins what will become the textile industry. Cotton production is rising fast because the combination of **cotton gins** and **cotton mills** has fundamentally changed the cotton marketplace. A process that used to take dozens of human hours to turn a pound of cotton into a pound of thread has been largely mechanized.

But the thread still needs to be woven into cloth, and that is being done by people sitting at manual looms. Things do get better in 1733 with the invention of the flying shuttle loom. But what the world needs is a completely mechanized loom to cut down on the human labor involved in making cloth.

The power loom, developed by Edmund Cartwright in 1784, was the invention that mechanized the production of cloth. When this important piece fell into place, the availability of cloth rose and the cost fell because factories could churn out much larger quantities. Engineer/inventors brought the power loom into existence, and then they rapidly improved the speed, reliability, and manufacturability of these machines.

Because of a campaign of secrecy in Britain, the invention of the power loom there did not make it to the United States until around 1814, when a man named Francis Lowell watched and memorized the operation of a power loom in the UK and then brought what he learned back to the US. This breakthrough allowed two things to happen: The first weaving factories sprang up, and engineers could see a working model of a power loom and start rapidly improving it.

How big of an effect did this industrialization have? One source notes that, right before the start of the Civil War, the state of Georgia—just one state—was producing something like 26 million yards of cloth per year from 33 factories. The North was producing far more because the textile industry started and was centered there. Between the time of the Revolutionary War and the Civil War, engineers and industrialists had brought textiles from a cottage product made completely by hand to a factory-made commodity manufactured completely by machines.

SEE ALSO Cotton Mill (1790), Cotton Gin (1794), Mass Production (1845), Sewing Machine (1846).

An 1836 plate depicting power-loom weaving at a mill.

1790

Cotton Mill

At the time of the American Revolution, every part of the production process for creating textiles was done by hand. People, in many cases slaves, planted the cotton, cultivated the cotton by hoeing it, and then picked the cotton one boll at a time. People pulled the seeds out of the cotton, spun the cotton on manual spinning wheels, and wove the cotton into cloth on hand looms. Then people cut and sewed the fabric by hand.

All of that began to change in America in the late 1700s. The first successful cotton mill opened in 1790 in Pawtucket, Rhode Island. At this site, the Industrial Revolution in the United States began.

If you have ever watched a human being spinning cotton into thread on an old-fashioned foot-operated spinning wheel, you know that it is a slow, tedious process. It starts with carding, which aligns the cotton fibers. Then the spinning wheel twists the fibers together to form thread or yarn.

To make the textile industry possible, this whole spinning process needed to become mechanized. The Pawtucket cotton mill handled the process of carding and spinning with machines.

The important machine was the spinning frame, which performed the actual act of spinning. But for this machine to work efficiently, it needed rovings, which were long, slightly twisted strands of fibers about the size of a fat pencil. To form the rovings, a series of machines carded the fibers, made them into slivers, combined the slivers, and sent them to a drawing machine. The basic idea was to create a consistent product rugged enough to feed continuously into the spinning frame without the rovings coming apart or jamming.

As you might imagine, this whole mechanization idea created a field day for inventors and mechanical engineers. In the same way that the **Wright brothers** demonstrated a working airplane and then aviation exploded, this first example of mechanization started a revolution in textiles.

SEE ALSO Power Loom (1784), Cotton Gin (1794), The Wright Brother's Airplane (1903).

Slater Mill, pictured, was the first water-powered cotton-spinning mill in America.

1794

Cotton Gin

Eli Whitney (1765–1825)

You can imagine what happened with the advent of the first American **cotton mill**—the demand for cotton skyrocketed. But the problem with cotton was the seeds.

If you were to walk into a cotton field at harvest time and pick a ripe cotton boll, it is nothing like the pure, white cotton balls you buy at the drug store. The reason a cotton plant exists is to make and spread its seeds, and the seeds are embedded in the cotton fibers. The cotton fibers are stuck to the seeds. Removing these seeds was something people did by hand at a rate of about one pound of cotton fibers per day. It was a common task for children and slaves, and it made cotton expensive.

That changed with the patented invention of the cotton gin in 1794 by Eli Whitney. The cotton gin is a perfect example of how one machine can make a huge difference to an entire industry. With the cotton gin, the raw cotton sits in a bin. A set of toothed wheels spins through the raw cotton, catching and tugging at the cotton fibers to separate them from the seeds. Then a set of spinning brushes pulls the seedless fibers off the teeth.

The original cotton gin was a small, hand-cranked box that made seed removal trivially easy and fast. Engineers scaled up the process, refined it, and mechanized it further to create a factory process that could feed clean cotton fiber to the proliferating cotton mills.

The cotton gin removed a major roadblock and expense in the cotton manufacturing process, and cotton use exploded. It is a great example of a major productivity breakthrough fostered by engineering. Once invented, the cotton gin rapidly evolved into an industrial-scale machine that removed human labor almost entirely from the ginning process. The same reduction in labor would happen in carding, spinning, and weaving, and also in planting, cultivating, and harvesting the plants. Today, through decades of engineering refinement, a handful of people can do what once took a crowd.

SEE ALSO Power Loom (1784), Cotton Mill (1790), Sewing Machine (1846), Drip Irrigation (1964).

Eli Whitney's patent for the cotton gin, March 14, 1794.

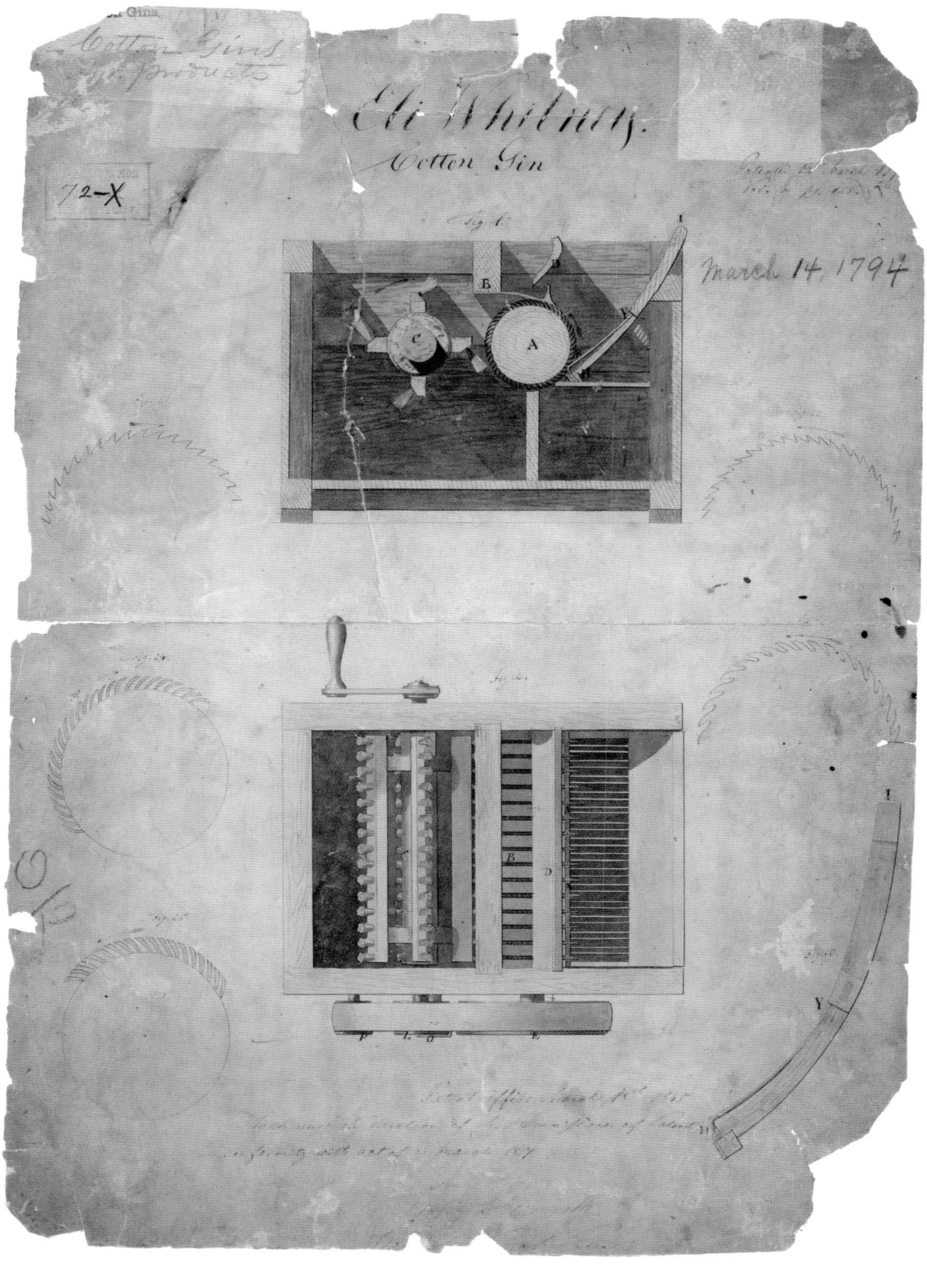

Cotton Gins

Eli Whitney.

Cotton Gin

72-X

March 14, 1794

1800

High-Pressure Steam Engine

Richard Trevithick (1771–1833)

There was a time in history when the human body was the only way to power things. Then we learned to harness horses and oxen. Then we figured out how to use water for power with **waterwheels**. But all these sources of power have their limitations. You cannot create a locomotive or a cruise ship like the ***Titanic*** with any of these power sources. And while you can create a power plant or a factory powered by water, you are severely limited as to where you can locate them. The world needed a better source of power.

The steam engine provided the transition to the industrial age. The first high-pressure steam engine was introduced in 1800 by British engineer Richard Trevithick. By 1850, engineers had incrementally improved steam engines and the Corliss steam engine became the state of the art for large stationary power needs. It was efficient and reliable, as well as large and heavy, making it a good engine for powering factories. The San Francisco **cable car** system used steam engines of this type.

The engine used to power the Centennial Exhibition in Philadelphia in 1876 is an example: a two-cylinder steam engine producing 1,400 horsepower (one million watts). Pistons more than a yard (one meter) in diameter moved 10 feet (3 meters) in their cylinders to spin a flywheel 30 feet (9 meters) across.

The *Titanic* used the next generation of steam engine, in which multiple cylinders captured energy from successive expansions of the same steam.

A key element for any high-pressure steam engine is the boiler, where boiling water creates the steam pressure. The problem with boilers is that, being under high pressure, they had some probability of exploding. One of the most horrific boiler explosions occurred aboard a steam-powered ship named the Sultana in 1865. It had four boilers, one of which had started leaking and had been hastily repaired. With roughly 2,000 people on board, the repaired area presumably failed, causing an immense boiler explosion that killed a total of about 1,800 people. Today engineers spec steam turbines instead. You find them in nearly every **power plant**.

SEE ALSO Waterwheel (100 BCE), Cable Cars (1873), Steam Turbine (1890), *Titanic* (1912), International Thermonuclear Experimental Reactor (ITER) (1985).

President Ulysses S. Grant and Don Pedro starting the Corliss engine at the Centennial celebration, Philadelphia, 1876.

1823

Truss Bridge

If we could get inside the brain of an engineer and look at the core values driving the thought process, one of the values near the top of the list would be *efficiency*. Engineers are interested in efficiency in everything they do. If an engineer is building something, then that value expresses itself in the efficiency of materials. Excess materials add weight and increase the cost.

A truss makes very efficient use of materials to span a gap. It is mostly air—a collection of open triangles engineered for great strength.

The oldest covered bridge in the United States, and therefore the oldest bridge truss, is the Hyde Hall Covered Bridge from 1823, located in upstate New York. The bridge is 53 feet (16 meters) long and it is made using two wooden trusses, one on either side of the bridge. A roof over the top protects the wood from the elements.

Trusses got their start in roof structures dating back to the Roman Empire, and it is easy to understand why. If you are trying to span the walls of a **cathedral** or a large room, the ceiling joists need to connect the two walls. But wooden joists, no matter how massive, start to sag under their own weight at about a 40-foot span. The solution is a kingpost truss, where a kingpost in the center of the joist ties into the peak of the rafters, supporting the joist. This is the basic idea behind any truss—using different pieces of the truss to support other members through tension or compression. It takes far less material compared to a solid beam of the same dimensions.

This is why we see trusses everywhere in the modern world: in bridges like the **Golden Gate Bridge**, buildings like the **World Trade Center**, **tower cranes**, power line towers, etc. Engineered properly, a truss is far less expensive and lighter than a solid beam of the same size.

SEE ALSO Basilica of Saint Denis (1144), Engineered Lumber (1905), Woolworth Building (1913), Golden Gate Bridge (1937), Tacoma Narrows Bridge (1940), Tower Crane (1949), World Trade Center (1973).

This photo, taken around 1862, depicts a train on a truss bridge.

1824

Rensselaer Polytechnic Institute

Stephen van Rensselaer (1764–1839), **Amos Eaton** (1776–1842)

In today's world, all engineers come from colleges and universities that offer bachelor's degrees in engineering. In almost all cases in the United States, these degree programs have been accredited by ABET, also known as the Accreditation Board for Engineering and Technology. If someone wishes to become a **professional engineer** then ABET accreditation is required.

There are thousands of engineering degree programs in the US offered by hundreds of institutions. But in 1823 there were none. The first school specifically designed to train engineers was Rensselaer Polytechnic Institute located in Troy, NY.

When the university opened, enrollment was tiny (only ten students in 1825). The first degrees, all in civil engineering, were granted in 1835. Even in 1850 the enrollment was only on the order of fifty students. But the school's civil engineering graduates were quite influential. For example, Theodore Judah, one of RPI's first graduates, served as the chief engineer for several railroads and was an important contributor on the **transcontinental railroad**.

Why was the first engineering school in the United States located in Troy, NY? There are several reasons, but one important factor was Troy's prominence as an early industrial center for the nation. In a way, Troy could be thought of as the Silicon Valley of its day. Located on the Hudson River, Troy served as an important transportation hub. When the **Erie Canal** connected into the Hudson River near Troy in 1825, the location's importance skyrocketed. Because of the steep cliffs along the Hudson River in the area and the many streams flowing over them, Troy and surrounding towns were great places to build water-powered factories. Troy was also an important center for **steel** production. All of this industrial activity and prosperity made Troy a good place to locate an engineering school. Technology-minded people, innovation-minded people, and industrial-minded people were attracted to Troy.

Today, engineering schools in the US graduate 80,000 to 90,000 new engineers each year.

SEE ALSO Gunter's Chain (1620), Cotton Mill (1790), Erie Canal (1825), Mass Production (1845), Professional Engineer Licensing (1907), Women's Engineering Society (1919).

An 1876 engraving of the original Rensselaer Polytechnic Institute. The building on the left, Winslow Chemical Laboratory, still stands, though none of the other buildings do.

POLYTECHNIC INSTITUTE, TROY, N.Y.

1825

Erie Canal

Benjamin Wright (1770–1842)

If you were a merchant in 1800, and you did not live near an easily navigable river, your transportation options were few. If there was a road, you could move your goods in a horse-drawn or ox-drawn wagon. If not, you strapped your goods onto a pack animal. There were no railroads yet so moving goods was a huge problem. Canals were coming into wide use in England and Holland. But America faced a problem—very few engineers. They trained in England, then returned to the States. A canal project definitely needs trained civil engineers for every aspect of design and construction.

A canal is a long, gently sloping waterway punctuated by locks that handle significant grade changes. The water level along the entire canal has to be maintained using surrounding water sources like rivers and lakes. Gravity does all the work of moving the water in early canals. If it's not done right, the canal dries up or floods.

The Erie Canal, overseen by principal engineer Benjamin Wright, was a monumental achievement for the time. It stretched from Albany, NY, all the way to Buffalo, NY, 360 miles (580 km) and 36 locks away. It connected the Hudson River, and therefore New York City, to Lake Erie. It was possible to get all the way to Toledo, Ohio, because Lake Erie is about 150 miles (240 km) long.

Once the canal was completed in 1825, it created a transformation. A canal boat could carry 60,000 pounds (27,000 kg) of freight. Therefore, the cost of moving a ton of freight fell rapidly. It might have originally cost $100 to $120 to move a ton of freight the distance of the canal. The Erie Canal dropped that price below five dollars. Moving tons of wheat or piles of logs to market suddenly became affordable. It was revolutionary.

The Erie Canal project had an even greater importance for the profession: it became a school of engineering for many people. It is no coincidence that **Rensselaer Polytechnic Institute**, the nation's first engineering college, opened just a few miles from the Albany, NY, end of the Erie Canal in 1824, one year before the completion of the project.

SEE ALSO Gunter's Chain (1620), Rensselaer Polytechnic Institute (1824), Steam Turbine (1890), Panama Canal (1914).

Erie Canal at Salina Street, Syracuse, circa 1904.

1830

Tom Thumb Steam Locomotive

The steam locomotive changed the course of civilization. For the first time, people could build efficient transportation systems without water and ships. Trains were land barges that could go anywhere tracks could take them. Compared to digging a canal, tracks were incredibly inexpensive and versatile. Tracks could go over mountains, across deserts, through tunnels—places where canals could never go.

The steam locomotive got a modest start in the United States in 1830 with the Tom Thumb. The steam-powered piston engine on the Tom Thumb generated only 1.4 horsepower (1,000 watts) by burning coal in a small boiler, but it was enough to carry a car full of passengers at 18 mph (29 kph) over 13 miles of track.

One hundred and eleven years later, engineers had brought the steam locomotive to its zenith with the Big Boy engine for the Union Pacific railroad. Big Boy was utterly gigantic, weighing over a million pounds with its required tender carrying coal and water, and measuring 85 feet (26 meters) long. It could develop 6,000 horsepower (4.5 million watts) and had two separate sets of drive wheels powered by their own cylinders. Twenty-five Big Boy locomotives were built and each one traveled an average of one million miles before **diesel locomotives** replaced them.

In between, there was the classic steam engine that you would see in old westerns—the kind with the cow pusher on the front, the big funnel smokestack, and the cab in the back for the engineers. This is the kind of steam engine you see in the golden spike photo for the **transcontinental railroad**. Jupiter was built in 1868 and used a wood-fired boiler. It remained in service for 41 years.

Steam locomotives made the coast-to-coast movement of freight possible. They also made it possible to travel from New York to San Francisco in just a week and for thousands of towns to spring up along the railroads. They played a huge role in the Civil War, moving soldiers and materiel. Engineers built a whole new transportation modality that changed the course of history.

SEE ALSO Square-Rigged Wooden Sailboats (1492), Truss Bridge (1823), Erie Canal (1825), Transcontinental Railroad (1869), Diesel Locomotive (1897), Panama Canal (1914), Bullet Train (1964).

Later steam locomotives like this one were developed based on the Tom Thumb, America's first.

1835

Combine Harvester

Hiram Moore (1801–1875)

If you look back at the US employment statistics for the late 1700s, about 90 percent of the workforce was employed in agriculture. Nearly the entire population spent their time growing food. Then engineers got involved in food production, and agriculture took off in terms of efficiency. The combine harvester, which was first developed by Hiram Moore in 1835, was one huge innovation that led the charge. Today, America produces all the food it needs (and then some) with about 1 percent of the workforce. Chemical engineers produce pesticides and fertilizers. Mechanical engineers produce machinery and tools. We produce far more food per acre of land and per person than ever before.

Harvesting grain used to be labor intensive and time consuming. People with scythes would cut a swath of wheat or oats and tie the stalks into sheaves. The sheaves would be transported to a threshing floor where feet or flails would separate the grain from the stalks. Then winnowing would separate the wheat from the chaff. Harvesting a big field could take dozens of people many days.

A modern combine does all this work quickly and efficiently in one step. The machine has a cutting bar up front. The stalks are ingested into the body of the machine on a conveyor. Inside the body, a threshing drum separates the grain, and a series of sieves and fans winnows the grain. The straw and chaff fall back to the field while the grain moves to a holding tank. A truck pulls up and offloads the grain from the tank periodically. A big combine harvester like this with a single driver can handle one hundred or more acres per day of grain, beans, or corn. A person with a scythe might do one acre a day, with several people following behind to make the sheaves. The work of transporting, threshing, and winnowing took even more people.

That improvement in human productivity shows the power of engineering. It drastically reduces costs. Similar improvements in plowing, planting, cultivating, and fertilizing, not to mention **center-pivot irrigation** and, later, **drip irrigation**, make today's farmers incredibly efficient.

SEE ALSO Hunter/Gatherer Tools (3300 BCE), Center-Pivot Irrigation (1952), Green Revolution (1961), Drip Irrigation (1964).

This modern combine owes its existence to Hiram Moore's original design.

1837

Telegraph System

Charles Wheatstone (1802–1875), **William Fothergill Cook** (1806–1879)

When we think of a telegraph system, we probably think of a person in an office tapping Morse code messages on a key, and receiving messages with a clicking metal bar. This arrangement, developed by English inventor William Fothergill Cook and scientist Charles Wheatstone in 1837, was the first telegraph implementation to be put into commercial service.

Many arrangements came before and after, but this one dominated for several reasons. Most importantly, it was incredibly simple. All you needed at each end was a key—essentially a switch—a sounder, an electromagnet that makes clicking noises, a single wire, and a battery. The earth was used as a second wire to complete the battery circuit. That simplicity meant it did not cost much to set up, and it was extremely reliable.

Once the basic system was in place, networks rapidly developed. The single wire had to go somewhere, and poles with glass insulators were the preferred place because they were inexpensive and easy to build. The poles went up along railroad tracks because it was an easy place to put them. Most train stations therefore had a telegraph office, and anyone in a town with a telegraph station could communicate with the rest of the world.

Imagine what would happen to a civilization that suddenly, for the first time, had the easy ability to communicate over long distances. A message that might have taken several days or a week to get through by letter on horseback could now get through in a minute.

During the Civil War, for example, the telegraph was a huge game changer for the North because messages could get to and from many battlefields almost instantly. President Lincoln himself could be found in the telegraph office getting instant information. It was much easier to move troops and supplies around with good communications in place.

Engineers found ways to insulate wires with gutta-percha and **undersea telegraph cables** soon followed. Engineers shrank the world.

SEE ALSO Transcontinental Railroad (1869), Telephone (1876), Radio Station (1920), TAT-1 Undersea Cable (1956), ARPANET (1969).

The telegraph represented an unprecedented ability to communicate over long distances.

1845

Mass Production

Cost reduction is one of the key improvements that engineers offer society. Engineers take complicated, expensive processes and devices and make them affordable. One way they do that is through mass production.

If you go to the Williamsburg historical area in Virginia, you can see how gunsmiths made guns prior to the Industrial Revolution. They made each part—components like triggers, barrels, springs, and frizzens—and then fit them together by hand. Even the screws were made by hand.

During the Industrial Revolution, engineers worked out three big changes in the status quo: 1) Making standardized, interchangeable parts, 2) separating tasks into repeatable stations along an assembly line, also known as the division of labor, and 3) using machines to do as much of the work as possible. One of the first places where these three innovations came together was in a gun factory at Harper's Ferry, Virginia, starting in 1845.

In this factory, machines did a great deal of the work. For example, one especially interesting type of machinery allowed for the automated production of gun stocks from wood. Instead of a person carving the stock by hand and then custom fitting it around the other parts of the gun, each stock was identical and interchangeable, and it took very little direct human effort to make. That same mentality was applied to barrels, triggers, springs, etc.

The cost of guns fell as the amount of human labor fell. In addition, it was no longer necessary to send gunsmiths into the field to repair handmade guns. Anyone could remove a broken piece and replace it with an interchangeable part.

Mass production of this type became known as the American system of manufacturing, and from the gun industry it spread to many other industries. It acted as the foundation upon which Fordism was built to produce the Model T.

Today nearly everything we use is mass-produced. Look around you—there may not be a single handmade object in your line of sight. Everything from your shoes to your **smart phone** comes from a factory. That's because factories bring down the cost of production to make more and more things affordable.

SEE ALSO Cotton Mill (1790), Top-Loading Washing Machine (1946), AK-47 (1947), Smart Phone (2007).

This photo shows mass airplane production at work.

240205
240209

1845

Tunnel Boring Machine

Henri-Joseph Maus (1808–1893)

The tunnel boring machine is one of the most sophisticated, largest machines engineers have developed. First introduced by Belgian engineer Henri-Joseph Maus in 1845, this early TBM, which was enormous and consisted of over 1,000 percussion drills, was used to bore through the Alps in order to connect France and Italy.

Modern TBMs are more streamlined than Maus's "Mountain Slicer," as it was called. At the front of the machine is a massive rotating disk the size of the tunnel's bore. This disk is called the cutting head. In the case of the two main tunnels for the **Channel Tunnel** between France and England, the bore was 25 feet (7.6 meters) in diameter. The rotating disk contains knives and scrapers that press against rock or compressed earth and chip it away. The debris created comes through holes in the disk to a collection area and then it moves out of the tunnel on a conveyor.

Behind this cutting head is a shield the diameter of the tunnel's bore. Inside the shield is a piece of equipment that installs the concrete wall segments. A new section of wall might be a ring that is 3 feet (1 meter) wide and consists of seven to ten semicircular segments. In one type of boring machine, a set of hydraulic rams pushes against the last ring installed in order to apply pressure to the cutting head. Periodically the cutting head stops, the rams retract, and a mechanized arm puts each wall segment in place. Once the complete ring is finished, the hydraulic rams move back into position against the new wall section so the boring machine can move forward again. A TBM typically injects grout between the concrete wall it forms and the bare tunnel wall to seal the whole tunnel.

The TBM lets engineers bore through the earth creating an extremely strong and durable tunnel as they go. The machine is steerable so that it can move in a left/right or up/down direction. All of the roads and buildings on the surface are completely undisturbed.

SEE ALSO Channel Tunnel (1994), Large Hadron Collider (1998).

Construction worker looking at a tunnel boring machine cutter head.

ARM 5
30
23
25
27
29
31
33

1846

Sewing Machine

Elias Howe (1819–1867)

By the 1850s, the textile industry had been revolutionized. The **cotton gin** plus **cotton mills** plus factories filled with **power looms** transformed the production of textiles—the process was thoroughly automated.

But the thing that had not changed was sewing. People with needles and thread still sewed garments together by hand. That was transformed with the advent of the lockstitch sewing machine, which appeared in the United States in an 1846 patent from inventor Elias Howe and then rapidly improved through the efforts of many inventors and engineers in the 1850s.

The key technology that had to be figured out was a machine that could produce a lockstitch—a seam that would not come undone if a person pulled on the thread. Singer sewing machines were the first to do this at a commercial scale in the 1850s. The mechanism is as ingenious now as it was then. A needle with thread pokes through the fabric from the top. This forms a loop of thread below. Underneath the fabric, a hook catches the loop and then wraps it around a bobbin thread. Once released by the hook, the upper needle cinches up the loop and locks the stitch. A sewing machine can produce long, straight, strong seams ten times or more faster than a seamstress or tailor could by hand.

All of this technology came together with one other innovation at the time of the Civil War—standardized clothing sizes, which allowed for standard items of apparel to be **mass-produced** in factories.

Here is a way to understand the change that occurred in the marketplace. In the Revolutionary War, the Americans did not have a standard uniform. Soldiers brought their own clothes to the battlefield. There was no way for the army to afford uniforms because clothing was rare and expensive. By the Civil War, a typical soldier would receive from the army three pairs of pants, three shirts, two coats, a great coat, and various undergarments and socks. This was a cornucopia of clothing, all in standardized sizing, and it was made possible by the industrial revolution in textiles. Engineers and inventors had a massive effect on this marketplace.

SEE ALSO Power Loom (1784), Cotton Mill (1790), Cotton Gin (1794), Mass Production (1845).

An 1892 advertisement for Singer sewing machines, showing people in national costume using treadle machines.

CARD No 2
ALL NATIONS
USE
SINGER SEWING
MACHINES.
ITALY.
SPAIN (VALENCIA)
TUNIS
SPAIN.
MANILA.
JAPAN.
BURMAH
SERVIA
BOSNIA (AUSTRIA-HUNGARY)
NORWAY
INDIA.
SWEDEN.
SPAIN
SPAIN.
PORTUGAL (VIANNA)
ITALY (ANCONA)
ROUMANIA.

America's Cup

Imagine a race in which engineers start with a blank sheet of paper and can let their imaginations run wild. This is the essence of the America's Cup yacht race, which first began in 1851. Yes, there are rules, but within those rules there is a great deal of latitude for engineers. And the designs that appear from such freedom have been revolutionary.

For example, the latest boats can exceed 45 mph (72 kph) on open water. Boat speeds can at times double wind speeds: if the wind is blowing at 20 mph, the boat can go 40 mph. Their sails can be 17 stories tall and rigid. The boats can be quite large—86 feet (26 meters) long by 46 feet (14 meters) wide—and able to hold crews of 11.

How is this even possible? How can a boat powered by the wind travel at 45 mph? The answer lies in multidisciplinary engineering. For example, to keep the boats light and strong they are composed almost entirely of **carbon fiber** and **titanium**. To reduce drag, the boats can hydrofoil using small carbon fiber underwater wings that lift the boat out of the water.

The "sails" represent a huge innovation. Imagine removing the wing from a 747 and mounting it vertically on the mast. That is the size of these rigid sails. They are split in the middle, hinged into essentially two wings that can flex in the center, something like a giant flap to change the full wing's shape. When the wind flows at the correct angle, it is not pushing as it would a traditional sail. Instead it is creating horizontal lift over the upright wing. The lift translates into forward motion.

Between the light weight of the boats, the winglike sails, the hydrofoils, and the well-trained crews, the speeds of these boats have surpassed even the most optimistic expectations.

The size of the sail and the stresses on the foils demand hydraulics. But electric pumps are against the rules. So you will see crew members furiously turning cranks throughout the race. They are the hydraulic pumps for the boat.

SEE ALSO Square-Rigged Wooden Sailboats (1492), Carbon Fiber (1879), Hall-Héroult Process (1889), Titanium (1940), Boeing 747 Jumbo Jet (1968).

The America's Cup, started in 1851, was one of the first international yacht races.

1854

Water Treatment

John Snow (1813–1858)

Millions of people wake up every morning, and one of the very first things they need is water. In America, consumption averages 80 gallons (300 liters) per person per day. Every municipality across the country has engineers who work to make that possible at the water treatment plant.

In many municipalities, the water starts in a reservoir or man-made lake. The municipality taps into the lake with a large inlet pipe, and the water contains two contaminants that need to be removed: 1) particles like soil, leaves, fish waste, etc., and 2) various types of bacteria and parasites. Thanks to English physician John Snow's study of the 1854 cholera outbreak, we now know that in order to prevent illness, we need to remove these contaminants. A quick burst of chlorine at the intake will kill the algae and bacteria and turn them into particles. The next step is to remove all the debris.

One common approach: add alum and other coagulants. Once in the water, alum molecules have a charge that attracts the particles. This process is called flocculation and produces flocs.

Engineers create large, still, settling areas for the water to pass through. This gives the flocs time to sink to the bottom. Next the water flows though large sand filters, like the sand filters found in swimming pools but at a much larger scale. These filters will catch any remaining particles, any remaining parasites, and many of the remaining bacteria. The final step is sterilization. Three techniques have become popular: chlorine, ozone, and ultraviolet light. Some people do not like the taste of chlorine, but the advantage is that it stays in the water as it moves through the pipes and it is easy to filter out once it arrives in the home. The study of water treatment has lead to other ways of making potable water out of potentially contaminated source material, such as **desalination**.

SEE ALSO Modern Sewer System (1859), Desalination (1959).

A pool for water treatment is pictured here.

1855

Bessemer Process

Henry Bessemer (1813–1898)

In the Iron Age, the prevalent use of iron implements changed the world. However, an equally revolutionary shift occurred after English engineer Henry Bessemer developed a process to refine iron and produce steel commercially, first patented in 1855.

Where does steel come from? Start with iron. Dig iron ore out of the ground and transform it into iron in a blast furnace. It comes out as pig iron with a carbon content of 5 percent or so. Put pig iron into a basic oxygen furnace to form steel. Pure oxygen blasts in under pressure and burns off much of the carbon, leaving behind 0.1 percent carbon (mild steel) to 1.25 percent carbon (high-carbon steel). The carbon content, plus any alloying metals, plus the quenching process, determine the properties of the steel in use.

Steel is a remarkable material: dependably strong and resistant to fatigue, it is also workable and highly mutable—it can take a number of different forms. For example, if you heat up steel and quench (cool) it one way, it is more ductile. Quench it another way and it is much harder and more brittle. It is even possible to get both effects at the same time with *case hardening*. The outer shell is hard and therefore difficult to cut, while the interior is softer to combat the brittleness.

Then there are the alloys. Add a little chromium to steel and it won't rust. Add extra carbon and it becomes much harder. Add tungsten or molybdenum and you get tool steels. Add vanadium and it holds up better to wear.

This combination of advantages explains why steel is so ubiquitous. Engineers use steel in car bodies and **engines** because of the combination of strength, cost, and durability. Engineers use steel in skyscrapers such as **Burj Khalifa** for the same reasons. Big bridges such as the **Millau Viaduct**, same thing. They reinforce **concrete** with steel to greatly improve its strength in tension. One place where steel isn't found is where weight is a factor—aluminum or carbon fiber is used there. Or where strength and durability isn't a big factor and cost is—**plastic** is used there.

SEE ALSO Concrete (1400 BCE), Plastic (1856), Hall-Héroult Process (1889), Nitrous Oxide Engine (1978), Millau Viaduct (2004), Burj Khalifa (2010).

White-hot steel pours like water from a 35-ton electric furnace, Allegheny Ludlum Steel Corp., Brackenridge, PA, circa 1941.

Plastic

Alexander Parkes (1813–1890)

Although humans have used naturally occurring plastics such as rubber and collagen for millennia, the first manmade plastic was Parkesine, patented by Alexander Parkes in 1856. Today, the amount of plastic that surrounds us is nearly indescribable. The affordability, malleability, and durability of plastic makes it an ideal material.

One reason for the widespread use of plastic is the work of chemical engineers, who have used mass-scale processes in factories to make the production of plastic so inexpensive. Another factor is the contribution of mechanical engineers and industrial engineers who design the parts and create the molding systems to produce shaped plastic objects. Plastic is also lightweight, strong for its weight, corrosion-free, easily molded, and it comes in so many forms with many different properties. While Parkesine, made of cellulose, was often used to create synthetic ivory, modern plastics are normally made of other, higher-quality components. Polyethylene is made of long chains of carbon and hydrogen atoms, so it is essentially solidified gasoline. The length of the chains, the amount of branching, and the amount of polymerization gives polyethylene its many different properties.

Engineers and scientists have worked together to create hundreds of other types of plastics as well. Some plastics form fibers that become soft cloth or pillow stuffing. Nylon fibers, produced by Wallace Carothers at Dupont in 1935, create a strong, abrasion-resistant fabric for parachutes, backpacks, and tents. **Kevlar** was developed later, and is strong enough for use in bulletproof vests. Some plastics are rubbery and they become seals, gaskets, O-rings, wheels, and grips. Some plastics are clear like glass; others are completely opaque. Some are so strong that they replicate the tensile strength of steel while being flexible and light. This diversity and versatility means that engineers can use plastics to make almost anything.

SEE ALSO Concrete (1400 BCE), Cotton Mill (1790), Wamsutta Oil Refinery (1861), Carbon Fiber (1879), Kevlar (1971), 3D Printer (1984).

Plastic toy blocks of this sort are often made out of ABS (acrylonitrile butadiene styrene) plastic.

1858

Big Ben

Benjamin Hall (1802–1867)

Big Ben is a clock tower in London that holds one of the largest chiming clocks in the world. In 1858, civil engineer Benjamin Hall was the Commissioner of Works, and he supervised the installation. What makes this clock so interesting from an engineering perspective is its massive scale, as well as, from a social standpoint, its function as a public fixture helping a rapidly industrializing city become conscious of time.

Big Ben's clock starts with a massive pendulum, 13 feet (4 meters) long, a scaled-up version of the original **mechanical pendulum clocks**. With this length, the period of the pendulum, or the time it takes to swing from right to left and back to the right, is 4 seconds. To precisely adjust this pendulum to account for temperature changes, pennies are added to the top of it. Each penny adjusts the clock by 0.4 seconds per day. The pendulum weighs 682 pounds (310 kg).

The pendulum controls the escapement. The escapement lets energy escape from a falling weight and allows the clock's gear train for the hands to move forward one increment. The escapement also supplies the energy that keeps the pendulum swinging. Big Ben was the first clock in the world to use the double three-legged gravity escapement, which was invented specifically for Big Ben and is particularly reliable.

The energy for the clock comes from a half-ton weight attached via a cable to a winding drum. Clock engineers wind the clock three times a week. The rotation from the escapement feeds into the clock's gear train for the hour and minute hands. The hands are immense; the minute hand is 14 feet (4.2 meters) long and weighs 220 pounds (100 kg). A single shaft rises from the clock's gear train to a gear that sends power down four shafts to the four clock faces.

There are two other weights and two other gear trains in the clock. One gear train controls the hammer for the main bell that counts the hours. The other gear train controls the hammers for the four smaller bells, which play the little song "Westminister Chimes" every quarter hour.

SEE ALSO Mechanical Pendulum Clock (1670), Atomic Clock (1949), Atomic Clock Radio Station (1962), LCD Screen (1970).

Big Ben is the third-tallest freestanding clock tower, and ushered in the Industrial Revolution by chiming for the entire city.

Domine Salvam fac Reginam nostram Victoriam Primam
Domine Salvam fac Reginam nostram Victoriam Primam

1859

Oil Well

Edwin Drake (1819–1880)

What would prompt someone to think that there were vast pools of oil underground? First, it was common knowledge that water sources existed underground, as people had been digging wells for centuries. Second, in some places, crude oil seeps to the surface. In 1859, near Titusville, PA, the first commercial attempt was made to drill for underground oil near an oil seep. Hired by the Seneca Oil Company, Edwin Drake used a rotating drill bit to bore down to bedrock. The hole was lined with cast-iron pipe to keep water from collapsing it. The drill went through bedrock and hit oil at 69 feet (22 meters) below ground. A pump brought the oil to the surface.

Today the process is similar, but with more environmental safeguards, regulations and engineering oversight. A drill bit on the end of a pipe, called the drill string, is rotated by a turntable on the drilling rig. A fluid known as drilling mud pumps down through the pipe to cool the drilling bit and force the debris out of the hole. When the drill hits a predetermined depth, or bedrock, steel casing is inserted into the hole. Then cement is pumped down the casing under high pressure so it exits the bottom of the casing and flows back up the hole on the outside of the casing toward the surface. Once set, this cement will lock the casing in place and protect any groundwater from the oil.

Then a smaller bit goes down inside this surface casing to reach the oil. More steel casing and then cement goes down the hole. At that point, the drilling equipment is removed, a wellhead is installed, and pumping can begin to extract the oil.

Today much of the economy depends on oil. Humans pump and refine about 85 million barrels a day, almost all of it sourced from oil wells, although in many cases **shipped** over long distances or conveyed via pipeline, as in the case of the **Trans-Alaska Pipeline**.

SEE ALSO Plastic (1856), Wamsutta Oil Refinery (1861), Trans-Alaska Pipeline (1977), Container Shipping (1984).

Oil rig in Titusville, Pennsylvania, circa 1900.

Joseph Bazalgette (1819–1891)

Since the time of the Roman Empire, there has been one key engineering marvel that has distinguished civilized cities from uncivilized ones: a sewer system to handle human wastewater. However, until the nineteenth century, many cities still dumped untreated sewage straight into waterways. This changed in 1858 because of the "Great Stink"—a citywide demand from residents of London that Parliament deal with widespread water contamination, and its resulting stench. Chief engineer Joseph Bazalgette, appointed by the Metropolitan Board of Works, proposed a sewerage system that was accepted and implemented beginning in 1859.

Bazalgette's sewer network has a lot in common with modern sewer systems. A complete sewer system consists of two parts: the pipes that carry sewage away from buildings, and the treatment plant that processes the sewage water before release into the environment.

While the freshwater supply in a city is almost always pressurized with pumps, sewage is almost always a gravity system. The sewage treatment plant is located as low as possible on the terrain, and all of the sewage falls toward it. When there is the occasional need to move sewage uphill, a lift station does the work.

There is a three-stage process at the sewage treatment plant. The first phase removes trash, lets heavy stuff settle out, and skims off grease and oil. The second stage uses an aerated pond or tank to let bacteria eat or precipitate everything it can out of the water. The water is allowed to settle again, and then it goes through a final filtration and disinfection step. In some cities, the effluent from the sewage treatment plant is so clean that it goes right back into the city's water supply.

Sewage treatment is one process in which engineers have a huge but hidden impact on every city. If the sewer system were to fail, the conditions would soon become atrocious. But because engineers build and keep the systems working almost flawlessly, we take them completely for granted.

SEE ALSO Pompeii (79), Water Treatment (1854), Frozen Pizza (1957), Drip Irrigation (1964).

1971

1860

Louisville Water Tower

Imagine that you are designing the water supply for a small city. You have to get water to every home and business. To do that, you might start by laying out water mains—the big pipes, many feet (meters) in diameter—that act as the backbone of the water system.

A set of pumps pressurizes the water from the treatment plant and injects it into the mains. On the other side of the mains is the distribution network that, through smaller and smaller pipes, eventually provides a connection to every building that needs water.

But what happens during a power failure or a pump failure? Every water system leaks, and during a power failure the system will depressurize. All of those leaks start letting pathogens into the water system, counteracting the effects of **water treatment**. And then a fire will start somewhere—it's part of Murphy's Law. But there will be no water for fire hydrants, which also connect into the water system.

The engineering solution to both of these problems is the water tower. It is simply a big storage tank raised into the air on some kind of tower structure. The tank might hold a million gallons or more. The operation is extremely simple—a large pipe connects from the water main into the bottom of the tower. Pressurized water rises up the pipe and fills the tank.

If a power failure occurs now, gravity keeps the system pressurized. If a water tower is 200 feet (61 meters) tall, gravity will create 86 psi (592 kPa) of pressure from the water in the tank. As long as the tank has water, the system will have pressure. The tank can be sized to provide the community with hours of water so that pumps can be serviced, pipes repaired, etc.

Water towers also help to handle surges, for instance, in the morning, when everyone in town is taking a shower at roughly the same time. The pumps that pressurize the system don't have to be quite as big. Louisville, KY, installed one of the country's first water towers in 1860, and thousands have been built since.

SEE ALSO Roman Aqueduct System (312 BCE), Water Treatment (1854), Modern Sewer System (1859), Drip Irrigation (1964).

Louisville water tower, one of the remaining standpipe water towers in the United States.

Wamsutta Oil Refinery

When crude oil comes out of the ground, it is a fairly disgusting and useless fluid. But once chemical engineers get involved and run crude oil through a refinery, a wide range of useful products appear. One of the first refineries in the United States was the Wamsutta Oil Refinery in Pennsylvania in 1861.

Crude oil is called a *hydrocarbon* because it is made primarily of hydrogen and carbon atoms. The carbon atoms have a tendency to chain together, with hydrogen atoms bonded to the carbon chains. The basic idea behind refining is that hydrocarbons behave differently depending on the length of the carbon chain. The longer the chain, the thicker the refined product. Butane (a gas at room temperature and pressure) has four carbons in its chain. Gasoline has eight. Kerosene has twelve. Paraffin wax has twenty.

To refine crude oil, start by heating it. The different chain lengths boil off at different temperatures. The vapor goes into a fractional distillation column to separate out all of the different products based on these temperatures.

If the primary goal of the refinery is to produce gasoline, then there are processes designed to produce gasoline-length chains. For example, a catalytic reformer can combine shorter carbon chains into gasoline-length chains. Catalytic cracking can take longer chains and break them down into gasoline-length chains. Different products from these different processes blend together to form gasoline at specific octane levels.

Another important part of the refinery process is the removal of contaminants. Sulfur is one of the more common. Sulfur in gasoline creates two problems. First, it forms sulfur dioxide in the combustion chamber. The sulfur dioxide mixes with rain to form sulfuric acid, also known as acid rain, a major ecological issue that lead to **automotive emissions controls** in the twentieth century. Second, sulfur destroys catalytic converters that clear up exhaust gases. A hydrotreater turns sulfur into hydrogen sulfide gas for easy removal.

The largest refineries can process nearly a million barrels of oil per day, making them gigantic collections of towers, pipelines, valves, and storage tanks covering many acres. Every part of the process involves engineering.

SEE ALSO Oil Well (1859), Automotive Emission Controls (1967), Trans-Alaska Pipeline (1977), *Seawise Giant* Supertanker (1979).

This is a modern oil, gas, and fuel refinery.

1861

Elevator

Elisha Otis (1811–1861), **Werner von Siemens** (1816–1892)

By the second half of the 1800s, architects and structural engineers were coming up with new ideas for buildings. These ideas would allow them to create taller and taller structures. But they faced the limitation of stairs. People will only climb so many flights of stairs before they give up.

Someone had to engineer something better than stairs to solve this impasse. The solution that eventually evolved is a room that moves vertically—an elevator. American inventor Elisha Otis patented a steam elevator in 1861, later founding the Otis Elevator Company.

For a steam elevator to function in a building, it had to have a boiler to create pressurized steam and a **steam engine** to turn a large drum for winding the cable. From there the elevator mechanism had the four key elements we are familiar with today: the car, a counterweight, the cables and their associated pulleys, and then the safety mechanism that made elevators from the Otis company famous. If a cable broke, it would release a braking mechanism that would immediately stop the car and keep it from falling back down the shaft. This braking system gave people the confidence to use elevators without fear.

In 1880, German inventor Werner von Siemens produced the first electric elevator. He is considered to be one of the founding fathers of electrical engineering for this and other contributions.

By the year 1900, there were over 3,000 elevators in New York City. They carried more than a million people every day. Engineers had created a safe device that was extremely useful to people living in cities, and it made the skyscraper era possible. Without elevators, buildings really would not have gotten much higher than four or five stories.

SEE ALSO High-Pressure Steam Engine (1800), Empire State Building (1931).

Elisha Otis's elevator (1861 patent shown here) was the first with a safety catch to protect passengers in the event of a broken cable.

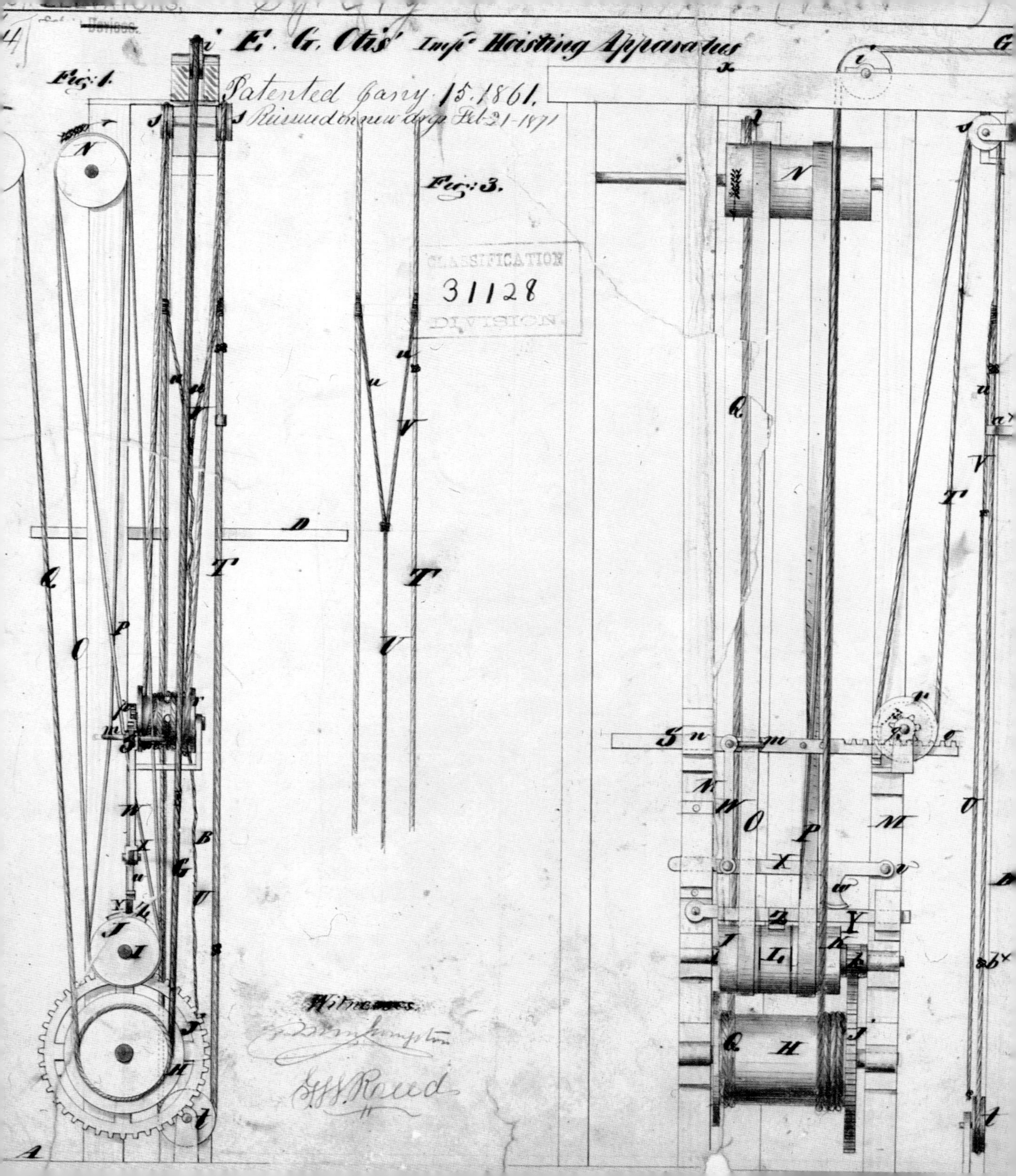
E. G. Otis' Impd Hoisting Apparatus
Patented Jany 15. 1861.
Reissued on new drgs Feb. 21. 1871
Fig: 1.
Fig: 3.
CLASSIFICATION
31128
DIVISION
Witnesses.
G. H. Reed

1869

Transcontinental Railroad

Before the transcontinental railroad was finished in 1869, the normal way for a person to get from a city like St. Louis or Omaha to San Francisco was on a wagon pulled by oxen. This journey took months and was fraught with danger. Once the track for the transcontinental railroad was complete, the trip from Omaha took less than four days. The train trip was far safer and more comfortable. You could take a lot of luggage. And it was reasonably priced, given the huge amount of time and trouble it saved. The fare from Omaha to San Francisco in 1869 was $81.50, or roughly $1,400 in today's dollars.

But how did the engineers do it? Sure, people had built lots of railroad lines by this time. But how could they start construction from two cities separated by nearly 2,000 miles (3,200 km) and then meet in the middle?

The answer lies in the surveyors. The first step was to pick the route. This was done by surveying five possible routes and comparing them, and it was a process authorized by Congress. The Corps of Topographical Engineers, a division of the army, surveyed the five different routes. Imagine working your way across 2,000 miles of rough terrain and mountains with devices like a **Gunter's chain** (to measure distance) and a transit (to measure angles). It took two years. But even with these simple tools, the accuracy was impressive.

Then the chosen route became a gigantic civil engineering project. Engineers had to design and build tunnels and bridges, cut and fill to create level roadbeds, blast passes through and around hillsides, and so on. The section of track through the Sierra Nevada mountains was particularly challenging. At the same time, a **telegraph** line was installed, making near-instantaneous communication possible across the entire continent. North America was the first continent humankind spanned.

When completed, the track at Omaha connected into the existing railroad network. A person could travel from New York to San Francisco in less than a week. It was an amazing accomplishment—one of the greatest engineering achievements of the nineteenth century.

SEE ALSO Gunter's Chain (1620), Erie Canal (1825), Telegraph System (1837), Professional Engineer Licensing (1907), Tacoma Narrows Bridge (1940).

Laying the last rail in the transcontinental railroad, 1869.

Cable Cars

Andrew Smith Hallidie (1836–1900), **William Eppelsheimer** (1842–unknown)

One thing that is true of all engineering endeavors is this constraint: engineers must work with the technology available at the time. Thus, an engineer working in the mid-1800s did not have access to things like **lasers**, **microprocessors**, or inexpensive aluminum.

So from 1873 through 1890, when promoter Andrew Smith Hallidie and engineer William Eppelsheimer set about to design a transportation system in San Francisco, one constraint they had to work around was the size of existing **steam engines**. Steam engines were big and they were heavy. They had to carry a water supply and a coal supply. That worked for a big steam locomotive pulling a long train with hundreds of passengers. It did not work at all for small streetcars running up steep hills on city streets.

The size of the era's steam engines therefore determined the architecture of the system. The steam engines would sit stationary in a building, and their power would be distributed with moving steel cables running in channels beneath the street. If the cable car line is two miles long, it requires at least four miles of steel cable running in a loop. Rollers and pulleys along the route minimize friction on the cable.

The cable car has a lever used by the operator. The lever manipulates the mechanism that grips the cable. When the car is gripping the cable, the car moves. When it is not gripping the cable, the car can coast or brake.

At the end of the line, the cable runs through a large pulley so it can return toward the steam engine. The car either has to turn around on a turntable, if the car has only one set of controls, or, if it has controls at both ends, it has to go through a switch that puts it on another track.

With a cable speed of 9.5 mph (15 kph), cable car systems seem like quaint anachronisms. But at the time, they were state of the art. Engineers had built a working system that made getting around the city much easier. A century later, the system is still working.

SEE ALSO High-Pressure Steam Engine (1800), Erie Canal (1825), Diesel Locomotive (1897), Internal Combustion Engine (1908).

San Francisco cable cars, shown here, are still in operation today.

VanNess Ave.. California
60
& Market
Streets

1876

Telephone

Alexander Graham Bell (1847–1922)

Imagine that the year is 1850 and you want to talk to someone. You have exactly one option: You can travel and meet with that person face-to-face, which could take days or weeks, depending on distance. Your alternative: a handwritten letter. Or, by 1850, the **telegraph system** has started to expand. But the simple act of talking to someone still requires a face-to-face meeting.

Enter the telephone. The patent for Alexander Graham Bell's telephone was issued in 1876. The device itself was incredibly simple—a microphone made of carbon granules and a speaker. To connect two telephones together, all you needed was copper wire and a small source of electric current, like a battery—with this innovation, two people talked to each other at a distance for the first time.

How did engineers scale this up? The first innovation was the central office. In a town, copper wires ran from each home or business to the central office. An operator could connect any line to any other line in town. Wires are added to connect the town to the next town over. At that point, anyone in the two towns could communicate. As additional towns connected, this led to the creation of regional central offices. Eventually trunk lines spanned the country, then the world, and now everyone could connect to everyone else.

Engineers developed mechanical switches to replace the human operators. The telephone dial told the switches what to do. As a result, the cost of calling fell. Engineers created much smaller computers to replace the mechanical switches, and touch-tone dialing became possible. The cost of calling fell again. Engineers turned voice signals into digital bits and sent the bits through **fiber optic** cables, drastically reducing costs and increasing capacity. Then engineers created voice over IP (VoIP) so calls were routed through the Internet. Internet telephony was born and calling became free on many VoIP networks. The success story for engineering: taking something that used to be impossible and eventually making it free!

SEE ALSO Telegraph System (1837), TAT-1 Undersea Cable (1956), T1 Line (1961), ARPANET (1969), Fiber Optic Communication (1970), Mobile Phone (1983), Smart Phone (2007).

Illustrations depicting the Bell Telephone.

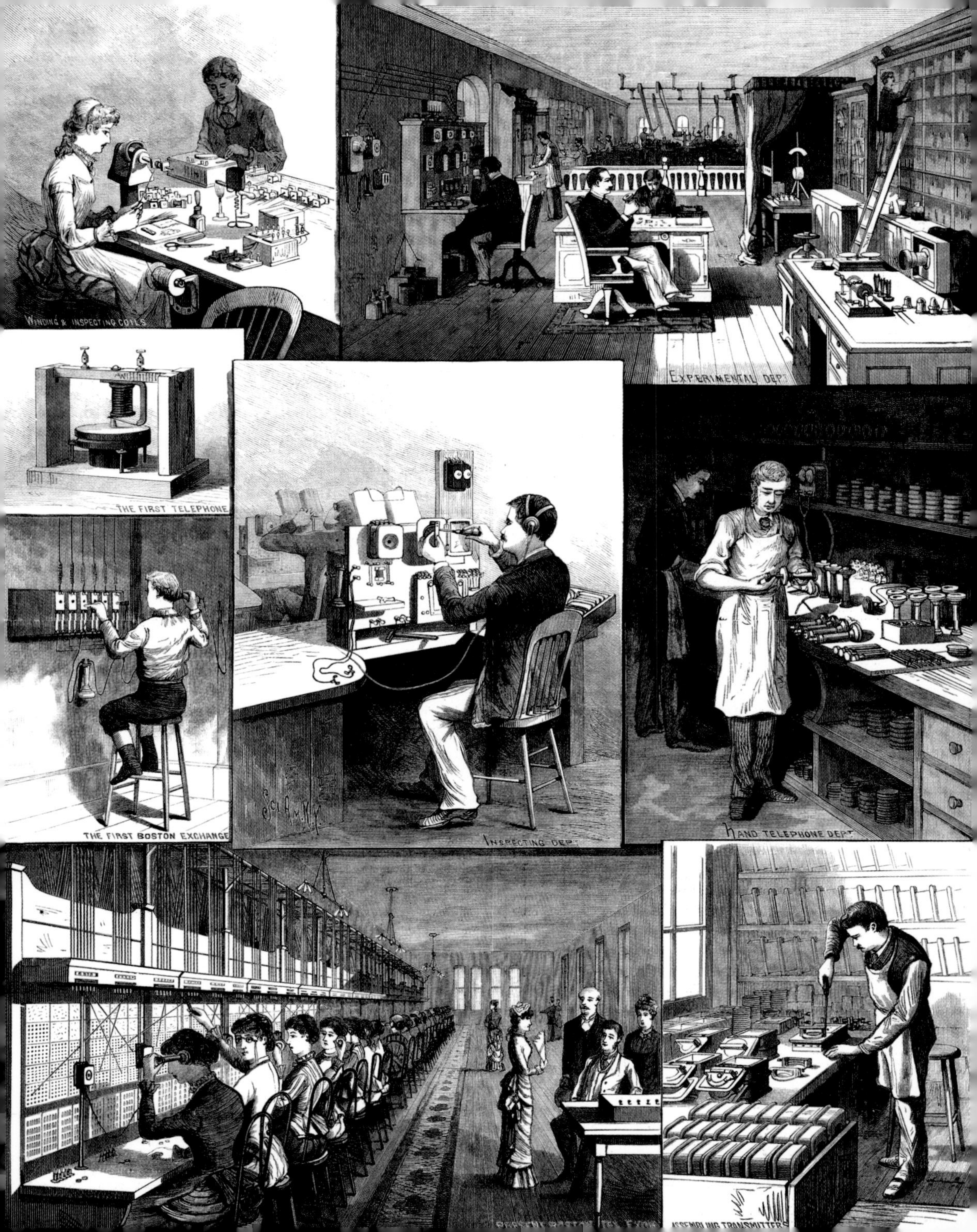

WINDING & INSPECTING COILS
EXPERIMENTAL DEPT.
THE FIRST TELEPHONE
INSPECTING DEPT.
HAND TELEPHONE DEPT.
THE FIRST BOSTON EXCHANGE
ASSEMBLING TRANSMITTERS

1878

Power Grid

In 1878, at the Paris World's Fair, visitors marveled at the Yablochov arc lamps (patented by Pavel Yablochov in 1876) powered by Zénobe Gramme dynamos. This was an example of an early commercial system of high-voltage power—the kind of power grid that exists invisibly behind the scenes around the world today.

It is possible to imagine a society where there is no power grid—where every home and business generates its own power on-site. But this approach has efficiency problems. A big power plant can realize economies of scale when purchasing fuel and can apply significant resources to **emission controls**. Advanced technologies like nuclear power are not possible without a big power plant. And site-specific power sources like hydropower, solar power, and wind turbines only really make sense if there is a grid. A power grid can also improve reliability. When a big power plant needs to go offline for maintenance, other power plants in the area use the grid to make up the load.

It is amazing to realize that the power grid has only two key components: wire and transformers. Transformers can multiply voltages up and down. For long distance transmission, transformers boost the voltage to 700,000 volts or more. Once the power arrives at its destination, transformers step down the voltage. It might travel at 40,000 volts in a community, and then 3,000 volts in a neighborhood. At your house, a final transformer brings it to 240 and 120 volts for use in your wall outlets and light switches.

The grid is not perfect, and occasionally we see widespread blackouts. On a sweltering summer day with the whole grid running at peak loading, a failure in a key transmission line can cause an irresolvable problem. Other lines try to pick up the load from the failed line, but they overload and fail. A ripple effect can leave several states in the dark. Engineers are working on new architectures to prevent this problem as well. Once perfected, the grid will be even more invisible.

SEE ALSO Hoover Dam (1936), Itaipu Dam (1984), Smart Grid (1998), Alta Wind Energy Center (2010), Ivanpah Solar Electric Generating Center (2014).

Without a power grid, energy would have to be generated on site.

1879

Carbon Fiber

Thomas Edison (1847–1931)

What if you want to engineer a structure that is both strong and light? And by strong we mean stronger than steel, and by light we mean lighter than aluminum. If that is what you need, and your budget is big enough to afford it, then your go-to material in today's world is carbon fiber, first developed by Thomas Edison in 1879, who used an all-carbon fiber filament to light the first incandescent light bulbs.

Carbon fiber reinforced plastic, which was developed from this initial material, uses hard plastic to encase the carbon fibers to stabilize them. The carbon fibers come from threads of high-carbon materials, the most common being polyacrylonitrile. By heating the fibers in oxygen and then without oxygen, everything but the carbon atoms boils off. These remaining carbon atoms are structured as long chains that have impressive tensile strength. They can form into threads, and the threads can be spiral wound or woven into a fabric.

The most common way to work with carbon fiber is to lay up the cloth in a mold and soak it with the plastic resin. Molding allows carbon fiber parts to be any shape, but it is an expensive, manual process. In items where cost is no object, like race cars, supercars, airplanes, and expensive bicycle frames, this is not a problem. But it has limited the spread of carbon fiber to a wider array of products. For example, your typical consumer automobile is not made of carbon fiber because of the cost.

How strong is carbon fiber? Imagine a piece of steel and a piece of carbon fiber both shaped like pencils. The carbon fiber piece might be three times stronger than the steel piece, but have one-third the weight. It is a huge difference. The first time you hold carbon fiber, it seems like it comes from another world because it is so light and strong compared to steel.

SEE ALSO Human-Powered Airplane (1977), V-22 Osprey (1981), Bugatti Veyron (2005), Human-Powered Helicopter (2012).

High-performance bicycle frames are often made of carbon fiber.

S-WORKS
ASTANA
SAMRUK
K A Z Y N A
ASTANA
WORKS
ASTANA
SRAM
SPECIALIZED

1885

Supercharger and Turbocharger

Gottlieb Daimler (1834–1900)

Think about what is happening inside the cylinder of a diesel or gasoline engine. During the intake stroke, the piston moves downward and sucks in a volume of air. Let's say the amount of air is one liter. Now the question is: How much fuel can burn using that one liter of air? The amount is limited by the number of oxygen atoms in the cylinder. The oxygen will combine with the carbon and hydrogen atoms in the fuel to create CO_2 and H_2O. Adding too much fuel is a waste—it can't burn because of the limited oxygen.

Engineers look at this situation and ask an obvious question: Is there a way to get more oxygen into the cylinder? One way is to boost the pressure of the incoming air. If incoming air arrives at double the normal pressure, then twice as much oxygen gets crammed in the cylinder and twice as much fuel can burn. The 2x boost roughly doubles the power available. Engineers can radically improve the power-to-weight ratio of the engine as long as the boosting equipment itself doesn't weigh too much.

A supercharger, patented by German industrial engineer Gottlieb Daimler in 1885, is the standard way to boost air pressure. It is an air pump attached to the engine's crankshaft with a belt. Three types of superchargers are in common use: centrifugal, screw, and roots types.

If you take a centrifugal supercharger and power it with an exhaust turbine instead of a belt attached to the crankshaft, you have created a turbosupercharger, or turbocharger. The advantage: a turbocharger uses less engine power. The disadvantage: increased complexity and less boost at low rpm.

In big engines, superchargers can be a no-brainer. Dragsters using four-stroke nitromethane engines, and locomotives using **two-stroke diesel engines** always use them. But in smaller engines, the size, weight, cost, and complexity of the supercharger, plus the beefier design requirements for the engine, may cancel the benefits. Engineers analyze the tradeoffs to decide if the benefits outweigh the disadvantages.

SEE ALSO Two-Stroke Diesel Engine (1893), Nitrous Oxide Engine (1978), Bugatti Veyron (2005).

Close up of supercharger on the engine in a customized 1968 AMX by American Motors Corporation (AMC), a two-seat GT-type car.

BDS
Racing
Edelbrock
Edelbrock
BDS
Edelbrock

1885

Washington Monument

In 1832, the first steps were being taken to build a monument for George Washington in Washington, DC. By 1835, the committee working on the monument could describe what they wanted: the monument would be "unparalleled in the world" and it "should blend stupendousness with elegance, and be of such magnitude and beauty as to be an object of pride to the American people." The design they eventually settled on was a gigantic obelisk, scheduled to be the tallest human-made object in the world. Now all they had to do was engineer it and build it.

If you go to Egypt and look at the obelisks there, they are solid rock, and that means that the tallest is about 100 feet (30 meters). The Washington Monument at 555 feet (170 meters) is gargantuan by comparison. So it is hollow and made of stacked blocks. Its hollow design allowed for a crane inside to bring building materials up as the column grew. Today the hollow core contains steps and an elevator.

The first component built was the foundation—all 37,000 tons of it. It measures 126 x 126 feet and is 37 feet thick, made of **concrete**. The base of the monument measures 55 feet (17 meters) square, centered on the foundation, with walls that are 16 feet thick. At the top of the column, where the pyramidal shape starts, it is 34.5 feet (10 meters) square, with walls only 1.5 feet (45 cm) thick. There are 36,500 blocks in the monument, and at the base there are inner and outer walls, with the gap in between filled with rubble.

Although it seems like a fairly simple project—it is just a big stack of blocks essentially—it took quite a while to build the monument. The cornerstone was laid in 1848, and the dedication occurred in 1885, 37 years later. What this meant is that the Washington Monument was the tallest structure in the world for only four years. The Eiffel Tower eclipsed it in 1889.

SEE ALSO Concrete (1400 BCE), Bessemer Process (1855), Statue of Liberty (1886), Eiffel Tower (1889), Gateway Arch in St. Louis (1965).

Although a proposal for a monument to George Washington was put forth in 1799, the structure wasn't completed until 1885.

SETTING THE CAP STONE.

CAPSTONE OF THE MONUMENT, SHOWING THE ALUMINIUM TIP.

HEIGHT 555 FEET.
WEIGHT 81,120 TONS.
COST # 1,187,710.

WASHINGTON MONUMENT.

CORNER STONE LAID JULY 4TH 1848.
CAP STONE SET DECEMBER 6TH 1884.
INAUGURATED FEBRUARY 22D 1885.

IN THE ELEVATOR.

ENTRANCE TO THE MONUMENT.

Statue of Liberty

Gustave Eiffel (1832–1923), **Frédéric Auguste Bartholdie** (1834–1904), **Maurice Koechlin** (1856–1946)

Art and engineering combine to create something as big and beautiful as the Statue of Liberty. Because of its size, the Statue of Liberty is highly engineered. In fact, with its skin removed, it is obvious. Under the skin, the Statue of Liberty looks a lot like the engineered skeleton of a skyscraper. And, in fact, this skeleton was created by the same firm that designed the **Eiffel Tower**.

Think about the problems faced by the sculptor/designer, Frédéric Auguste Bartholdi, as well as Gustave Eiffel, who assisted in its design, and his structural engineer, Maurice Koechlin. First, he is creating a sculpture 150 feet (45 meters) tall. He wants to be able to build the sculpture in France and mail it to America on a ship. So a giant marble sculpture is out. In marble the uplifted arm would be difficult as well. He decides to make it out of a thin copper skin instead. But there will be 160,000 pounds (72,600 kg) of copper skin when he is done. And the sculpture, once assembled, has to be able to handle hurricane force winds.

So inside the sculpture there is a huge metal frame. Four vertical beams a hundred feet (30 meters) tall anchor the sculpture to its pedestal and provide support for the internal staircase. From those beams, a metal truss framework extends out toward the 300 sheets of riveted copper that form the skin. The copper sheets attach to a lattice of custom-bent iron bars that provide rigidity and structure. The curtain-wall architecture of a modern skyscraper like **Burj Khalifa** manages the building's load in a similar way: the Statue of Liberty actually is a prototype of today's skyscrapers.

What about the uplifted arm? It has its own truss and ladder to climb to the torch.

The statue was built in France, then disassembled, crated, and shipped to America for reassembly, which took about a year. When it was dedicated in 1886, it stood both as an important work of art and an important work of engineering.

SEE ALSO Washington Monument (1885), Woolworth Building (1913), Burj Khalifa (2010).

Interior view of the Statue of Liberty, looking up from the inside.

Eiffel Tower

Émile Nouguier (1840–1898), **Maurice Koechlin** (1856–1946)

If you have ever gotten close to the Eiffel Tower, there is no denying that it is an engineered structure. It is immense, it is made of thousands of pieces of metal, and it is intricate. There are many places in the tower where the complexity is impressive. And it has stood the test of time at well over 100 years old, opening in 1889.

The size is surprising, especially given the age. With a roof height right at 300 meters (nearly 1,000 feet), it was the tallest object in the world for four decades until it was surpassed by the Chrysler Building and then the **Empire State Building** in New York shortly after. The **Millau Viaduct** is the only thing surpassing it in France today.

The fact that something this intricate could be conceived, designed, engineered, fabricated, and erected in that era is also impressive. You can get a sense of this if you look at high-definition images of the tower or visit it in person. Look at the four columns that stretch from the first floor (187 feet or 57 meters off the ground) to the second floor (377 feet or 115 meters off the ground). This is the conceptually simplest part of the tower, yet the complexity is impressive.

Those four columns are approximately 58 meters (190 feet) tall. Over that distance they gently curve and gently taper. For each column there are four thick steel beams riveted together from plates, and then lattice girders that turn each column into its own lattice girder. All of that iron was prefabricated in a factory, brought to the site on horse-drawn wagons, and then assembled with rivets, and it all needed to fit perfectly. The structural engineers, Maurice Koechlin and Émile Nouguier, made thousands of precise drawings to tell the factory what to manufacture and the construction workers what to assemble.

It is said that the Eiffel Tower contains 18,035 pieces of metal and 2.5 million rivets. When you consider that an engineer placed each piece of metal and each rivet on a drawing, and those drawings were then manifested in reality with tenth of a millimeter precision, you get a sense of the engineering achievement.

SEE ALSO Washington Monument (1885), Statue of Liberty (1886), Woolworth Building (1913), Empire State Building (1931), Millau Viaduct (2004).

The Eiffel Tower is an example of extremely precise engineering that has resulted in monuments that have stood the test of time.

Hall-Héroult Process

Charles Martin Hall (1863–1914), **Paul Héroult** (1863–1914)

Think about how common aluminum is today, and how important it is to engineers as a structural material. Out in the garage we find aluminum in bicycle frames and rims, and some cars now have aluminum parts to lighten the vehicle. But the place where we see aluminum the most is in aircraft. Every airliner that flew up until 2010 was made primarily out of aluminum. The same goes for just about every rocket and spacecraft, including the **International Space Station**. The Boeing 787 is the first commercial airliner to be made primarily of carbon fiber instead of aluminum.

The reason aluminum is so popular, especially in airplanes, is because it is relatively inexpensive, easy to work with, durable, and light. A piece of aluminum that has the same strength as a steel piece might weigh half as much. If you take a cube of aluminum that weighs 1 pound, the same size cube of steel weighs about 2.8 pounds.

If you were to double the weight of most airplanes by using steel instead of aluminum, the airplane would not be able to fly, or it could only fly with zero payload. Aluminum makes it possible for engineers to design airplanes that can get off the ground. Before aluminum became common, airplanes were made of wood and cloth.

Aluminum was first purified in 1825. But it was extremely expensive—roughly equivalent to gold in price. In 1889 a new patent for a system called the Hall-Héroult process, which was developed by American chemist Charles Martin Hall and Frenchman Paul Héroult, demonstrated how to make inexpensive aluminum, and the rest is history. Once aluminum became inexpensive, and then abundant, its use exploded. Today, over 30 million tons of aluminum are used every year.

Just about every vehicle is better when it is lighter, so engineers use aluminum everywhere they can. With new mass production techniques, aluminum is replacing steel in more and more cars.

SEE ALSO Bessemer Process (1855), Carbon Fiber (1879), Boeing 747 Jumbo Jet (1968), The Concorde (1976), Space Shuttle Orbiter (1981), Apache Helicopter (1986), International Space Station (1998).

The interior of a bomber, created with aluminum.

1890

Steam Turbine

Sir Charles Parsons (1854–1931)

If you go to any large power plant today, one of the landmarks will be a huge steam turbine bigger than a bus. You find steam turbines on aircraft carriers and nuclear submarines, too. With the steam turbine, engineers were able to reconceptualize the extraction of power from steam and thus abandon pistons.

Let's get in our time machine and go back to the engine room of the ***Titanic*** in 1912. Here they are using steam drawn from over one hundred massive coal-fired boilers and it is going into three **steam engines** driving three propellers. Two of these steam engines are gigantic piston machines that produce 30,000 hp (22 million watts) each, and the third is a steam turbine producing about half that. What we witness here is a period of transition. Steam turbines, first invented by Sir Charles Parsons in 1890, had not yet been perfected, but they would soon replace pistons to extract rotational energy from steam.

The basic idea behind a steam turbine is extremely simple. The expanding steam turns a series of vanes attached to a shaft. The vanes get progressively larger, so that the steam's energy can be captured as it expands. Compare that process to the *Titanic*'s piston engines; the piston engines use three cylinders of increasing size. The steam first expands in the smallest cylinder. Then it flows to the next cylinder, somewhat larger in size to extract more power from the less dense exhaust of cylinder one. Then to the third even larger cylinder. This worked but made for a large and heavy piece of equipment. One steam piston engine on the *Titanic* weighed 1,000 tons.

A steam turbine does the same job, but is much smaller, lighter, and more efficient than an equivalent steam piston engine. Modern steam turbines appear in almost every major coal-fired and **nuclear power plant** today because of these advantages. Instead of just three expansion chambers, the steam turbine can have many stages of vanes of increasing size to extract as much power as possible. This shows how engineers switch to completely new concepts to get better results.

SEE ALSO High-Pressure Steam Engine (1800), *Titanic* (1912), CANDU Reactor (1971).

Contemporary turbines are so precisely made that they can only be constructed with computers.

B-539

1891

Carnegie Hall

Dankmar Adler (1844–1900), **William Burnett Tuthill** (1855–1929)

Carnegie Hall in New York City, built in 1891, is considered to be an excellent venue. It seats 2,800, yet an orchestra can perform with no amplification at all. Its architect, William Burnett Tuthill, was a cellist and studied European concert halls for their acoustics, in addition to consulting with acoustic specialist Dankmar Adler.

It is easy to understand some of the problems they faced by thinking about two different situations. First, imagine a person giving a lecture in a big open field. As the sound moves away from the speaker's mouth, it has to fill more and more volume, so the available energy dissipates quickly. However, there is no echo whatsoever.

Now imagine that you are in a closed space, like a racquetball court. All the walls are flat, smooth, and solid. The sound from the lecturer is contained within a finite space in this case, so you can definitely hear it. But the echo problem can make speech difficult or impossible to understand.

To create Carnegie Hall, acoustical engineers blended features from these two cases. Listeners benefit from "depth" to the sound. Depth is created when the sound from the speaker's mouth or the musical instrument arrives via multiple paths, first via a direct line, and then other versions arriving quickly from reflections off the ceiling and side walls. Actual echoing is canceled out by the audience itself, heavy drapes, or acoustical panels that absorb sound at the back of the room. This translated to a long, narrow room with the orchestra at one end, rather than a wide room or a room that widened from the stage. In wide rooms, the sound energy dissipates like it does in a field and people in the back cannot hear.

To maximize the quality of a performance or event, engineers continue to make improvements, from giant **stadium TV screens** to **retractable roofs**.

SEE ALSO Parthenon (438 BCE), Tape Recording (1935), Retractable Stadium Roof (1963), Stadium TV Screen (1980).

Muziekgebouw Concert Hall, Amsterdam, uses some of the same principles of acoustic engineering as Carnegie Hall in New York City.

1891

Zuiderzee Works

Cornelius Lely (1854–1929)

From an engineering perspective, the Netherlands is a fascinating country. It has extremely high human density (over 1,200 people per square mile), about 15 times greater than the density of the United States. Because of this density, combined with its location, the country has been reclaiming new land from the sea with dikes for many centuries. Approximately a quarter of the country lies below sea level—some as much as 23 feet (7 meters) below. And approximately half of the country is barely above sea level. To drive the point home, the word Netherlands means "lowlands."

On a typical day, the sea remains on its side of the dikes. But when storm surges occur, problems arise. And in a country where much of the land is below or at sea level, storm surges can be catastrophic.

So the Netherlands has invested heavily in protection against storm surges. The Zuiderzee Works consists of dozens of different projects: **dams**, dikes, gates, and so on. Built based on an original plan by civil engineer Cornelius Lely in 1891, construction did not begin until 1920, and continued through 1986.

Of those works, the most impressively engineered by far is a gigantic, movable storm surge barrier across the mouth of the Rhine River as it exits to the sea. Most of the time, the two parts of the barrier sit on land in dry docks on both sides of the river. The river there is 1,180 feet (360 meters) wide. When a storm surge is expected, the dry docks flood. The two barriers, which float, come out into the river and meet in the middle. Then they submerge to seal the mouth of the river against storm surge arriving from the sea.

Given that the barriers are closing off a river, won't water back up behind them? It can happen. So the gates can partially refloat to let excess river water flow underneath.

If you think about the size of these barriers, you realize they are some of the biggest moving objects on Earth. The fact that they can move both horizontally and vertically means they have two of the largest ball and socket joints on Earth as well.

Storm surge is a huge problem for the Netherlands, and engineers have risen to the occasion with this moveable barrier.

SEE ALSO Hoover Dam (1936), Venice Flood System (2016).

Northern half of the Maeslantkering, a storm surge barrier in the Nieuwe Waterweg near Rotterdam and Hook of Holland in the Netherlands.

Two-Stroke Diesel Engine

Rudolf Diesel (1858–1913)

Four-stroke diesel engines work well and they are extremely common in road applications that require hundreds of horsepower. Most tractor-trailer rigs, school buses, and passenger buses use a four-stroke diesel engine. But in some applications, diesel engines need to scale up to thousands of horsepower, as in a **diesel locomotive**, or tens of thousands of horsepower, as in a **container ship** or **supertanker**. In these cases, engineers use two-stroke diesel engines, patented by German engineer Rudolf Diesel in 1893, instead.

Why? One big advantage of a two-stroke diesel is that it produces twice as much power from an engine of any given size. In an eight-cylinder four-stroke diesel engine running at 1,000 rpm, there are 4,000 power strokes per minute. If the same engine is two-stroke instead, there are 8,000 power strokes.

In a two-stroke diesel engine, as the piston moves downward in the power stroke, it bottoms out and uncovers intake ports cut into the cylinder wall. Right before that happens, exhaust valves open at the top of the cylinder to release exhaust gases. The air coming in through the intake ports is pressurized by a supercharger, so it forces its way into the cylinder at 2x atmospheric pressure or more. As the piston finishes bottoming out, it starts moving back upward and compresses the air in the cylinder. Right before top dead center the diesel fuel is injected. It ignites spontaneously in the hot, compressed air, providing power to push the cylinder back downward in the cylinder.

The two-stroke diesel engine in a typical diesel locomotive is already massive, weighing 30,000 pounds (13,600 kg) or more. If it were a four-stroke design instead, with the same power, it would need to be twice as big. This is why two-stroke diesel engines are so common in large applications like locomotives and cargo ships.

In smaller applications, the two-stroke's requirement for a supercharger often offsets the power-to-weight advantage of the two-stroke approach. Engineers have to balance the parameters to achieve optimal results in any given application—the weight, power draw, cost, and complexity of a supercharger may outweigh the other advantages of the two-stroke approach when only 100 hp is needed.

SEE ALSO Supercharger and Turbocharger (1887), Diesel Locomotive (1897), *Seawise Giant* Supertanker (1979), Container Shipping (1984).

Pictured: The Wärtsilä X72 and the Wärtsilä X62 two-stroke engines in Doosan Engine Co., Ltd, in Chagwon, South Korea.

Ferris Wheel

George Washington Gale Ferris, Jr. (1859–1896)

People will climb mountains and ride **elevators** to the top of skyscrapers like the **Empire State Building** to take in a good view.

So say that you wanted to engineer a ride that gives people a mountaintop view without the climbing—how might you do it? The Ferris wheel is one answer to the question. The first Ferris wheel opened in 1893 in Chicago, designed by George Washington Gale Ferris, Jr.

If you look at today's tallest Ferris wheel—the High Roller in Las Vegas—you can see one reason why Ferris wheels are a favorite of engineers: they can be extremely efficient with materials. Like a bicycle wheel, which can be both strong and light because of its tensioned spokes, the High Roller looks remarkably sparse, with just a rim, the spokes, the hub, and the base. The cabins for riders attach to the rim and rotate with electric motors to stay level.

Because the wheel is so large at 560 feet (167 meters) in diameter, and so sparse in terms of materials, it is easy to lose track of how much weight it is carrying. Looks can be deceiving. First, there can be more than 1,000 people riding the wheel simultaneously. They ride in 28 cabins that hold up to 40 people each. Every cabin connects to a section of rim that is 53 feet (16 meters) long. Four steel cables (the spokes for this wheel) tie each rim section to the hub. A cabin weighs approximately 45,000 pounds (20,500 kg) empty, or up to 55,000 pounds (25,000 kg) fully loaded. So there is something like 1.5 million pounds (680,000 kg) hanging off the rim. It is an immense amount of weight for the structure to support and be able to rotate. In addition, there are a dozen **tuned mass dampers** in the base to even out any vibrations and create an extremely smooth ride.

The wheel turns at two revolutions per hour, meaning that a passenger stays on the wheel for about 30 minutes, allowing 2,000 or so people per hour to take in the view.

SEE ALSO Under Friction Roller Coaster (1919), Empire State Building (1931), CN Tower (1976), Tuned Mass Damper (1977), Harry Potter Forbidden Journey Ride (2010).

Ferris wheel at The World's Columbian Exposition, 1893.

1897

Diesel Locomotive

Rudolf Diesel (1858–1913)

Imagine that you are an engineer in the 1930s and you want to create a new locomotive based on a diesel engine, first operated successfully by Rudolf Diesel in 1897. Steam engines have been operating successfully for more than a century, but a diesel engine is a completely different animal. If it is going to pull a train that is one hundred cars long, then the load could weigh thousands of tons. Pulling that up a hill requires immense power from the engine. You also want it to be as efficient as possible.

Your first instinct might be to engineer the locomotive like a diesel truck. These trucks pull heavy loads. But the problem is the transmission. Where a car needs a four- or five-speed transmission, a truck might need nine, thirteen, even eighteen gears because of the much greater load. Therefore, if we were to go from truck to locomotive and needed to add even more gears, the size and complexity of the transmission starts to get ridiculous. To eliminate the transmission, locomotives therefore start with a **two-stroke diesel engine**. The engine is huge, with twelve to twenty cylinders, and a **supercharger** improving efficiency.

This engine then connects to an electric generator or alternator rather than a transmission. If the engine is rated at 4,000 horsepower, the alternator can produce up to three megawatts. In other words, a diesel locomotive is actually a small electrical power plant. The electricity then flows into electric motors. If the locomotive has six axles, there can be six motors.

The advantages of this engineering innovation are numerous. First, the use of electric motors eliminates the need for a large, complicated transmission. Second, it is much easier to drive the six axles with six electric motors than to connect them all together with drive shafts from the transmission. Third, if one electric motor fails, the train can still function. Fourth, the engine runs at discrete speeds, improving efficiency and making it possible to gang multiple locomotives together.

When miniaturized and combined with batteries, this same kind of approach makes **hybrid cars** possible. Engineers developed an important efficiency breakthrough.

SEE ALSO Supercharger and Turbocharger (1885), Two-Stroke Diesel Engine (1893), Prius Hybrid Car (1997).

Santa Fe's Super Chief diesel streamliner rounding a curve on the Los Angeles to Chicago route.

1899

Defibrillator

Jean-Louis Prévost (1838–1927), **Frédéric Batelli** (Dates Unavailable), **Frank Pantridge** (1916–2004)

Imagine that you are walking down an airport concourse when a man in front of you suddenly clutches his chest and falls to the floor. He is experiencing sudden cardiac arrest. The good news is that many public buildings now have portable defibrillators. The fact that engineers have been able to shrink, simplify, and cut the cost of these lifesaving machines is a testimony to the engineering ethos. The concept of defibrillation dates back to 1899, when Jean-Louis Prévost and Frédéric Batelli, two physiologists at the University of Geneva, Switzerland, demonstrated the effect of electricity on the hearts of dogs. Portable defibrillators were not available until 1965, when Irish professor Frank Pantridge developed an early model weighing over 150 pounds (70 kg).

The portable units that we see at **airports** and malls are called AEDs or Automatic External Defibrillators, first used outside of a hospital in 1980. The goal of a defibrillator is to stop the fibrillation and reestablish a regular heartbeat. A defibrillator does this with a strong electric shock.

In sudden cardiac arrest, the heart has stopped its regular beating action for some reason. The human heart has its own independent nerve network that causes its regular beating, and if this network gets disrupted (for example by a suddenly clogged artery on the heart muscle), the heart can enter a cycle of fibrillation, where the nerve network goes off script and starts beating irregularly and rapidly.

After placing the self-stick electrodes on the patient's chest, a **microprocessor** in the device analyzes the heart rhythm and detects if fibrillation is occurring. If it is, the defibrillator uses its internal batteries to load a capacitor with a high-voltage charge (e.g., 700 volts) and then delivers it to the patient. In many cases, a single charge is enough to start a normal heartbeat, but if not, the device can try multiple times.

The best way to treat sudden cardiac arrest is to use a defibrillator as soon as possible. What electrical engineers, software engineers, and industrial engineers have been able to do is create an inexpensive, lightweight, mass-produced package that can help treat a leading cause of death.

SEE ALSO Microprocessor (1971), Artificial Heart (1982), Kansai International Airport (1994).

While the first defibrillator was prohibitively massive, today lighter portable models are available in most public buildings.

304
SANTA FE
SANTA FE

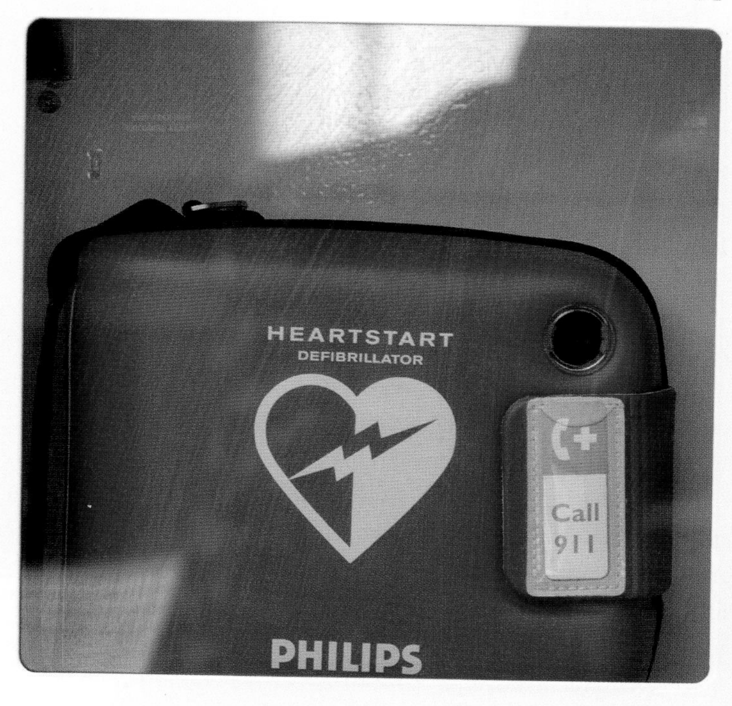
DEFIBRILLATOR
HEARTSTART
DEFIBRILLATOR
Call
911
PHILIPS

TRAINED
RESPONDERS
ONLY
ALARM
WILL
SOUND

1902

Air Conditioning

Willis Carrier (1876–1950)

Some of the technologies that engineers create are essential for making life bearable in certain parts of the world. Air conditioning is a great example of this.

Invented by American engineer Willis Carrier in 1902, the idea for air conditioning is surprisingly simple. Take a liquid that boils at something like -20°F (-29°C). Squirt it at low pressure into a pipe. Inside that pipe the liquid will boil, leaving the pipe at -20°F. Then, take a pump to suck out the vapor. Compress the vapor at high pressure and stuff it into another pipe. That pipe will be hot with the heat of compression. As the vapor cools, it condenses back to a liquid. The liquid squirts back into the cold pipe and the cycle repeats. Put the cold pipe in the building with a fan, and the hot pipe outside with another fan, and voila—air conditioning.

Once this simple cycle was invented, engineers went to work making efficient, inexpensive motors and compressors, and creating systems in all different sizes. Today you can buy a small room air conditioner for $100. It is also possible to buy a complete building-wide air conditioning system big enough to cool off an entire **stadium** or a sprawling **airport**. Cars, planes, and boats have air conditioners, too.

Once perfected in buildings, air conditioning technology morphed to create the refrigerator/freezer. The box that makes up the refrigerator is essentially a very small, well-insulated room.

Several things that we take for granted today would not be possible without air conditioning and refrigeration: **frozen foods** and the modern server farm come to mind.

With the development of inexpensive, reliable air conditioning systems, areas around the globe that had been inhibited by extremely hot weather became bearable, and they grew rapidly. One key technology changed the entire economic landscape in these areas.

SEE ALSO Electric Refrigeration (1927), Frozen Pizza (1957), Irradiated Food (1963), *Toy Story* Animated Movie (1995).

This rooftop air conditioning system is located in Hong Kong.

1903

EKG/ECG

William Einthoven (1860–1927)

Imagine how useful it would be if engineers could develop an inexpensive, reliable device to monitor the heart. It would let doctors see how a patient's heart is performing simply by attaching a few electrodes to the patient's skin. It is a completely non-invasive procedure that helps a doctor monitor normal heart function as well as a number of abnormalities. The ECG machine, also known as an EKG machine, is the diagnostic device that makes this dream a reality.

The first reliable EKG machine was invented by William Einthoven, a doctor working in the Netherlands, in 1903. Because of the heart's very low signal strength, it takes a great deal of sensitivity to sense and record it. The system he invented did not use electrodes—instead patients placed their limbs in buckets of salt water to get a good connection. An extremely thin glass fiber coated in silver and placed in a strong magnetic field would move in response to the heart's faint signals. Why this strange approach? Because electrical engineers had not yet created reliable, flexible amplifier circuits.

By projecting an image of the string's vibrations on the screen, the EKG could be seen and interpreted. Einthoven then developed a standard naming scheme to identify the parts of the signal.

The development of the machine and the interpretation of the signals was such an important advancement that the doctor won the Nobel Prize for his work.

But this original machine weighed more than the patient and required several people to make it function. This is where engineers come in—after the scientific groundwork has been laid, they find ways to shrink the machine, better capture and display the signals, make things more reliable, and eliminate the buckets. And they succeeded spectacularly so that today an EKG machine is inexpensive and ubiquitous.

And in today's world they have taken things a step further. The computers found in portable **defibrillator** machines can now read EKGs, decide if a patient is experiencing a heart attack, and administer a life-saving shock, all completely automatically. This laid the groundwork for later artificial medical products like the **artificial heart**.

SEE ALSO Defibrillator (1899), Radio Station (1920), Heart-Lung Machine (1926), Artificial Heart (1982).

The modern EKG is considerably more efficient than Einthoven's original design.

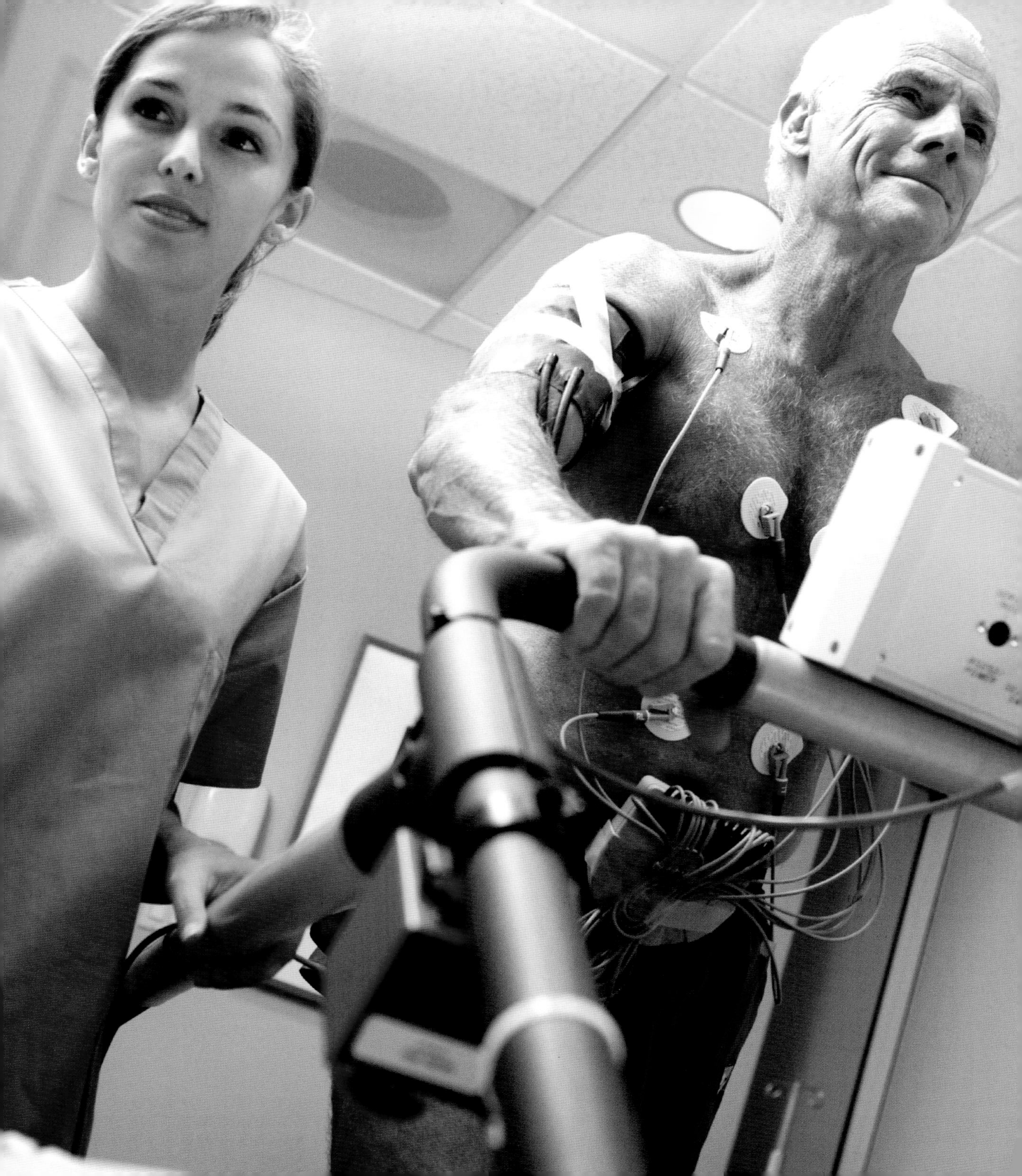

1903

The Wright Brothers' Airplane

Wilbur Wright (1867–1912), **Orville Wright** (1871–1948)

Airplanes are so familiar today that it's hard to imagine life without them. But at the start of the twentieth century, there was not a single airplane in existence. Many believed that humans would never fly.

The Wright brothers made the dream of flight a reality in North Carolina in 1903. They certainly were engineers, but they were also scientists and inventors. There were so many problems and fundamental questions that they needed to resolve: How to create lift? How to generate sufficient thrust? How to control flight? How to make the plane light enough? How to combine it all together?

For example, they built a wind tunnel and did fundamental research to discover wing shapes that provided maximum lift. Then they had to render those shapes as strong, lightweight structures—the bi-wing arrangement of the original Wright flyer using wood, fabric, and wire. Then they had to bend those structures during flight to control the plane. We look at their solution today as slightly bizarre—the entire wing warped, and they controlled warping with their hips. We consider their forward-mounted control surfaces to be strange as well. That's because the Wright Brothers started with a blank sheet of paper, with everything unprecedented and unknown. The conventions of rudder, elevator, and ailerons would evolve quickly once the brothers unlocked the core secrets of flight.

The entire aircraft weighed 605 pounds (275 kg) empty. How to get it off the ground? The **engine** seems primitive by today's standards. At 200 cubic inches (3.3 liters) and roughly 200 pounds (91 kg), it produced just 12 horsepower (9,000 watts). A small pan of gasoline in the engine's air intake served as a carburetor, contact breakers in the cylinders created the spark, and evaporating water cooled the engine. A man named Charles Taylor built it from scratch from a three-way dialog with the brothers. But it reliably produced its 12 horsepower to spin two counter-rotating, hand-carved wooden propellers.

It seems nearly impossible that three people could bring so many ideas and engineering disciplines together to create a flying machine that worked. Inspiration, curiosity, persistence, and the thrill of discovery powered them through.

SEE ALSO Two-Stroke Diesel Engine (1893), Boeing 747 Jumbo Jet (1968), Human-Powered Airplane (1977), Space Shuttle Orbiter (1981).

Pictured: The first sustained flight of the Wright Brothers' airplane.

1905

Engineered Lumber

Standard lumber comes from cutting up a tree trunk with a saw. Something like a 2x4 or a 2x6 is not engineered at all—it is wood in the exact form that nature made it, knots and all. The problem is that people often need something other than 2x4s. This is where engineering comes in. For example, while it is possible to make a floor out of 2x4s, it is wasteful in most cases. An engineered product like plywood, which was first produced by the Portland Manufacturing Company in 1905, is a much better option—lighter, thinner, cheaper.

It would be possible to span a 20-foot (6 meter) distance with a solid wood beam, but this too is wasteful. A wooden truss or a laminated product are two engineered forms that can do the same thing with much less wood. In other cases, engineers can take less desirable pieces of wood and turn them into something usable. This is the idea behind oriented strand board (OSB), in which small flakes of wood combine together with glue to create a sheet-like product similar to plywood. While lumber requires whole pieces of wood, and plywood requires full-size veneers, OSB can turn small flakes into a useful product rather than wasting them.

Another example is the wooden I beam. The two flanges and the web of the wooden I beam come from different types of laminates that are then glued together. The wooden I beam is significantly stronger and lighter than a solid beam of wood, and it eliminates problems like cracking, warping, and shrinkage over time. I beams used as floor joists can even help prevent squeaky floors.

Another common place to apply engineering is in roof trusses. **Trusses** are created on factory jigs, so they all match perfectly. They are brought to a construction site ready to install, and they go up very quickly, reducing construction cost. Because they use less wood, and less expensive wood, roof trusses cost less than a stick-built structure.

Engineered lumber shows the advantages of engineering in general: a product that provides the same functionality in a form that is lighter, stronger, and less expensive.

SEE ALSO Truss Bridge (1823), Instant Skyscraper (2011).

A wooden bridge containing engineered lumber in Montmorency forest crossing Montmorency River, Quebec, Canada.

1907

Professional Engineer Licensing

Imagine that your city wants to **dam** a river to create a new reservoir, but a million people live along the river below the dam. If the dam were to break, those million people would all die in the flood. Would you want twenty-two-year-old Joe Engineer with his freshly printed college degree in civil engineering designing that dam? Probably not. And there are lots of other things that you probably don't want twenty-two-year-old Joe Engineer being chief engineer on: just about any structure, any skyscraper, any vehicle falls into this category. If the object being engineered is important, chances are you would prefer a highly competent, highly experienced, highly disciplined engineer in charge of the project.

This is the purpose of Professional Engineer licensing, which began in the US in 1907. In the United States, there are four requirements that an engineer must meet in order to get a professional engineering license in his or her area of expertise:

- Get a four-year engineering degree from an ABET-accredited college. ABET is the Accreditation Board for Engineering and Technology. It ensures that colleges are delivering high-quality engineering degrees.
- Take the Fundamentals of Engineering (FE) exam and pass it. It is a five-hour-long exam broken down by specialty. For example, chemical engineers and civil engineers take different exams. Usually this exam is administered not long after graduation.
- Work for four years under a professional engineer, something like an apprenticeship in order to gain valuable experience.
- Take the Principles and Practice of Engineering (PE) exam and pass it. This is an eight-hour exam taken after serving for four years under a professional engineer and it is highly specialized. For example, there is one test for structural civil engineering and a different test for transportation civil engineering.

At that point an engineer can apply for a Professional Engineer license in a specific state. Once licensed, an engineer can sign and seal engineering plans that require a Professional Engineer. These would include all public works projects and many large-scale private projects.

SEE ALSO Gunter's Chain (1620), Rensselaer Polytechnic Institute (1824), Erie Canal (1825), Women's Engineering Society (1919), Tacoma Narrows Bridge (1940).

Engineering failures like the Quebec Bridge, which collapsed in 1907, prompted Professional Engineer licensing.

1908

Internal Combustion Engine

The first widely adopted internal combustion engine was found in the Model T Ford starting in 1908. The Model T engine was based on the Otto cycle engine, also known as the four-stroke engine, patented in 1861 by Alphonse Beau de Rochas. An incredible amount of engineering went into making the Model T engine cheap, reliable, and long-lasting, given the materials and manufacturing processes available at the time. Over fifteen million Model Ts had been manufactured when production ended in 1927.

The engine contained a number of engineering marvels. Materials engineers improved on the **Bessemer Process** to create vanadium steel, which is so strong that some Model T engines still run today. Electrical engineers created the trembler coil ignition system that helped the engine run on gasoline, kerosene, or ethanol. Engineers also created the thermosiphon system, which moved water through the radiator without a water pump. But the real heroes were the manufacturing engineers, who made it possible to eventually produce two million cars per year with amazing efficiency, keeping prices low.

But if you compare the Model T engine with today's engines, you can see that engineers since have been able to achieve a galaxy of improvements. The Model T engine had four cylinders displacing 2.9 liters, yet it produced only 20 horsepower. There are motorcycles today that produce 200 horsepower from one-liter engines. How is this possible? Engineers created overhead valve trains and high compression ratios to replace the flathead design on the Model T. They increased the redline and created fuel injection systems to replace carburetors. They designed much more powerful and precise ignition systems. They created tuned intake and exhaust systems.

Few technologies are in such widespread use and have been so highly refined without a reconceptualization. After over one hundred years of change, all of the core principles of the Model T engine—pistons, valves, spark plugs, water cooling, gasoline—are still in place. But each one has been fine-tuned and highly optimized by engineers to create the compact, reliable, high-power engines of today.

SEE ALSO Bessemer Process (1855), Wamsutta Oil Refinery (1861), Turbojet Engine (1937), Formula One Car (1938), Bugatti Veyron (2005).

Technician working on an internal combustion engine, possibly at Ford Motor Company, in 1949.

1910

Laparoscopic Surgery

Hans Christian Jacobaeus (1879–1937)

For hundreds of years, the process of surgery was pretty straightforward. A surgeon uses a scalpel to cut open a big hole in the patient—big enough so the surgeon has a clear field of view and room to fit his or her hands. The procedure is performed, and then the hole is sewn up.

The problem is that big holes in people create a lot of side effects. They may get infected. They cause more pain. Surgeons may need to cut through a significant amount of muscle. They take a long time to heal. They leave big scars. Recovery times increase. So doctors, scientists, and engineers contemplated the idea of reducing the size of the hole. Laparoscopic surgery is the result of this thinking. In 1910, Swedish internist Hans Christian Jacobaeus performed the first thorascopic diagnosis using a cytoscope—a complete reconceptualization of the surgical process. Although laparoscopic surgery cannot replace all types of operations, when it can be used, it eliminates the big hole and all of the problems that go with it.

The fundamental idea at the core of laparoscopic surgery is to eliminate the surgeon's need to directly see and directly touch the site of the operation. This means that several small holes (e.g., 1 cm or smaller) are cut in the patient. Two of the holes allow long, small, remotely operated tools to be inserted. One allows the insertion of a small camera and light. And one allows gas to flow in to keep the abdominal wall inflated and away from the surgical site. These small holes heal much more quickly than a single big hole would.

The surgeon now sees everything on a video monitor and does the entire procedure with long, small tools. There are gripping tools, cutting tools, and sewing tools that the surgeon inserts through the ports. She manipulates the tools with a scissor-like grip.

Laparoscopic surgery is a great example of how scientists and engineers can look at a problem and then find completely new ways of doing things that are significantly better than the obvious or traditional approach. When it happens, the new solution is often quite unexpected. Without developments such as this, refined medical processes such as the implantation of an **artificial heart** would not be possible.

SEE ALSO Defibrillator (1899), CT Scan (1971), MRI (1977), Artificial Heart (1982), Surgical Robot (1984).

Gastric bypass surgery being performed in a hospital.

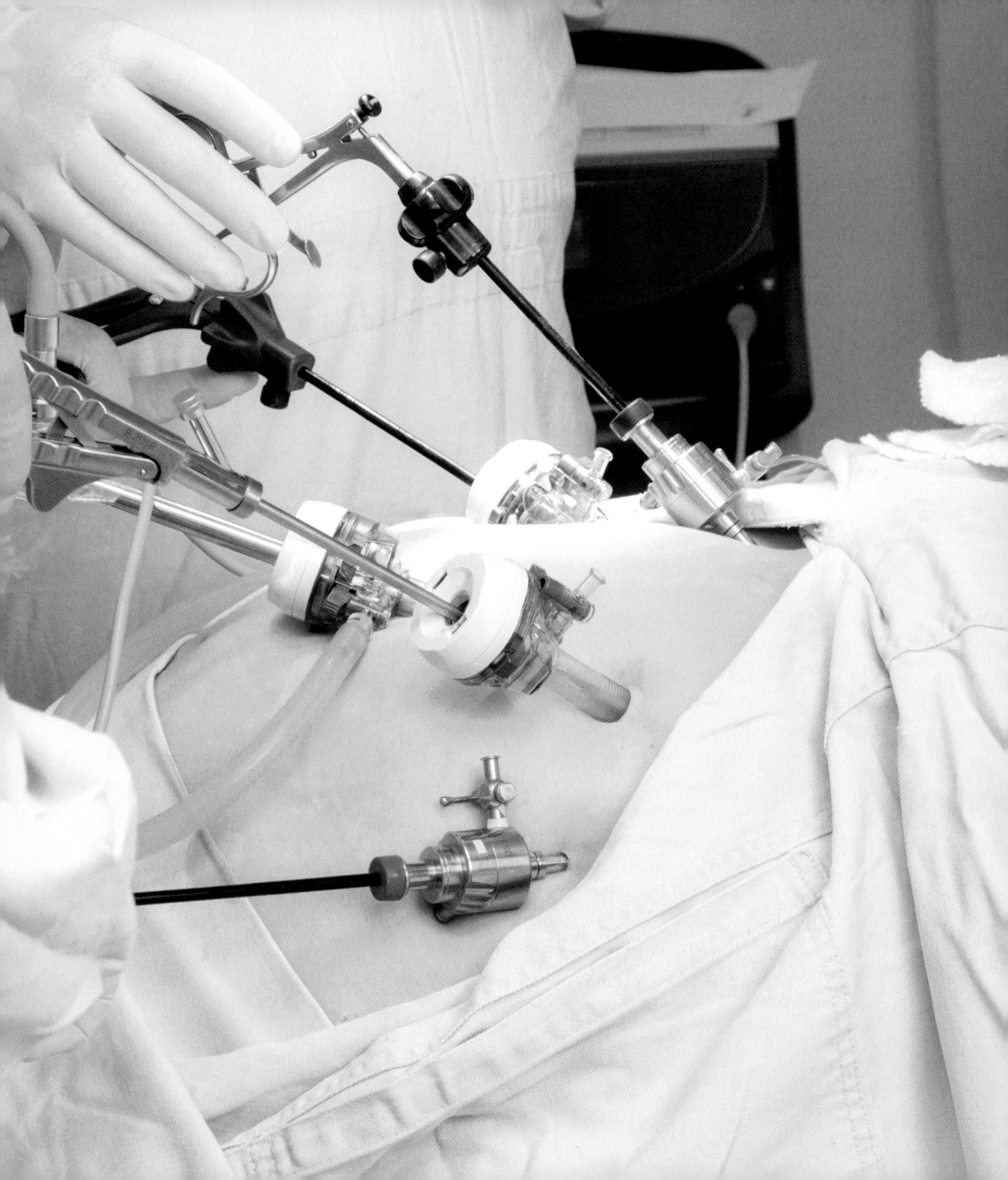

1912

Titanic

Thomas Andrews (1837–1912)

Think about the world's famous engineering failures: ***Apollo 13***, the **Tacoma Narrows Bridge**, **Fukushima**, and so on. The *Titanic* is the most famous of all. It had been claimed to be unsinkable, yet it sank on its maiden voyage. Media hype, new technology, the death toll, and famous passengers propelled *Titanic* to the top.

The obvious question is: How does an unsinkable ship sink? What did naval architect Thomas Andrews, as well as the engineers who assisted him, do wrong?

The RMS *Titanic* was 883 feet (269 meters) long and 92 feet (28 meters) wide. So imagine that we made a big bathtub with those dimensions and put it in the water. It would float just fine. Now we shoot a hole in the side of the bathtub with a cannon. Water will pour in though the hole and eventually the bathtub will sink. A good bilge pump solves the problem. If it can pump faster than the water comes in, then the bathtub can float indefinitely even with a hole in its side.

The engineers of the *Titanic* came up with the following scheme to make an unsinkable ship: They turned their boat into a series of 16 compartments by adding 15 sealable bulkheads. They also had a bilge pump system that could move approximately 4 million pounds (1.7 million kg) of water per hour. The idea was that if the *Titanic* sprung a leak, one compartment would flood. The bilge pumps would easily handle the water. But even if the pumps failed, one flooded compartment was fine. The ship was in fact unsinkable if that happened.

The problem is that when the *Titanic* sideswiped its iceberg, it compromised six compartments. Far more water came in than the pumps could handle. Six compartments started filling. Unfortunately, the bulkhead walls did not extend to the top of the hull. So when the water in one compartment reached the top of its bulkhead, the water spilled over into the next compartment. Thus the unsinkable ship filled with water.

In retrospect, there are a number of things that the engineers might have done to solve this problem. But apparently they never considered the possibility that one accident could breach six compartments. The *Titanic* sank in the North Atlantic Ocean on April 15, 1912 due to that failure of imagination.

SEE ALSO *Hindenburg* (1937), Tacoma Narrows Bridge (1940), *Apollo 13* (1970), Fukushima Disaster (2011).

RMS Titanic *departing Southampton, April 10, 1912.*

TITANIC

1913

Woolworth Building

Gunvald Aus (1851–1950), **Cass Gilbert** (1859–1934), **Korte Berle** (Dates Unavailable)

An architect or engineer working in the 1700s was using wood, stone, and brick for construction. Exterior walls and interior columns carried all the load. With stone, making a multi-story building taller means that the walls, especially at the bottom, have to be thick. Windows weaken the structure, so they have to be small.

In the 1800s, cast iron appeared. Columns and walls could be thinner and windows could be bigger, but building height was still severely limited. Seven or eight stories was the maximum height.

All that changed with the introduction of cheap steel created by the **Bessemer Process**. Because of steel's strength, frameworks of steel I beams could rise to amazing heights and less of the building's load had to be carried by exterior walls. The exterior walls of modern skyscrapers are called curtain walls to reflect how much their role has changed—the exterior walls can hang off the steel structure rather than vice versa. If engineers want the exterior of a building to be entirely glass, that is possible with modern structures.

Other technologies helped to enable the rise of skyscrapers. Skyscrapers would be impossible without elevators—no one would be willing to walk 50 flights of stairs. Auxiliary pumps and storage tanks make it possible to have water pressure in tall buildings. All of these elements came together in the late 1800s.

New York's Flatiron Building had 21 floors and used all of these techniques in 1903. But was it a "skyscraper"? Engineers upped the ante very quickly. Just 10 years later the Woolworth Building with 60 floors became the world's tallest building for the next 17 years. Its height and appearance make it undeniably a skyscraper in the modern sense. The Woolworth Building, designed by architect Cass Gilbert and constructed with the assistance of structural engineers Gunvald Aus and his partner Korte Berle, was an amazing engineering achievement at the time, with 34 high-speed elevators, 5,000 windows, and nearly a million square feet (93,000 square meters) of floor space.

Skyscrapers demonstrate a consistent engineering theme—when new technologies and materials become available, engineers can create new things.

SEE ALSO Basilica of Saint Denis (1144), Bessemer Process (1855), Statue of Liberty (1886), Empire State Building (1931), Burj Khalifa (2010).

Woolworth Building under construction, 1912.

ENERAL PHO TOGRAPHE
WOOLWORTH BLDG. IRON WORK

1914

Panama Canal

John Findlay Wallace (1852–1951), **John Frank Stevens** (1853–1943), **George Washington Goethals** (1858–1928)

To travel from New York to San Francisco by boat in the early 1900s was a major undertaking, consisting of two months and 13,000 miles (21,000 km) of travel around Cape Horn at the bottom of South America.

The Panama Canal became a much-needed shortcut in 1914, trimming 7,000 miles (11,300 km) off the journey by traveling across Panama. And the canal is a stunning engineering achievement. Engineers had to create a water pathway almost 50 miles (80 km) long through jungles and mountainous terrain.

The architecture of the Panama Canal is interesting because engineers could have solved the problem in several different ways. The principal engineers—John Findlay Wallace (from 1904–1905), John Frank Stevens (1905–1907), and George Washington Goethals (1907–1914)—chose to solve it this way: They built the world's largest dam at the time to form Lake Gatun, then the largest manmade lake. The Gatun locks on the Atlantic side raise ships up from sea level to the lake in three stages. When built, these locks were the largest concrete project ever conceived. Ships travel 20 miles (33 km) on the lake. They then pass through the Culebra Cut, an immense 7.8-mile (12.6 km) artificial channel created by blasting through a mountain. A lock lowers ships to Miraflores Lake, and two more locks lower ships to the Pacific Ocean.

After the canal was completed, operators noticed a problem with the water supply. Each time the locks cycle to raise or lower a ship, an immense amount of water flows from Lake Gatun into the ocean. In the dry season, there isn't enough water. In the wet season, too much rain arrives. Madden Dam and Alajuela Lake solved this problem by storing water and allowing operators to release it when needed.

When the canal opened, it was celebrated as a gigantic engineering achievement. The shortcut it created has saved billions of dollars. The process used to create it has also been refined in later projects such as the **Three Gorges Dam** in China.

SEE ALSO Great Wall of China (1600), Erie Canal (1825), Hoover Dam (1936), Container Shipping (1984), Three Gorges Dam (2008).

Seen here: The Panama Canal in 2009.

1917

Laser

Albert Einstein (1879–1955)

Laser pointers, laser scanners, laser printers, giant laser weapons . . . Why are lasers so common? A good part of it comes in the name: Light Amplification by Stimulated Emission of Radiation, or LASER. The Stimulated Emission part is important: laser light is usually a single, pure color with a single phase and a narrow beam that stays that way over long distances. As light goes, this is about as organized as it gets. Light from a normal bulb has photons with all different colors, all different directions, and all different phases.

When engineers got hold of organized laser light, they were able to do things they could not do with disorganized light. For example, the tight beam of the laser lets engineers shoot it at a tiny spot on the mirror surface of a DVD or CD and see whether the light reflects off cleanly. If it does, that represents a 1, if not, a 0. The tight beam and concentrated light also lets a laser cut paper, wood, or metal very precisely. The beam can contain hundreds of watts of power all concentrated into a pinprick of light.

Lasers make **fiber optics** possible. The basic idea is simple—a light at one end of the fiber turns on and off to send 1s and 0s to the other end. Lasers can switch on and off very quickly. The focused, powerful beam can travel dozens of miles through modern fiber before needing a repeater. And because of the precise color, it is possible to put multiple colors of laser light through the same fiber to multiply capacity. Massive amounts of data can flow through a single fiber—something that would be impossible without lasers.

Laser light shows us something interesting about engineers. Once the fundamental discovery is made (in this case by Albert Einstein, who set down the foundations for lasers and masers in a paper published in 1917), engineers often find many different ways to exploit it. The original laser sources were quite crude. Engineers find ways to make them smaller, faster, brighter, cheaper. These improvements open up new uses for lasers, and soon we find that engineers have built lasers into hundreds of products. It's the engineering way.

SEE ALSO Compass (1040), Trinity Nuclear Bomb (1945), Fiber Optic Communication (1970).

The laser guide star (LGS) at Yepun, one of the four individual telescopes that comprise the Very Big Telescope (VBT) in the Atacama Desert in Chile.

1917

Hooker Telescope

George Ellery Hale (1868–1938)

Back in the early 1900s, scientists had not yet discovered galaxies because they did not yet have a telescope big enough to resolve the stars in distant galaxies. The Hooker telescope, which saw "first light" after being commissioned by American solar astronomer George Ellery Hale in 1917, has a 100-inch (2.5 meter) mirror and was the first telescope big enough to do it. Using the Hooker telescope, Edwin Hubble was able to announce the existence of galaxies in 1929.

The Hooker telescope was the largest telescope in the world for three decades and the second largest for three more. It was an amazing engineering achievement at the time because of the many colossal, precise pieces that had to come together for it to work so flawlessly.

The first piece was the 100-inch mirror, made of a solid glass disk 12 inches thick and weighing 9,000 pounds (4,100 kg). A huge grinding machine was built to grind and polish it to its precise shape. This glass disk was silvered and then mounted in a framework able to hold it steady, keep it from flexing, and swivel and tilt it smoothly and precisely.

The framework and trusses that hold the mirror came from a battleship factory. Combined with the mirror, they weigh 100 tons (91 metric tons). The assembly sits on a mercury bearing to assure smooth operation. The mirror is backed by a network of water-filled pipes to maintain its consistent temperature. And the entire structure moves using a clockwork that keeps the telescope pointing at the same point in the sky for long periods of time as the planet rotates, allowing long-exposure photography of faint celestial objects. A 4,000-pound (1,800 kg) mass pulls on cables as it falls, and turns the whole telescope and dome very slowly in synch with the earth's rotation.

The engineering and precision of this telescope allowed it to take the photos in the 1920s that for the first time confirmed the existence of galaxies. Amazingly, the same telescope is still used today over one hundred years after the massive glass disk was cast. Future telescopes like **Hubble** and **Keck** were developed based on some of the fundamental ideas behind the Hooker telescope.

SEE ALSO The Hubble Space Telescope (1990), Keck Telescope (1993).

Plans for the 100-inch reflecting telescope at Mount Wilson Observatory.

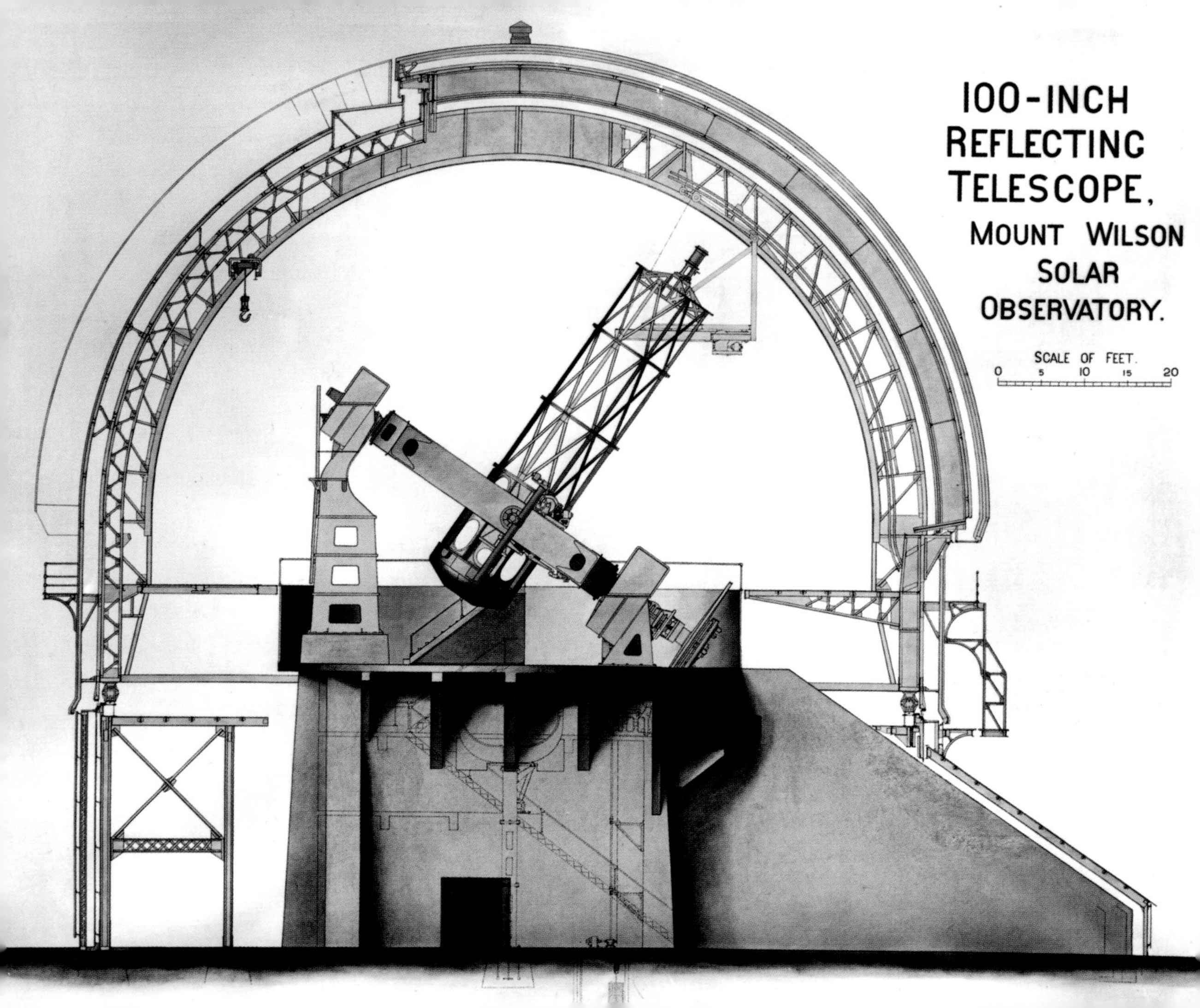
100-INCH
REFLECTING
TELESCOPE,
MOUNT WILSON
SOLAR
OBSERVATORY.
SCALE OF FEET.
0
5
10
15
20

1919

Women's Engineering Society

Verena Holmes (1889–1964)

By 1818, the Institution for Civil Engineers was formed in London, making it the oldest society for engineers. By the First World War, several other organizations had been formed, such as the Institution of Mechanical Engineers (established in London in 1847), the Association of German Engineers (1856), the Canadian Society for Civil Engineering (1887), the American Society of Mechanical Engineers (1880), and many others. Technical **colleges** specializing in engineering also proliferated.

However, an unprecedented number of women entered the workforce during World War I, and many continued to work once the war was over. England lost over 800,000 men, and so-called "surplus women" were left unmarried and often had to support themselves through gainful employment. Professions that had been closed off to them in the past were now opening up, in spite of the fact that some employers were inclined to send female workers home after the surviving soldiers returned from the front.

Verena Holmes had wanted to be an engineer all her life. After graduating from Oxford High School for Girls, she took night classes at a technical college and apprenticed as a drafter. During this time, she established the Women's Engineering Society to promote education and enable job training for women interested in engineering. She obtained her degree from the University of London in 1922, and in 1924, she was the first woman to be inducted as an Associate Member of the Institution of Mechanical Engineers.

The Society's journal, *Woman Engineer*, performed a number of useful services. For example, in 1924, editor Caroline Haslett held a contest to determine which imaginary home improvement would be most useful in freeing women from domestic drudgery: the winner was a **dishwasher** operated by a hand pump, followed by a thermostatic **oven**. Driven by the possibilities of electrical engineering for improvement within the home, she founded the Electrical Association for Women.

The Women's Engineering Society was followed by other organizations for female engineers, including the Society of Women Engineers (SWE) in the United States, which was established in 1950. Both groups offer scholarships, lectures, programs, and support for women around the globe who are interested in studying and practicing engineering as a profession.

SEE ALSO Rensselaer Polytechnic Institute (1824), Professional Engineer Licensing (1907), Top-Loading Washing Machine (1946), Microwave Oven (1946).

A 1920 cover of Woman Engineer.

THE WOMAN ENGINEER

THE ORGAN OF THE WOMEN'S ENGINEERING SOCIETY (Incorporated 1920).

VOL. I., NO. 10. MARCH, 1922. PRICE 6D.

PHOTO: KADEL & HERBERT, N.Y.

MRS. CARLIA S. WESTCOTT,
THE FIRST AMERICAN WOMAN MARINE ENGINEER.

(*See page* 138.)

THE "WOMAN ENGINEER" IS ISSUED QUARTERLY—PRICE 6d.

1919

Under Friction Roller Coaster

John Miller (1872–1941)

If you think about a basic roller coaster, it is very simple. You take a free-rolling car up to the top of first hill and then let it roll down the hill's other side. In the process, the car loses all of the stored potential energy and turns it into velocity. The next hill is a bit shorter to accommodate the losses from friction and drag. The process repeats.

If the builders are working with wood, a roller coaster will often be that simple.

But with the advent of the under friction roller coaster, patented by American entrepreneur John Miller in 1919, wheels under the cars facilitated even more creative routes. With the "Miller Under Friction Wheel," the coaster could go through twists and spirals. Now, with the advent of steel tubular track, the only real barriers are the g-force limits that the human body imposes on the design.

The traditional way to get up the first hill is by using a chain. It allows the coaster to gather its potential energy slowly over a minute or more. But roller coasters like the Hulk in Orlando take a completely different approach. Riders launch from the bottom of the first hill using 230 large electric motors attached to rubber tires. The car goes from zero to 40 mph (64 kph) in less than two seconds and comes flying off the top of the first hill straight into an inversion and the first drop.

To take a fully laden car with 32 passengers up a hill and accelerate it that quickly requires something like 11,000 horsepower (8,000,000 watts). Engineers can't just say to the average **power grid**, "I'd like eight megawatts for the next two seconds please" without some serious ramifications of the brownout variety. So the system uses flywheels to accumulate the needed power in the time between launches. Then the flywheels dump their accumulated power into the two-second launch. This evens out the load on the grid to a steady draw.

SEE ALSO Power Grid (1878), Ferris Wheel (1893), Harry Potter Forbidden Journey Ride (2010).

Ohio's "Son of Beast" roller coaster at King's Island.

1920

Kinsol Trestle Bridge

It's a classic shot in the old western movies—the wooden trestle bridge used to carry **train** tracks over a valley. The reason it is so classic is because these bridges were extremely common in the nineteenth century, especially for railroads, especially in the western United States. If you were building a railroad and it needed to cross a valley, chances are the engineers used a trestle bridge to do it. It was the easy, accessible, inexpensive thing to do.

The Kinsol Trestle in British Columbia, completed in 1920, is an example of a large wooden trestle bridge that survives today. It is 617 feet long (188 meters) and 144 feet (44 meters) high at the deepest part of the valley. It also has a graceful curve built into the structure. It is made of wooden timbers and bolts, with concrete footings to separate the timbers from ground contact.

The basic engineering idea behind a trestle bridge is pretty simple. The railroad track is sitting on top of a series of A-shaped frames. The outer two legs of the A shape provide side-to-side stability as well as the verticality. There may be one or more vertical center posts as well between those outer legs. Then a significant amount of horizontal and diagonal cross bracing helps to keep this structure rigid.

It is also possible to make a trestle bridge out of iron or steel, and there are many modern examples. The Tulip Trestle in Southern Indiana is a 2,307-foot-long (703 meters) bridge made of steel lattice beams, and is the longest trestle bridge in the United States. Another place you can see a trestle type of design today is in classic wooden **roller coasters** at amusement parks and in long wooden piers at the beach.

SEE ALSO Truss Bridge (1823), Diesel Locomotive (1897), Under Friction Roller Coaster (1919).

Train of the Tanana Valley Railroad, Alaska, USA, crossing a trestle bridge at the head of Fox Gulch.

1920

Radio Station

If we could get in our time machine, go back to 1912, and stand on the deck of the sinking ***Titanic***, there is something we would see overhead that marked the beginning of a new era in communication. The *Titanic* had two masts, one at either end of the ship, and a long wire stretching between them. This was the antenna for a 5,000-watt spark-gap radio, and the *Titanic* was using it to send out Morse code distress signals.

The *Titanic* put radio on the map. Because of that disaster, the Radio Act of 1912 required ships to monitor for distress calls 24 hours a day and set up a system for the US government to license radio stations.

By 1920, the first AM radio station was broadcasting in the United States: KDKA in Pittsburgh, PA. What had happened between 1912 and 1920 was the mass production of vacuum tubes, accelerated by World War I. Vacuum tubes gave electrical engineers the ability to create amplifiers for radio transmitters and receivers. Once engineers created the equipment, radio exploded in popularity. Everyone had to have a radio. By 1922 there were more than a million radio receivers in the US. Hundreds of organizations—newspapers, colleges, department stores, and individuals—had created radio stations. The Golden Age of Radio was born.

NBC started in 1926 and CBS started in 1927. Government regulation changed to make the advertising model possible in radio. With a revenue stream in place, there was a good reason for broadcasters to expand and plenty of money to pay for content.

This whole story is fascinating. The war led to tubes, which led to radios. The result was an entirely new way of thinking—instantaneous, electronic, free mass media to millions of people through nationwide networks funded by advertising. None of that existed in 1920. By 1930 nearly half of United States homes had radios. With the Great Depression starting, radio provided an inexpensive form of news and entertainment. Electrical engineering had created a massive societal change.

SEE ALSO Telegraph System (1837), *Titanic* (1912), Tape Recording (1935).

A young woman turns on an early-model radio.

1921

Robot

The first use of the word "robot" was in Czech writer Karel Capek's 1921 play, *RUR*. But what is a robot? When a **chess computer** beats a human, is that a robot? When the **Curiosity Rover** touches down on Mars, is that a robot? When an ATM gives you money, is that a robot? *Merriam-Webster's* definition of a robot is, "a machine that can do the work of a person and that works automatically or is controlled by a computer." By that definition, all three of these examples are certainly robots. Engineers create robots to replace the work of people. The goal of that robotic replacement may have to do with money, safety, convenience, monotony, disgust, or some combination. The reason we send a robot to Mars, for example, is because it is cheaper and safer than sending a person, and the trip there and back would be quite monotonous.

We see robots in many different roles today. Most manufacturing plants in the US are highly automated, with robots doing all of the welding, painting, machining, and molding. Many warehouses and shipping hubs are becoming completely robotic as well. Hundreds of autonomous robots can move across a warehouse floor simultaneously. **Self-driving cars** and trucks represent a robotic growth area. Every vehicle will soon be robotically controlled.

One thing that is missing right now is a good, general vision capability. A human being can straighten a shelf full of products, gather all of the shopping carts in a parking lot, sort laundry, and empty a dishwasher because a human being has a great vision system. As soon as such a system exists for robots, there will be an explosion of robotic functionality. It is quite possible that every job in retail, restaurants, construction, etc., could be eliminated once robots can see and have a little more dexterity.

What has made this possible is a combination across a variety of engineering disciplines. Motors, structures, sensors, computers, batteries, and power management systems all come together to make robots possible, and engineers have brought the capabilities up and the prices down. It is only a matter of time before engineers have robots doing all the work that people do today.

SEE ALSO Chess Computer (1950), Surgical Robot (1984), Watson (2011), Curiosity Rover (2012), Mars Colony (c. 2030).

Atlas is a bipedal humanoid robot primarily developed by the American robotics company Boston Dynamics.

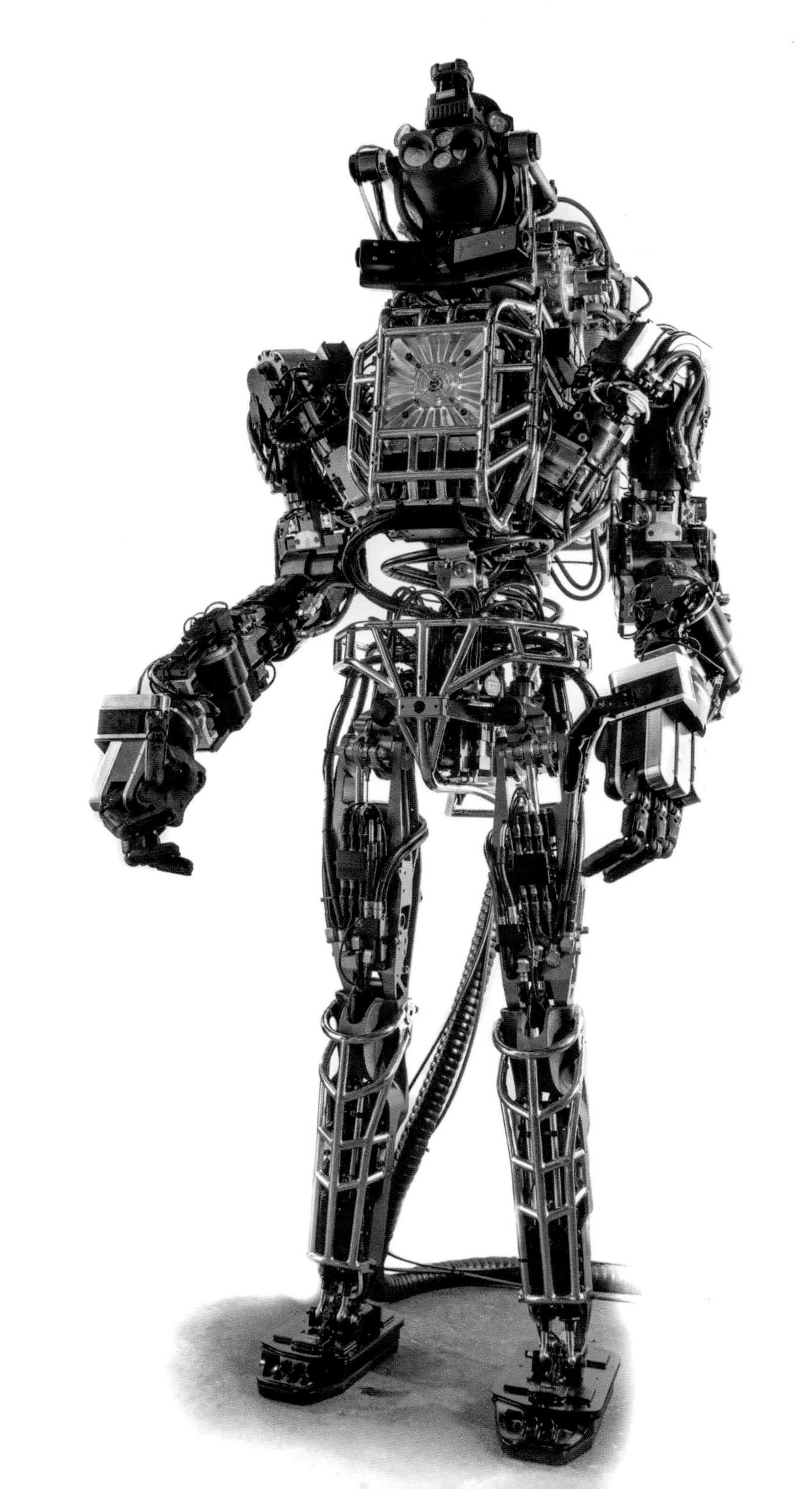

1926

Heart-Lung Machine

Sergei Brukhonenko (1890–1960)

The idea of replacing someone's heart with a donor heart seems nearly impossible. And so does heart repair, for example to fix a valve. The reason is because we all believe that when a person's heart stops beating, the person dies. In order to allow heart surgery to occur, engineers had to find a way to create a bridge over death, to keep the person alive while his or her heart is not beating. So engineers created the reliable heart-lung machine.

A heart-lung machine, developed by Soviet scientist Sergei Brukhonenko in 1926, is a device that takes over the pumping and oxygenating functions normally performed by a person's heart and lungs. The machine has to be able to do this, often for several hours, without in any way damaging the person's blood cells. There is also one other twist—the device needs to be able to connect in quickly and easily.

Engineers have solved all of these problems and heart-lung machines are used in hospitals hundreds of times each day.

The basic idea is simple and has four parts. There is a connection—a tube—usually into one of the heart's arteries or a femoral artery, to take blood from the body. A blood-friendly pump pulls the blood into the machine. A membrane system allows oxygen to flow into the blood and carbon dioxide to leave the blood. The blood is often cooled, which lowers the person's metabolism and therefore the need for oxygen. And then the freshly oxygenated blood flows back into the person's body through another tube.

Once the heart-lung machine has been installed, the person's heart can be stopped for repair, or even removed and replaced with an **artificial heart**. This is the truly impressive thing about the engineering of the heart-lung machine—it is able to take over a task that is vitally essential to human life for a period of time, giving surgeons the window they need to do their work.

SEE ALSO EKG/ECG (1903), Dialysis Machine (1943), Artificial Heart (1982), Surgical Robot (1984).

Heart-lung machine in a cardiac surgery.

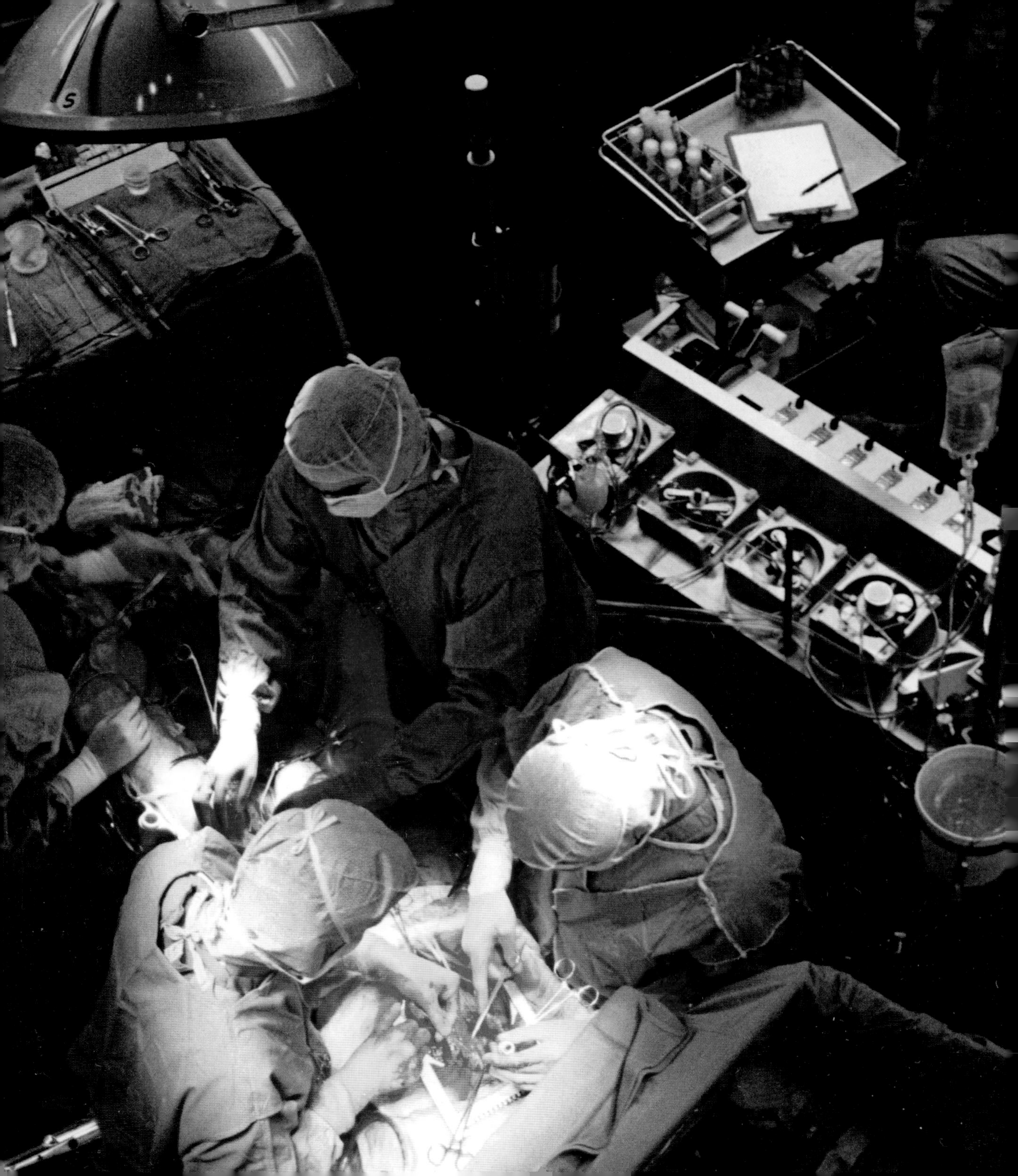
S

1927

Electric Refrigeration

Just about every home in the developed world has a refrigerator/freezer. Why? Because it helps food last longer. A gallon of milk sitting on the kitchen counter turns sour in just a few hours. Bacteria in the milk produce acid that curdles it. Put a piece of raw meat on the counter and it starts to go putrid as bacteria live their lives there.

By putting food in a refrigerator, the bacteria slow down. Now milk, meat, vegetables, leftovers, and more can last a week before bacteria can spoil them. In the freezer, bacteria stop cold, so **frozen foods** can last for months.

The first widespread technology for refrigeration came in the form of the icebox. The iceman would come by every few days and deliver the ice for it. Two problems: home ice delivery could get expensive, and the icebox was not quite cold enough.

To create a real refrigerator/freezer, engineers needed to take the same vapor compression idea used in air conditioning and make it smaller, more efficient, and safer. Early refrigerators were huge and they used poisonous liquids like ammonia. One key development was the miniaturization of motors and compressors, leading to GE's first electric home refrigerator in 1927. The invention of Freon as a refrigerant was also key. It was safe, inexpensive, and made things easier.; Refrigerator prices fell significantly. Soon just about every household had a refrigerator. Ice in the door arrived in 1965.

There was one fly in the ointment, however, that engineers did not anticipate. It turns out that Freon molecules can leak out of a refrigerator, and in the air they last a hundred years or more. They are not toxic to humans but the Freon molecules drift into the stratosphere and start destroying ozone. A ban on Freon and the development of new, safer refrigerants has reversed this problem, although it will be decades before it is completely solved.

SEE ALSO Water Treatment (1854), Air Conditioning (1902), Frozen Pizza (1957), Green Revolution (1961), Irradiated Food (1963).

Electric refrigerators were huge improvements over the icebox.

TIP-TOP
BREAD
CRISPER
CRISPER

1931

Empire State Building

Homer G. Balcom (1870–1938)

For 40 years the Empire State Building in New York City held the record as the tallest building in the world. This was just one of many records it set.

Since the time of its opening in 1931, the Empire State Building has been considered to be one of the world's great engineering achievements. It was originally designed by the architecture firm Shreve, Lamb, and Harmon; developed by John J. Raskob; and its chief structural engineer was Homer G. Balcom. The assembly process was highly tuned and impressively rapid—the elapsed time from first foundation work to completion was just 405 days. To make this kind of construction rate possible, the 58,000 tons of steel used in the building were all prefabricated into columns and beams in a factory, then trucked to the site—one of the early examples of just-in-time manufacturing. Workers used bolts and hot rivets to connect the steel pieces together. The rate of construction was 4.5 floors per week, with steel arriving at the site at a rate of approximately 10,000 tons per month.

Another key to quick construction was parallelizing the work. As workers finished each floor's steel framework, it was covered in heavy wood planking to create a platform for the construction of the floor above. Plumbers and electricians followed the steel workers in close succession, and the construction of the outer cladding started just weeks after the steel framework began rising.

The engineering of the outer cladding itself contributed significantly to the construction speed. If you look at the building above the first few floors, you see a regular pattern of chrome trim pieces rising vertically. Then there are the 6,400 windows. Between the windows vertically are cast aluminum panels, and beside the windows are limestone panels. Behind the aluminum and limestone are eight-inch-thick brick walls made of ten million bricks resting on steel beams at each floor. Plaster covering the bricks forms the interior walls. Each floor is a concrete slab poured in place. Engineers have developed innovative modular building practices, setting new records many decades later, and making unimaginably tall buildings like **Burj Khalifa** possible. But given the technology available in 1931, the construction speed and height of the Empire State Building is still impressive even today.

SEE ALSO Tuned Mass Damper (1977), Burj Khalifa (2010), Instant Skyscraper (2011).

Aerial view of New York, focusing on the Empire State Building.

1935

Tape Recording

Fritz Pfleumer (1881–1945)

Think how useful it is to be able to record sound. Today we make audio and video recordings at the touch of a button on our **smart phones**, and we do it constantly.

The popularity of recording began nearly the instant the technology was brought to market in 1888. The original technology—wax cylinders and a needle—made it possible for people to record and play back sounds. What really took off was prerecorded music, and it has been popular ever since.

Tape recording started in 1935 in Germany with the invention of the first practical tape recorder, the AEG Magnetophon K1, which was developed by German electronics company AEG and based on the invention of magnetic tape by engineer Fritz Pfleumer. A coating of ferric oxide on a thin plastic backing formed the tape, thousands of feet of which could fit on a real. A motor pulled the tape across a read-write head to another reel. A varying current in the head created a magnetic field to record onto the ferric oxide. Or the recording in the ferric oxide would create a small current in the head, which could then be amplified and heard. The initial systems were incredibly simple like this, yet they yielded very realistic recordings.

Once engineers had the basic technology of recording and playback nailed down, they morphed the technology in several different directions. The 8-track cartridge and then the compact cassette made tape much easier to use. Helical recording heads made **video recording** possible, and this morphed into VHS systems that allowed people to record TV shows and play movies at home. Computer tape drives were incredibly important in the 60s and 70s for data storage. Watch any older sci-fi film like *Dr. Strangelove* and spinning tape drives add a high-tech feel. Engineers improved the form factor to create floppy disks for computers as well. It is the same technology rendered onto a disk to make random access quicker. We even find short pieces of magnetic tape on the back of our credit cards.

Tape recording is a great demonstration of how engineers can take one technological idea and spin it off in many different directions.

SEE ALSO Hard Disk (1956), VHS Video Tape (1976).

The first practical tape recorder was developed in 1935; this model is from the late 1970s.

PLAY REC
PAUSE
POWER
TRACK SELECTOR
STEREO
REC
MONITOR
S-O-S
EQ
TAPE
SELECTOR
LINE — REC LEVEL — MIC/DIN
MIC
OUTPUT
PHONES
VU
VU
TAPE
SOURCE
OFF
ON
LOW NOISE
WIDE RANGE

1936

Hoover Dam

John L. Savage (1879–1967)

It is easy to take water for granted—until it is taken away. If you don't live near a water source, or there is a drought and the source goes dry, there is a big problem. Without water, a human being dies in a couple of days.

The Hoover Dam, completed in 1936, is an amazing engineering achievement for two reasons. The first is because it formed a giant manmade lake in the middle of a desert. When full, holding a total of 7,700 cubic miles of water (32,000 cubic kilometers), Lake Mead is the biggest manmade lake in the US. Its water keeps cities and farms in Arizona, Nevada, and California from drying up.

It is also an engineering achievement because of the dam itself. There are three basic dam types. A gravity dam is a big pile of dirt, rocks, and/or concrete that holds the water back through its sheer weight. An arch dam is a thin concrete arch, usually between two rock walls. The Hoover dam is an arch-gravity dam that combines both techniques, because the lake behind it is so huge.

John L. Savage, the chief design engineer appointed by the Bureau of Reclamation to oversee completion of the dam, faced dozens of significant challenges when designing and building it. Where would all the concrete come from? How to move so much concrete before it cured? How and where to divert such a big river? How to create electricity from the water flowing through such a large dam? Perhaps the most interesting problem was the heat generated by the curing **concrete**. The dam contains 3.25 million cubic yards (2.5 million cubic meters) of concrete. If poured all at one time, the concrete at the core would heat up to hundreds of degrees, destroying the dam. So the concrete was poured in blocks, with pipes carrying refrigerated water through each block to cool it. Later the pipes were filled with grout.

The dam is so massive, so tightly constructed, and so perfectly fitted into its canyon that it should last, approximately, forever.

SEE ALSO Concrete (1400 BCE), Desalination (1959), Itaipu Dam (1984), Three Gorges Dam (2008).

Hoover Dam is a concrete arch-gravity dam in the Black Canyon of the Colorado River, on the border between the US states of Arizona and Nevada.

1937

Golden Gate Bridge

Joseph Strauss (1870–1938), **Leon Moisseiff** (1872–1943), **Charles Alton Ellis** (1876–1949)

At 4,200 feet (1,280 meters) long in its main span, the Golden Gate Bridge, opened in 1937, is still one of the top ten longest suspension bridges on earth today. Amazingly, engineers and hundreds of construction workers built this massive structure in just four years.

The original designer and chief engineer of the bridge was Joseph Strauss, assisted by Leo Moisseiff (who later designed the **Tacoma Narrows Bridge**). In collaboration with Moisseiff , senior engineer Charles Alton Ellis was the principal engineer for the project. The engineers judged that a suspension bridge with just two towers was optimal because the water in San Francisco Bay is so deep in the main channel, and many large ships enter San Francisco's harbor. A multiple tower approach could not work on this site. Spanning the 4,200 feet between the two towers could not possibly be done with an arch, a truss, or cantilevered sections in any practical way.

The towers sit on massive concrete foundations poured on top of bedrock. The south tower is in the water, and 40-foot–thick (12 meters) fender walls made of steel-reinforced concrete surround its base. These walls prevent ships from crashing into the tower. The towers themselves are made of steel and rise to a height of 746 feet (227 meters) above the water.

The two main suspension cables are 3 feet (1 meter) in diameter. However, they are not solid steel. Instead, they were "spun" in place out of 27,000 lengths of steel wire. Two reasons why: first, it was the only practical way to do it, and second, the cables need to flex—in strong winds, the bridge moves up to 27 feet (8.2 meters) side to side. Suspender cables hang down to hold the deck for the road. Trusses support the deck between the suspender cables.

The last pieces of the puzzle are the two anchorages at the ends of the bridge. The towers are being pulled toward each other by the massive weight of the cables, deck, and roadway. To keep the towers standing, opposing cables pull them outward by attaching to anchorage blocks weighing 130 million pounds (60 million kg) each.

SEE ALSO Truss Bridge (1823), Tacoma Narrows Bridge (1940), Millau Viaduct (2004).

View from the southern tower of the Golden Gate Bridge.

1937

Hindenburg

Every now and again, engineers will create something, put it out in public, and people will look at it and say, "That does not seem like a very good idea." Or they might say, "That looks like a recipe for disaster." Or perhaps, "There must be a better way." Sometimes engineers do these things out of ignorance, sometimes because of economics, sometimes pure hubris or delusion. Single-hull **supertankers** are one example. Anyone could look at that idea and say, "Well, what if you run into something? Then there's going to be a million gallons of crude oil in a marine ecosystem." The idea of dumping gigatons of carbon dioxide into the atmosphere falls into the same category.

And then there was the *Hindenburg*. The *Hindenburg* was a rigid airship that derived its lift from a lighter-than-air gas. The original idea was to use helium as the lifting gas, but the German company Zepplin that made the *Hindenburg* could not secure enough helium. So they switched over to hydrogen as the lifting gas. Hydrogen is cheap, easy to make, and is the lightest lifting gas available. Its big problem is that it is easy to ignite and extremely flammable. The Zepplin company felt they had taken all the needed steps to guard against an explosion, and the *Hindenburg* operated without incident over tens of thousands of miles.

While it was operating, the *Hindenburg* was an engineering marvel—a luxury cruise for 50 passengers in the air. The ship was just over 800 feet (245 meters) long, with the passenger compartment being tiny by comparison. The passenger accommodations might be compared to those of a cruise ship of the era, but the *Hindenburg* was quite a bit faster with speeds up to 75 mph (125 kph).

But then, on May 6, 1937, something went wrong. The hydrogen ignited and the entire load of gas—something like 7 million cubic feet (200,000 cubic meters) of hydrogen—went up in flames in just a few seconds.

Since then, there have been many efforts to resurrect both blimps and rigid airships, using helium obviously, rather than hydrogen. The idea is so tantalizing because the lift provided by the lighter-than-air gas requires no fuel and is therefore free. The problem is that the ships have to be so gigantic to get that lift that they can't go very fast and have problems in severe weather. Compared to jet airplanes going 500 mph in any kind of weather, dirigibles have trouble competing.

SEE ALSO *Titanic* (1912), *Seawise Giant* Supertanker (1979).

This photo, taken during the initial explosion of the Hindenburg, *shows the 804-foot German zeppelin just before subsequent explosions sent the ship crashing to the ground at Lakehurst Naval Air Station in Lakehurst, NJ, USA, May 6, 1937.*

1937

Turbojet Engine

Frank Whittle (1907–1996)

The ramjet engine is about as simple as engines get. It is basically a tube traveling through the air like an arrow. At the front opening, air rams into the tube because of the tube's velocity. An inlet cone aids the process, providing compression. In the middle of the tube, fuel sprays into the air stream and burns in the combustion area, assisted by a flame holder. The burning fuel creates heat, expanding the air significantly. Since air is ramming in the front of the tube, the expanding gas rushes out the back, creating thrust. The problem is that a ramjet engine does not work when it is standing still. Frank Whittle, a British engineer air officer, produced a solution to this in the late 1930s, which lead to the first successful turbojet engine appearing in 1937.

A turbojet engine has a powerful fan at the front inlet. The fan moves air through the engine even when standing still. Engineers developing Whittle's initial concept evolved this idea to a multistage compressor—a series of fans and blades. By compressing incoming air by a factor of 10 or more, the engine can burn additional fuel and create more thrust. An electric motor could turn the compressor, but engineers realized that another turbine in the exhaust stream is a better approach. The exhaust turbine connects to the compressor with a shaft running through the center of the engine: a very efficient design.

However, the exhaust turbine is sitting in an extremely hot airstream. Materials scientists and engineers solved this problem by creating new alloys and techniques to prevent meltdown. Second, the shaft is turning at high rpm, so it is important for the system to be balanced and lubricated.

By using the shaft's rotation to additionally turn a large bypass fan, the efficient turbofan engine was born in 1943. With the turbofan's efficiency, engineers make air travel affordable for millions. By adding a power takeoff shaft, the gas turbine engine seen in the **Apache Helicopter** and **M1 tank** is born.

SEE ALSO Jet Engine Testing (1951), M1 Tank (1980), Apache Helicopter (1986).

Pictured: Close-up of a turbojet engine of a plane.

1937

Magnetically Levitated Train

When a **train** is moving at speeds below 100 mph (160 kph), steel wheels work well. They are low cost and efficient without any major downsides. Above 100 mph, wheels start to have problems that become increasingly difficult for engineers to solve. One of those problems is vibration. Another is acceleration and **braking**. At some level of braking, steel wheels lose their grip on the track and skid. The simple act of keeping the train aligned on the track with wheel flanges causes friction, especially on curves.

The solution to all of these problems is to eliminate the wheels. The best replacement available is magnetic levitation. The concept of a "maglev" train has existed since German engineer Hermann Kemper received a number of patents for the device from 1937 through 1941.

A magnetic levitation train actually implements a package of three magnetic effects: One set of magnets lifts the train off the ground so it is floating. Another set of magnets keeps the train on the track in a left and right sense, especially as the train goes around curves. A third set of magnets (often implemented in conjunction with the lifting magnets) creates a linear motor to accelerate and decelerate the train.

By modulating a set of electromagnets, the train can control its precise height above the ground and its rates of speed, acceleration, and deceleration.

In addition to the benefits of ride smoothness, acceleration, and deceleration, getting rid of the wheels also eliminates the friction that they cause. However, at speeds over 200 mph, most of the energy is consumed overcoming aerodynamic drag. Therefore, engineers pay close attention to aerodynamics, which are as important as they would be in a jet aircraft.

The Shanghai Maglev Train in China, which began service in 2004, was the first public system running at over 200 mph (320 kph). It has seen a top speed of 310 mph (501 kph) and does regular runs (over 100 per day) of its 19-mile (30 km) track in seven minutes.

SEE ALSO Tom Thumb Steam Locomotive (1830), Diesel Locomotive (1897), Anti-Lock Brakes (1971), Vactrains (c. 2020).

Shanghai Maglev Train or Shanghai Transrapid is the first commercially operated high-speed magnetic levitation line in the world.

SMT

Formula One Car

Gioacchino Colombo (1903–1988)

A Formula One car is shockingly fast—so fast that the car pushes the limits of a human's ability to handle g-forces.

How do engineers create a car this fast, with a top speed over 200 mph (322 kph) and the ability to accelerate and decelerate so quickly? It all began in 1938 with the Alfa Romeo 158. Its principal engineer was the Italian automobile designer Gioacchino Colombo. Incredibly, the car won every race it entered in the first season of Formula One in 1950. Since then, Formula One cars have advanced considerably.

A Formula One car's performance starts with the **turbocharged** engine, and one key is the engine's maximum rpm. A typical passenger car might top out at 6,000 rpm (revolutions per minute). Formula One (F1) engines can exceed that by three times or more. It means that an F1 engine can process three times more fuel per unit of time than a normal engine of the same size. The engine's power-to-weight ratio is stellar. It can rev so high because a team of mechanics tune it to perfection and it only has to last two races.

Weight is important—it controls how fast the car can accelerate and decelerate. Formula One cars weigh less than 1,100 pounds (500 kg). One lightening technique: the engine replaces part of the frame. The engine block serves as a block, but also as the stressed member connecting the rear wheels and transmission to the rest of the car. The car is also lightened by the **carbon fiber chassis** and lightweight alloys used throughout.

And then there are the aerodynamics. The front and rear wings on an F1 car provide downforce instead of lift—so much downforce that an F1 car could drive inverted, sticking to the ceiling of a tunnel. It glues the tires to the track for better cornering and braking.

Formula One cars are a playground for engineers, with the F1 rules acting as their constraints. F1 cars contain maximum engineering for maximum performance.

SEE ALSO Supercharger and Turbocharger (1885), Internal Combustion Engine (1908), Bugatti Veyron (2005).

Modern Formula One cars can reach speeds of over 200 mph (322 kph).

Santander
V-Power
Santander
Santander
7
KASPERSKY
infor
Santander

1939

Norden Bombsight

Carl Norden (1880–1965)

If the Air Force wants to drop a bomb precisely onto a target today, the bomb is equipped with a set of fins that can be adjusted by a computer connected to a **GPS** receiver and inertial guidance system. Before the bomb is dropped, the computer receives the latitude and longitude of the target. As long as the bomb releases anywhere in the general vicinity of the target, the bomb will hit that target precisely.

Now let's get into our time machine and travel back to the 1930s. There are no GPS satellites. Yet, with World War II being fought in Europe, there is a need to drop bombs from airplanes and hit targets as accurately as possible. To solve this problem, Dutch-American engineer Carl Norden developed the Norden bombsight, which was introduced in 1939.

Before the bombsight's inception, airplanes dropped gravity bombs with no guidance systems. The gravity bomb followed an arc once the airplane released it. The arc's exact shape was controlled by the plane's speed and heading, the wind's speed and heading, and the altitude of release. Therefore, there is an exact place and time when the airplane needed to drop the bomb during the bombing run. The Norden bombsight gets the plane to the right place and calculates the right time so the bomb arc precisely intersects with the target.

Given the lack of **microprocessors**, this was not a trivial problem. The Norden bombsight used two systems: a gyro-stabilized platform to keep the bombsight level during the bombing run, and a mechanical computer to do the calculations. The key to the effectiveness of the mechanical computer was the user interface. After initial setup, the bombardier would put the target in the crosshairs and then adjust knobs so that the target stayed in the crosshairs without any drift. The mechanical computer would then release the bombs at the right time.

The Norden bombsight had a great advertising campaign associated with it once its top-secret classification was removed. It was never as accurate in real-world situations as claimed publicly. But given the technology available, it was a remarkable piece of engineering.

SEE ALSO Trinity Nuclear Bomb (1945), Cluster Munition (1965), Microprocessor (1971), Global Positioning System (GPS) (1994).

Pictured: A view through the Norden bombsight.

1939

Color Television

Peter Carl Goldmark (1906–1977)

In the 1950s, there were millions of black and white TVs in use, receiving free broadcast television channels. How did engineers bring color TV to the masses without orphaning all of those existing TV sets? This change from black and white to color TV represents a decision that engineers must frequently make: Is there value in being backward compatible with existing equipment, or is it time to abandon all of the old hardware?

In order to make a backward-compatible system for color TV, engineers would have to develop a new color signal that contains all of the necessary information for color pictures. At the same time, the signal must impersonate an older black and white signal so that it works in old B&W TVs that were built with no intention or forethought of understanding the new color signal.

This problem was solved in a remarkable way: the color information became encoded as a sine wave rippling along the normal, existing B&W signal. Older black and white TVs would ignore this new sine wave, while new color TVs could decode it.

The other problem engineers had to solve was also challenging: how to create a device that could display this new color signal. A first attempt, developed by German-Hungarian engineer Peter Goldmark for RCA in 1939, used a mechanical color wheel rotating in front of the display. This had a number of problems: it suffered from flicker and the mechanical color wheel was three times bigger than the screen itself. It felt like a kluge. Another system involved multiple CRTs (cathode ray tubes) and mirrors.

The ultimate solution was perfected and popularized by RCA in 1953—engineers developed a new form of CRT using three electron guns, three colors of phosphor in the screen, and a new shadow mask between the guns and the phosphors. With the backward-compatible color signal in place to avoid obsoleting millions of B&W sets and a new, all-electronic picture tube to display it, color TV could take off. Engineers had created an elegant system. The first national color broadcast was the Tournament of Roses Parade in 1954. The shift from analog to **HDTV** in 2006 echoed this original innovation.

SEE ALSO Radio Station (1920), Stadium TV Screen (1980), HDTV (1996)

Display of color television sets for sale in a department store, circa 1960s.

COLOR
COLOR

1940

Tacoma Narrows Bridge

Leon Moisseiff (1872–1943)

Imagine that you are an engineer designing a bridge. You follow the best practices of the day, but then the bridge collapses in spectacular fashion. When this happens, engineers learn that their equations are not perfect. The result: best practices are updated, equations are modified, and the world of engineered structures becomes a little bit safer. This is exactly what happened with the Tacoma Narrows Suspension Bridge, designed by Leon Moisseiff, when it disintegrated in 1940.

Suspension bridges were quite common in 1940. Moisseiff himself had designed the Manhattan Bridge in 1909 and the **Golden Gate Bridge** in 1937. These were gigantic bridges with gigantic spans, and Moisseiff became a highly regarded expert on designing suspension bridges.

So when Moisseiff designed the Tacoma Narrows Bridge across the Puget Sound near Seattle, it probably seemed straightforward. The bridge would need towers, the suspension cables, anchors for the cables, the suspenders, and the deck. There were budget constraints, so when Moisseiff proposed a less expensive design using smaller girders to support the deck instead of the 3x larger trusses that engineers had used in the past, his design won out for two reasons: 1) it was less expensive to build because it contained less steel, and 2) the thinner deck cross section was aesthetically pleasing. Moisseiff himself called the Tacoma Narrows "the most beautiful bridge in the world."

The problem: engineers did not fully understand aeroelastic flutter. If you have ever experienced a taught rope vibrating in strong wind, you have seen the phenomenon. The bridge's girders were not stiff enough, and in a strong wind it began to flutter too. In this case, the deck acted as an immensely long and heavy rope, so it fluttered with huge up and down motions that caused it to rip itself apart. If you watch the videos of the bridge before the collapse, the amount of motion seems impossible.

Although fortunately there were no human casualties, engineers analyzed the failure. Since then, the decks of suspension bridges must be stiff enough to prevent this problem.

SEE ALSO Truss Bridge (1823), *Titanic* (1912), The Golden Gate Bridge (1937), *Apollo 13* (1970), Fukushima Disaster (2011).

Pictured: The Tacoma Narrows Bridge collapsing.

1940

Radar

Heinrich Hertz (1857–1894), **Robert Watson-Watt** (1892–1973), **Arnold Frederic Wilkins** (1907–1985), **Edward George Bowen** (1912–1991)

In World War I, airplanes were a recent invention, and a major threat. While soldiers could shoot down airplanes dropping bombs, they needed plenty of advanced warning, because airplanes are fast. The problem: there was no good way to sense airplanes from a distance. The best technology at the time used gigantic ear horns to hear airplanes as they approached. Today, these contraptions seem ridiculous, and they did not work particularly well. However, by World War II, the technological landscape had changed quite a bit.

The word "radar" stands for Radio Detection And Ranging. Research for what would become radar began as early as 1886, when German physicist Heinrich Hertz conclusively proved the existence of electromagnetic waves. However, it wasn't until the Second World War that radar was used to detect airplanes at a distance using radio waves. The Chain Home system was deployed by Britain in 1940. The idea—developed by Robert Watson-Watt (descendent of James Watt, who invented the **steam engine**), Arnold Frederic Wilkins, and physicist Edward George Bowen—was to have a radar system that could look out from Britain and detect airplanes approximately 100 miles (160 km) away. The system was extremely important in the Battle of Britain, where Germany tried to use an aerial campaign to force a British surrender.

Radar is based on a simple idea: Send out a pulse of radio waves, then look for radio waves that get reflected back. By measuring the length of time taken for the reflection to return, a good estimate of distance is possible. An oscilloscope-like display showed blips of reflected energy on a screen. Interestingly, the Chain Home system contained the engineering philosophy known as incremental improvement. The first iteration was nowhere near perfect or ideal, but it was good enough to receive reflections and provide some sense of distance and direction. Improved systems were developed and installed over time. The power of the transmitters increased, giving better range. New systems detected airplanes closer to the ground. Rotating antennas and much better displays arrived.

Electrical engineers had developed a completely new sense for human beings. Today we still use radar for many applications, including **self-driving cars**.

SEE ALSO High-Pressure Steam Engine (1800), Cluster Munition (1965), F-117 Stealth Fighter (1983), Self-Driving Car (2011).

A Japanese sound detector, made obsolete by the discovery of radar.

國
七一
愛

1940

Titanium

William Justin Kroll (1889–1973)

Imagine that you are an engineer and you are commonly working with both aluminum and steel. Steel has an advantage because it is inexpensive and strong, but it is relatively heavy. Aluminum has an advantage because it is lighter than steel by about 50 percent for the same strength. But it melts at a much lower temperature and it suffers from fatigue faster.

As an engineer you would ask yourself, "Is there a metal that is light like aluminum and strong like steel and keeps its strength at higher temperatures?" The answer would be "titanium." This is why the **SR-71** aircraft is made mostly of titanium. There is no way the SR-71 could get off the ground if it were made of steel—it would be too heavy. There is no way the SR-71 could fly at Mach 3 if it were made of aluminum—the skin of the SR-71 gets too hot and an aluminum SR-71 would disintegrate. Titanium solves the problem. In addition, there are no rust, corrosion, or fatigue problems.

So what's not to like about titanium? Why aren't all airplanes made of this metal? The big thing is cost. In 1940, the Kroll process, developed by metallurgist William Justin Kroll, made titanium cheap enough to consider using it outside the lab, but no one has yet come up with a truly inexpensive process for producing titanium. What this means in round numbers is that if a pound of steel costs X, then a pound of aluminum costs 2X and a pound of titanium costs 20X.

The other problem with titanium is its workability. There is not really a way to make cast titanium parts. Machining titanium is slower and more problematic than steel. And welding titanium requires quite a bit of care to avoid faulty or contaminated welds. Sometimes you see titanium jewelry with pretty colors. The cause of those colors is oxygen contamination, which is bad structurally. Any kind of oil (including fingerprints) on the metal can also cause contamination.

Besides high-speed airplanes and rockets, engineers use titanium inside the human body because it is strong, light, and inert. If you have an artificial hip, it is probably titanium.

SEE ALSO Bessemer Process (1855), Hall-Héroult Process (1889), SR-71 (1962).

Pictured: X-ray of a hip prosthetic made of titanium.

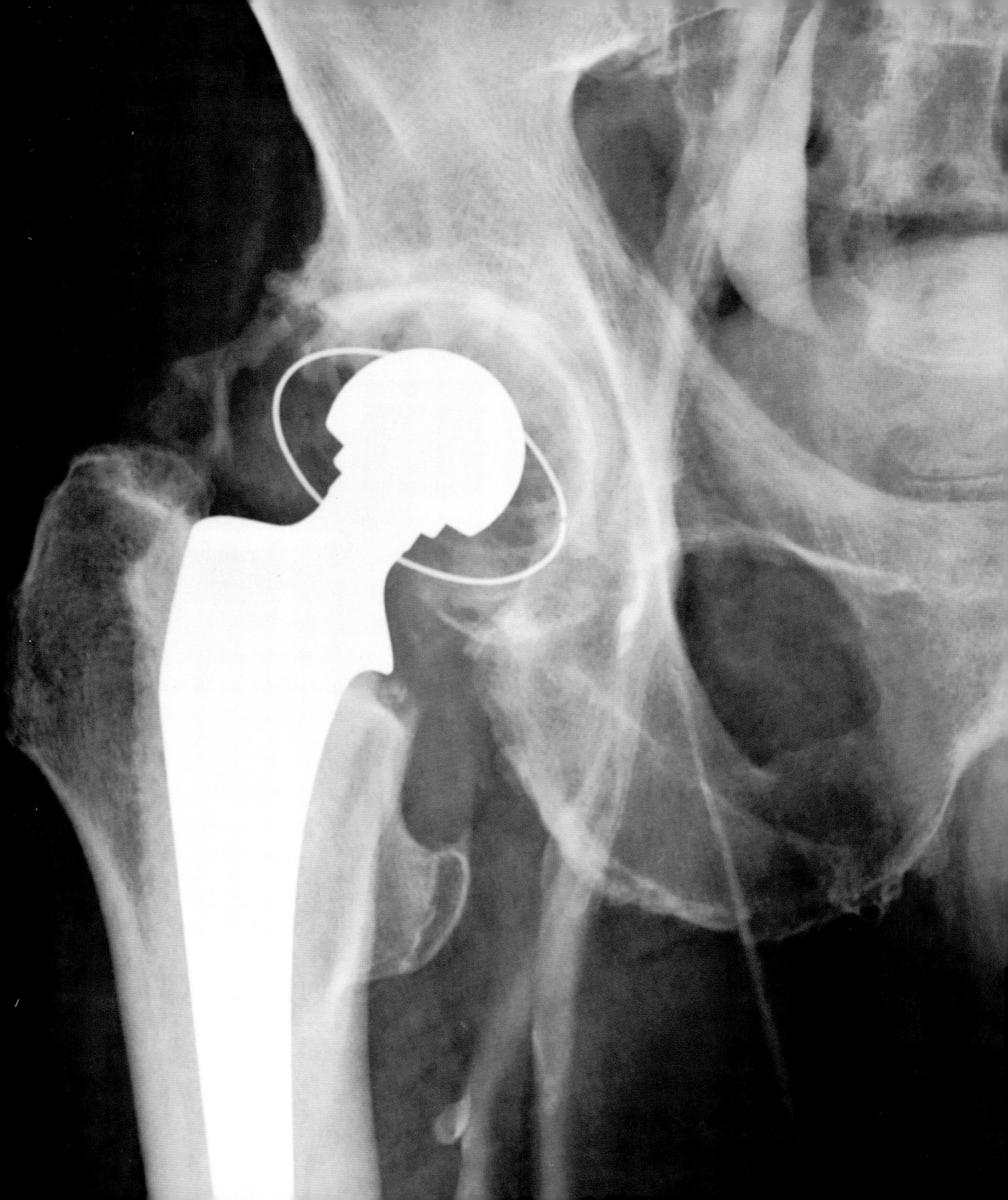

1941

Doped Silicon

John Robert Woodyard (1904–1981)

If we had to pick a substance that engineers have used to impart the biggest effect on humanity, what might that substance be? Maybe it is gunpowder, which engineers use in guns, cannons, and bombs. By killing untold millions of people, gunpowder has certainly had an effect, although not a particularly happy one. Maybe it is uranium, which engineers use in both **nuclear bombs** and **nuclear power plants**. Or **asphalt**, which billions of people use every day for transportation, or the **concrete** used in so many structures. What about gasoline, powering most of our vehicles?

The award for the most influential material might best go to . . . drumroll please . . . doped silicon. Doped silicon was developed by physicist John Robert Woodyard while in the service of the Sperry Gyroscope Company in 1941. This material, the foundation of the **transistor**, has transformed our society in a thousand different ways. Look around you and count how many objects use computers in one form or another. Think about how much time you spend using a laptop, **tablet**, or **smart phone**. Think about the billions of computers connected to the Internet.

And think about where we are headed. The "Internet of things" is the next big thing. It is predicted that, in just a decade or two, there will be 100 trillion objects connected on the Internet. They will be everywhere: home appliances, cameras, sensors, cars, tracking devices, drones, our homes and their security systems. Doped silicon has made computers so inexpensive, so power-efficient, and so intelligent, that computers are embedding in everything and connecting together on the Internet. And then there are **robots**, which will be arriving in massive numbers in the not-too-distant future.

The doping process is conceptually simple. Start with a pure silicon crystal. Add various dopants, like boron to create an area of holes, or phosphorous to create an area with free electrons. Combining these doped areas properly, an engineer can create diodes and transistors. With transistors, engineers can create amplifiers, receivers, and computers. Our computer and electronics industries are built on top of doped silicon.

SEE ALSO Concrete (1400 BCE), Asphalt (625 BCE), Wamsutta Oil Refinery (1861), AK-47 (1947), Transistor (1947), Cluster Munition (1965), Microprocessor (1971), HDTV (1996).

A silicon wafer, pictured here, is a thin slice of semiconductor material.

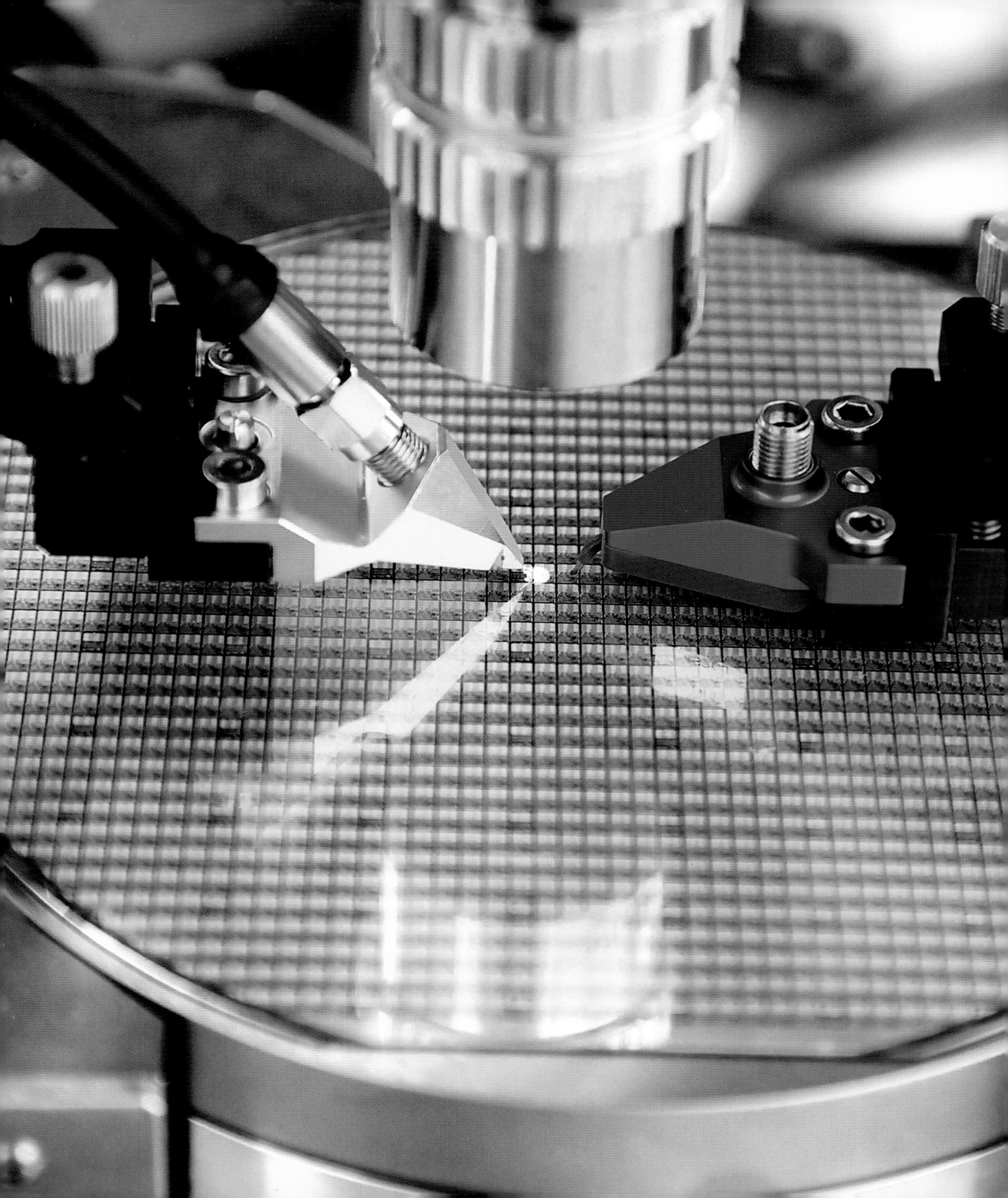

1942

Spread Spectrum

George Antheil (1900–1959), **Hedy Lamarr** (1914–2000)

Imagine you are creating a military radio station to broadcast commands into a battlefield. If it transmits on a fixed frequency, then it is easy for the enemy to find and intercept the broadcasts. Even if encoded, the enemy may break the code. The enemy can also jam that frequency by setting up a high-power transmitter and saturating the frequency with garbage.

Engineers first used spread spectrum technology and frequency hopping to avoid these problems. In 1942, American avant-garde composer George Antheil and his movie-star wife Hedy Lamarr were granted a patent for a "Secret Communications System," a spread spectrum technique that eventually lead to the electronic spread spectrum system used by the US military during the Cuban Missile Crisis in 1962.

In the electronic spread spectrum system, the radio transmitter sends a short burst on one frequency, hops to a new frequency and transmits some more, hops to a new frequency . . . and so on. Instead of using one frequency, the transmission crosses many. The receiver must be able to follow the hopping. This is usually done by using the same pseudo-random number generator on both sides. Now it is harder for the enemy to find the transmission, and very much harder to jam it.

It turns out that this same technology has great application today because of a different problem. Say you have many people all over the world who want to set up their own little transmitters. It simply is not possible for them to all get specific assigned frequencies. Spread spectrum really helps here. As long as all of the transmitters use different hopping patterns, they can coexist without interfering with each other very much. Why would someone want to set up a transmitter? Every **Wi-Fi** hotspot and Bluetooth source is a transmitter. Hundreds of these devices are able to coexist in a dense place like an apartment complex or an airport because of spread spectrum. Engineers took the solution to military problems and used it to solve civilian problems as well.

SEE ALSO Women's Engineering Society (1919), Radio Station (1920), Radar (1940), Wi-Fi (1999).

Pictured: The patent for a "Secret Communications System" developed by Hedy Lamarr and her husband, George Antheil. Their innovations eventually lead to spread spectrum technology as we know it today.

Aug. 11, 1942. H. K. MARKEY ET AL 2,292,387

SECRET COMMUNICATION SYSTEM

Filed June 10, 1941 2 Sheets-Sheet 2

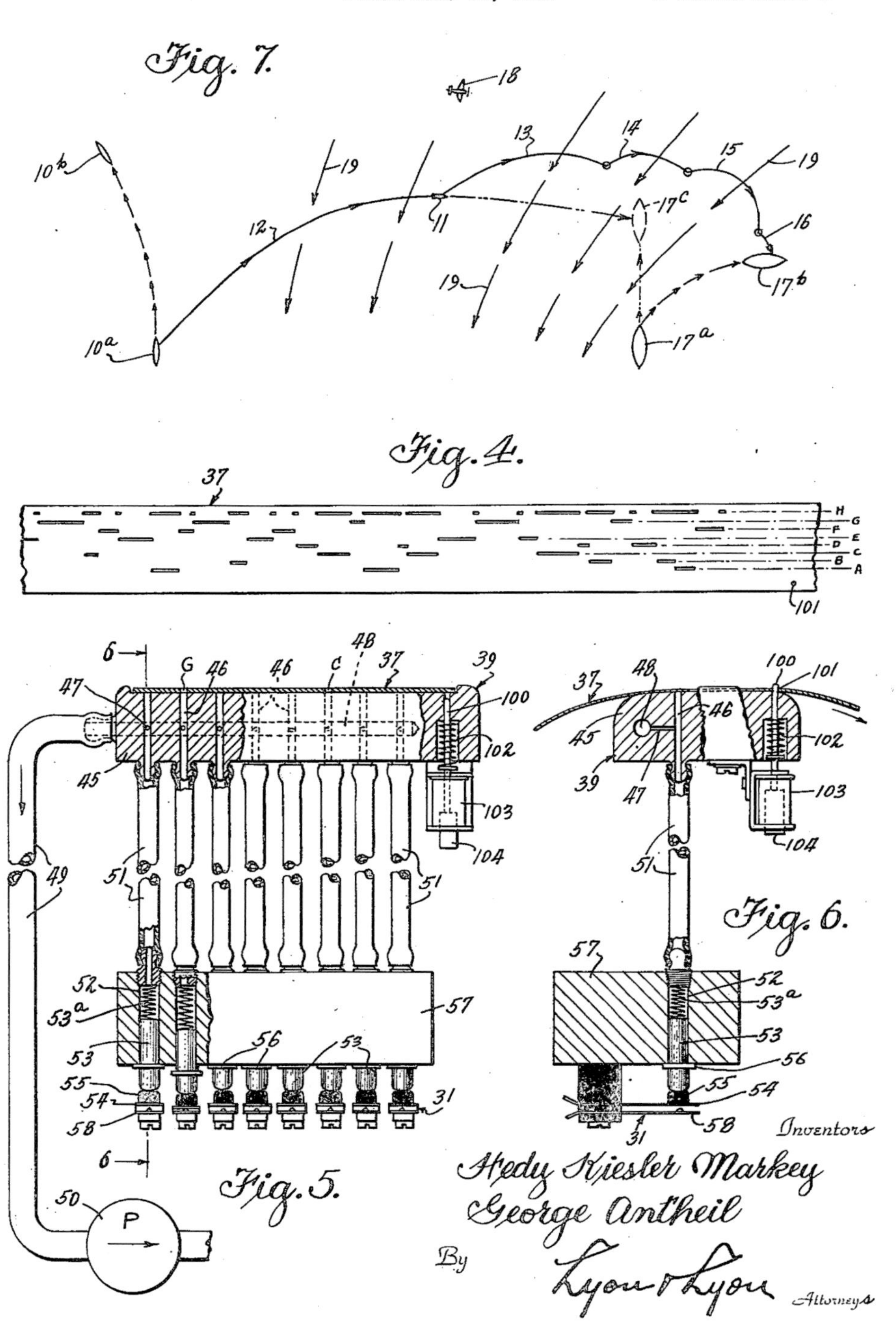

1943

Dialysis Machine

Willem Johan Kolff (1911–2009)

In the United States, something like one in ten of the adult population has kidney disease in one form or another. Approximately 400,000 people have kidney failure, requiring dialysis or a transplant. Without dialysis, a person with failed kidneys will die within a few weeks as toxins that would normally be removed by the kidneys build up in the blood.

Physician and inventor Willem Johan Kolff looked at this situation and asked if a machine could perform the same tasks as a kidney. The basic principles of the result of this inquiry—the very first dialysis machine in 1943—are easy to understand.

Dialysis started with two basic elements—a membrane and a liquid. The membrane came in the form of a natural sausage casing, which is simply a long piece of intestine from an animal. Imagine that blood is flowing inside the casing, and pure, sterile water is flowing outside. Certain undesirable chemicals in the blood, like urea, will pass through the membrane and be absorbed in the water. And so will many other useful chemicals

In a dialysis machine, the water is called dialysate and it is carefully mixed with chemicals so that harmful chemicals transfer from the blood to the dialysate while useful ones stay in the blood. By setting concentrations of certain chemicals high in the dialysate,those chemicals stay in the blood.

A typical patient whose kidneys have completely failed will need dialysis three times a week at a treatment center, or more frequently with a home unit. A fairly large flow rate is required for the blood into and out of the machine. To make regular dialysis possible, the patient needs to have surgery of some sort in order to facilitate frequent tapping, usually in a forearm.

Could engineers create an artificial kidney, as they have an **artificial heart**? Could this engineered kidney create urine that flows naturally to the bladder? There is quite a bit of research in this direction. Devices will probably start outside the body—a so-called "wearable" version. As engineers shrink them even more, we can expect the devices to move inside the body.

SEE ALSO Defibrillator (1899), Laparoscopic Surgery (1910), Artificial Heart (1982), Surgical Robot (1984).

A dialysis machine can be used for home or hospital sessions.

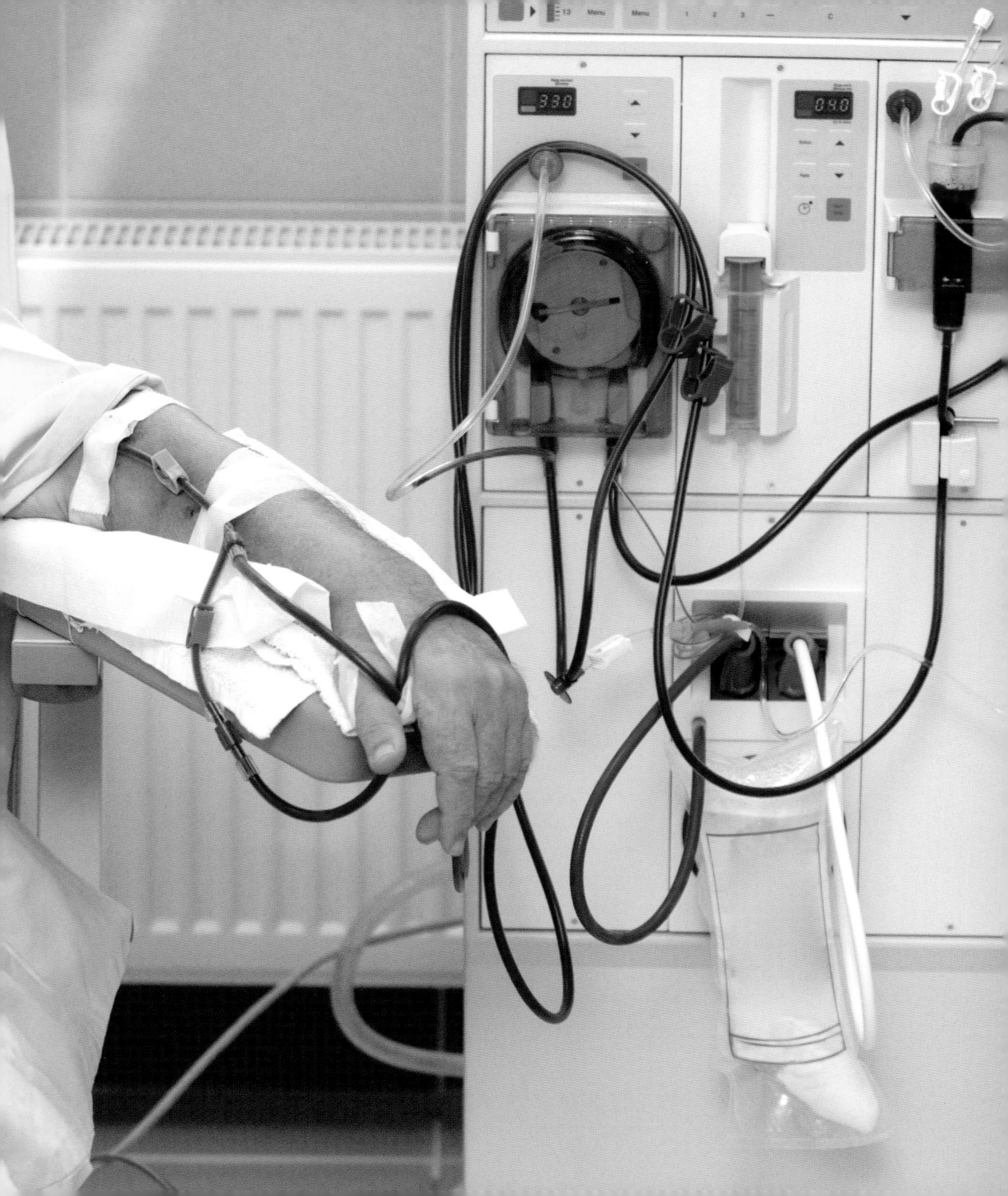
13
Menu
Menu
1
2
3
C
33.0
04.0

1943

SCUBA

Emile Gagnan (1900–1979), **Jacques-Yves Cousteau** (1910–1997)

Just about anyone who has ever been swimming has had this simple dream: Wouldn't it be great if I could breathe underwater? Following generations of inventors attempting to answer this question with human innovation, Emile Gagnan and Jacques Cousteau were able to create what they called the "Aqualung"—the first SCUBA or the Self Contained Underwater Breathing Apparatus, also known as an air supply in a backpack, in 1943.

The main component of a SCUBA system is about as simple and obvious as it gets: a tank full of compressed air. The air is highly pressurized to squeeze as much in as possible, so the tank is made of steel or aluminum, with 3,000 psi (207 bar) being a typical pressure. If the tank has a volume of 15 liters, it means that the tank holds approximately 207x that amount of air at standard atmospheric pressure, or 3,100 liters of air. If a person were to walk around on dry ground and breathe normally, at a rate of 30 liters of air per minute, the tank would hold about 100 minutes of air. (For comparison, firefighters estimate their air consumption at 40 liters per minute.)

But a person who is SCUBA diving is underwater, and the depth has a big effect on the amount of air per breath. The job of the regulator attached to the SCUBA tank is to **regulate the pressure** down from tank pressure (e.g., 3,000 psi) to the pressure necessary to breathe. The deeper a diver goes in the water, the more pressure the surrounding water is placing on the diver's lungs, thus more pressure is required from the regulator to push air into the lungs so the diver can take a breath. This means that, deep underwater, a tank of air gets used up much more quickly than at the surface.

The other way to breathe underwater is a closed-circuit system, also known as a rebreather. A rebreather is engineered to use soda lime granules to absorb carbon dioxide and a small tank of pure oxygen to replenish the oxygen consumed by the diver.

SCUBA is a great example of how engineers can create reliable, inexpensive solutions that turn dreams into reality.

SEE ALSO Carbon Fiber (1879), Hall-Héroult Process (1889), Cabin Pressurization (1958), Mars Colony (c. 2030).

Modern SCUBA gear, pictured here, is considerably more streamlined than its original incarnation.

1944

Helicopter

Airplanes are fantastic. They let us fly through the air like birds. Modern airlines let us fly over the Atlantic and Pacific oceans in hours for reasonable prices.

But airplanes have one big problem: they require runways for takeoff and landing, and runways take up a lot of space. We are never going to be able to commute to work every morning in private airplanes because there can never be enough runways for this to be convenient.

What we need is a flying machine that can take off and land vertically, without a runway. And so the idea of the helicopter was born and first **mass-produced** in 1944 by Sikorsky.

From a layman's standpoint, a helicopter seems simple enough—you just take an airplane propeller and spin it vertically rather than horizontally, right? But from an engineer's standpoint, it is not nearly that simple.

If we take a propeller and spin it vertically, the engine wants to spin in the opposite direction just as fast as the propeller does. To counteract the engine's rotation, engineers attach a long tail boom with its own propeller. The tail propeller needs power, so a long driveshaft transmits it from the engine. And the pilot needs to be able to control this boom propeller, so she or he does that with foot pedals.

Now the machine can rise straight into the air. How do engineers control its direction? And the rate that it rises or falls? A device called a swash plate can change the angle of the rotor's wings as they are rotating. So it is possible to create more lift on the back side of the rotor than the front, tilting the helicopter forward and causing it to fly forward. Doing the opposite lets a helicopter fly backwards just as easily.

The idea of a heavy aircraft hovering in midair seems almost impossible. Yet mechanical engineers had made it happen.

SEE ALSO Mass Production (1845), *Hindenburg* (1937), V-22 Osprey (1981), Apache Helicopter (1986), Human-Powered Helicopter (2012).

The world's first mass-produced helicopter, the Sikorsky R-4, returns from survey of South Pole waters.

BEWARE
ROTORS
U.S.COAST GUARD
043

1945

Uranium Enrichment

Imagine the following situation faced by engineers working on the Manhattan Project in 1942. The uranium that comes out of the ground is almost entirely U-238. But mixed in with the U-238 atoms is the occasional U-235 atom (less than 1 percent). The U-235 atoms are what engineers need to build a nuclear bomb. How is it possible to separate the U-235 atoms from the U-238 atoms?

There are lots of processes that engineers use in factories to separate one thing from another. **Oil refineries** use different boiling and condensation temperatures. Quarries use sieves to separate different sizes of gravel. If salt and sand mix together, water can chemically dissolve the salt to separate it. But separating U-235 from U-238 is difficult because the atoms are nearly identical.

People came up with many different ideas for performing the separation: thermal, magnetic, centrifuge, etc. The method they settled on as the best means of separation at the time is called gaseous diffusion and it involves two steps: Turn solid uranium into a gas called uranium hexafluoride, and let the gas diffuse though hundreds of micro-porous membranes, which have a slight preference for letting U-235 atoms through instead of U-238 atoms.

While this sounds simple, engineering a structure to perform the operation reliably turned out to be a gigantic engineering challenge. The K-25 building—the first full-scale gaseous diffusion plant, at Oak Ridge, Tennessee—came online in 1945, cost $500 million at the time ($8 billion today) and used a noticeable percentage of the nation's electricity. The building was enormous—one of the largest in the world, with something like fifty enclosed acres holding thousands of diffusion chambers along with their pumps, seals, valves, temperature controls, etc. One of the biggest problems was the highly corrosive nature of uranium hexafluoride. Newly developed materials like Teflon helped block its action.

With an unprecedented level of secrecy and a speed that boggles the mind, engineers built K-25 (and other plants) and brought them online to purify the uranium for the first atomic bombs. After World II, the gaseous diffusion process kept purifying uranium until it was replaced by more efficient centrifuges.

SEE ALSO Wamsutta Oil Refinery (1861), Trinity Nuclear Bomb (1945), Light Water Reactor (1946), CANDU Reactor (1971).

Gas centrifuges used to produce enriched uranium. This photograph is of the US gas centrifuge plant in Piketon, Ohio, from 1984.

1945

Trinity Nuclear Bomb

Robert Oppenheimer (1904–1967)

At its heart, a nuclear bomb exploits this simple principle: if 115 pounds (52 kg) or more of U-235 (uranium) comes together in a roughly spherical shape, the sphere will explode with amazing force. It works because a U-235 atom has this surprising property: If it absorbs a neutron traveling past it, the atom will split into two smaller pieces and emit three new neutrons, which fly off to trigger other U-235 atoms to do the same. The two smaller pieces plus the three neutrons weigh less than the original U-235 atom. The missing weight converts to energy at the rate $E = mc^2$. In other words, a very large amount of energy becomes available through direct mass-to-energy conversion. Hence the tremendous explosive energy seen in nuclear bombs. The absorption and splitting process happens nearly instantaneously.

To create a nuclear bomb that exploits this natural property, the engineer's job is to figure out an efficient way to keep masses of U-235 separate until the bomb needs to explode, and then bring the masses together. The first design was almost too simple to believe. A critical mass of U-235 is separated into two parts. One part is stationary. The other part is loaded into an artillery barrel. When the artillery round fires, the two masses combine, become supercritical, and explode.

The problem with this design is inefficiency. As the bomb explodes, the supercriticality is lost, so perhaps only 1 percent of the uranium has a chance to fission before the mass shatters. Engineers therefore worked to increase the efficiency. One way is to shape the mass of U-235 (or plutonium) into a hollow, broken sphere, and then use a uniform conventional explosion around the sphere to create a solid, supercritical sphere. The initial explosive force along with momentum will keep the sphere together for just a bit longer, improving efficiency. This design was employed in the first nuclear explosion in history—the Trinity bomb, created in 1945 by scientists affiliated with the Manhattan Project, including Robert Oppenheimer, who gave the bomb its name. Other engineering techniques include neutron reflectors and tampers—strong, heavy containers that keep the exploding mass together longer. Further tests in fusion would yield a successful **hydrogen bomb**.

SEE ALSO Bow and Arrow (30,000 BCE), Ivy Mike Hydrogen Bomb (1952), Cluster Munition (1965), Fukushima Disaster (2011).

This graphic illustrates the two methods of assembling a fission bomb.

Conventional chemical explosive

Sub-critical pieces of uranium-235 combined

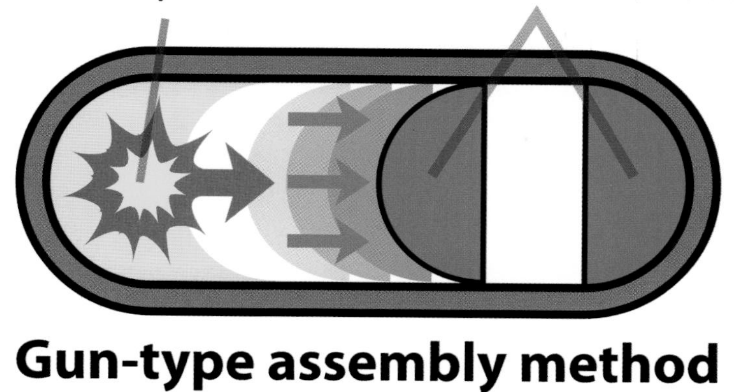

Gun-type assembly method

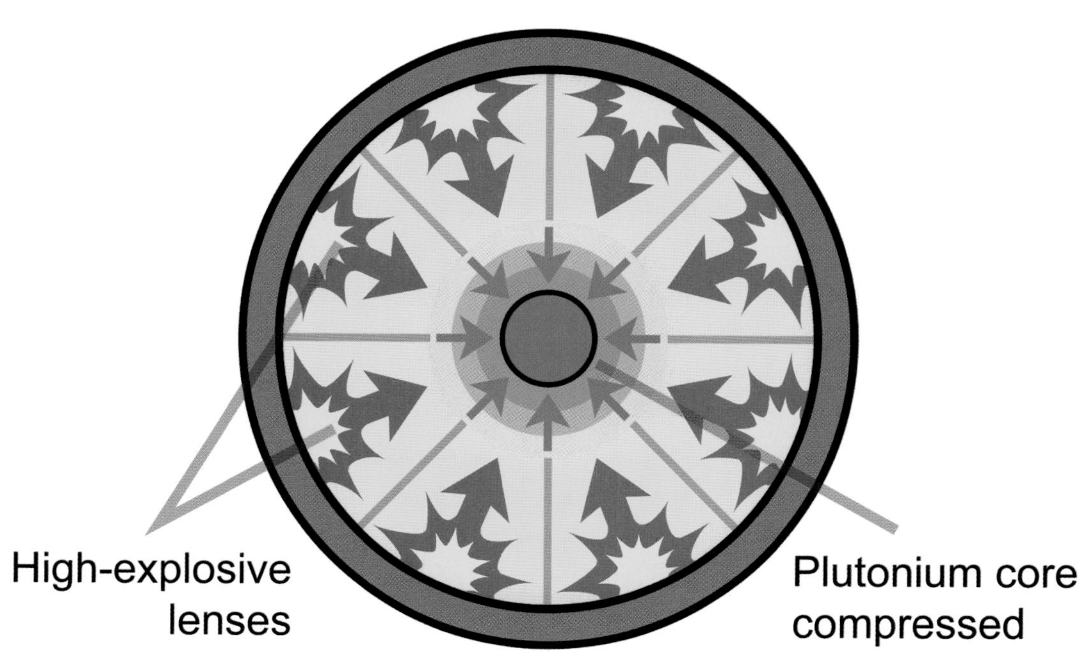

Implosion assembly method

1946

ENIAC—The First Digital Computer

John Mauchly (1907–1980), **J. Presper Eckert Jr.** (1919–1995)

There once was a time when computers did not exist. If the Army wanted to calculate range tables for its artillery guns, it turned to a room full of human beings who did the work by hand with paper and pen or with a mechanical adding machine. You would do the same thing to calculate the orbit of a comet or the forces on a structural beam. When you think that it might take a human being five to ten seconds to add two complex numbers together, you realize how long it took to do any sort of real computation.

And then, in 1946, with the christening of the ENIAC computer, engineers John Mauchly and J. Presper Eckert Jr., along with a team of design engineers, created the first machine that would start changing everything.

ENIAC was the first general-purpose programmable computer. By today's standards it was incredibly primitive. It used 18,000 vacuum tubes. This meant that the computer was as big as a house, weighed 60,000 pounds (27,000 kg) and needed 150 kilowatts of electricity to operate. It could perform 5,000 additions per second.

ENIAC was nothing like the computers we use today, in the way that the **Wright Brothers' first airplane** looks nothing like airplanes of today. ENIAC worked on decimal numbers rather than binary numbers and it handled 10 digits at a time. Data came in through a card reader and out through a card punch. A programmer would have to set up the computer by configuring switches and wires, and the process took multiple days.

But ENIAC could do general-purpose calculating, and this is what engineers needed. ENIAC made it possible to do calculations for the first **hydrogen bomb** being developed by the Manhattan Project. It took 500,000 punched cards to input the data for the problem into the machine. It is easy to imagine how difficult it would be to do those calculations with human beings and adding machines.

Computers became enablers for engineers. Problems that would have been impossible without computers suddenly became possible. Today, nearly every aspect of engineering leverages computers.

SEE ALSO Transistor (1947), Ivy Mike Hydrogen Bomb (1952), Microprocessor (1971), Watson (2011).

From left to right: Patsy Simmers, Gail Taylor, and Milly Beck holding computer boards.

A
C
A
B
C
F
G
H
J
K

1946

Top-Loading Washing Machine

There is a famous lecture by Hans Rosling, a well-known Swedish doctor and speaker, that discusses the washing machine. He makes a number of great points, but two that are relevant here are: 1) Everyone wants a washing machine—even the most diehard of green advocates washes their clothes in a washing machine, and 2) thank goodness for the engineers and systems that make inexpensive washing machines possible.

Mass-market washing machines require inexpensive steel, inexpensive **plastics**, and inexpensive motors, gears, and controls. And to use a washing machine we need a water system, a **sewer system**, a **power plant**, and a **grid**. Only when all of these things are available in a society can the masses afford and use washing machines. The fact that two billion people on the planet own and use washing machines is a triumph. The fact that five billion people do not tells us how far we have to go in terms of economic equality and infrastructure development.

The automatic washing machine that we know today came into its own shortly after World War II. Because of rationing, no one was making washers during the war, but in the postwar boom the sales of washing machines in the United States were impressive. It was helped by the Rural Electrification Act (1935, 1944), which made electricity available to just about everyone in America. Europe followed the same kind of trendline.

From an engineering perspective, the top-loading washing machine is a fairly simple device. It needs the tub that holds the water, and an inner basket that holds the clothes. This basket is important during the spin cycle to extract most of the water. An agitator twists back and forth during the wash cycle. A single heavy-duty motor and gearbox often handles the agitation, the spinning, and the pumping needed to drain the tub. An electromechanical or electronic control system tells the motor, gearbox, and the solenoid valves what to do.

The washing machine is one engineering achievement that has made life easier for billions of people. In an ideal world, there will come a day where every human being has access to one.

SEE ALSO Roman Aqueduct System (312 BCE), Plastic (1856), Modern Sewer System (1859), Power Grid (1878), Women's Engineering Society (1919), Light Water Reactor (1946).

General Electric introduced its first top-loading washing machine in 1946.

1946

Microwave Oven

Percy Spencer (1894–1970)

Every now and then discoveries happen by accident. Once they do, engineers get hold of the discovery and incrementally improve it to the point where just about everyone can afford it. This is the story of the microwave oven.

The cooking power of microwaves was discovered by Percy Spencer, an engineer working near radar equipment, in 1945. **Radar** uses microwave radio pulses at thousands of watts to detect aircraft. The engineer noticed the chocolate bar in his pocket melting from the microwaves. Of course if the chocolate bar was melting, he was also cooking himself to some degree, and perhaps this discovery also led to some basic safety precautions.

If you have radar equipment handy, creating a crude microwave oven is easy. You take the radio waves from the radar transmitter and send them into a metal box. The radio waves will reflect off the walls of the box. Any food placed in the box heats up as the radio waves induce molecular motion in water and fat molecules.

The first commercial microwave ovens were sold in 1946. Called Radaranges, they were made by Raytheon, which also made the candy-bar-melting radar set. The Radaranges were basically a radar transmitter and a metal box with a door, packaged in the size and weight of a **refrigerator**. In today's dollars this oven cost perhaps $50,000.

Microwave ovens were too expensive for most people to afford until engineers simplified things and brought the cost down in the 1970s. Those new designs also solved a big problem—protection against the empty state. Prior to the 1970s, it was catastrophic to run a microwave oven with nothing inside it.

Today we take microwave ovens completely for granted and prices have fallen as low as fifty dollars for simple ovens. We see microwave ovens everywhere—in hotel rooms, break rooms, dorm rooms. They are far faster than normal ovens and far more energy efficient—almost every bit of energy the oven uses goes into heating the food. The microwave oven is a great example of the power of engineering to bring a technology to millions of people.

SEE ALSO Women's Engineering Society (1919), Electric Refrigeration (1927), Radar (1940), Frozen Pizza (1957).

A chef using a Raytheon RadaRange III, an early commercial microwave oven, circa 1958.

RadaRange

1946

Light Water Reactor

Eugene Wigner (1902–1995), **Alvin Weinberg** (1915–2006)

Nuclear fuel was revolutionary. A very small amount of fuel—less than a hundred tons—could power a whole city for a year or more with no airborne pollutants. It would take 20,000 times more coal to generate the same power.

The basic idea is incredibly simple, and is the same principle used in the development of the **nuclear bomb**. U-235 atoms will split and generate heat when a neutron hits them. Since a splitting U-235 atom creates three new neutrons, a chain reaction can occur. Allowed to proceed uncontrolled, the chain reaction creates a nuclear bomb. If controlled, however, the U-235 becomes a powerful and consistent source of heat.

All engineers had to do was design a reactor that could safely extract the heat from the fuel without melting down or blowing up. The most popular way to solve the problem turned out to be the light water reactor. In 1946, Hungarian-American theoretical physicist Eugene Wigner and nuclear physicist Alvin Weinberg proposed and developed what would become the light water reactor as we know it.

First, engineers needed a way to hold the fuel safely, and then a way to control the rate of heat production. Enriched uranium pellets are stacked into metal tubes, which are arranged in bundles. To control the heat, control rods can be lowered between the fuel rods. The material in control rods absorbs neutrons. To shut down the reactor, all of the control rods are fully inserted.

The engineers also needed a moderator—something to slow down the neutrons enough to cause U-235 to split. They chose water for this purpose because water also extracts the heat from the reactor in the form of steam.

It seems simple enough—fuel rods, control rods, and water are the essential elements. So why are nuclear reactors so complex and expensive? It comes from the fact that if the water stops flowing, the reactor overheats and melts, probably releasing radioactivity into the environment in the process. Engineers try to build a huge amount of redundancy and safety into a nuclear power plant.

SEE ALSO Trinity Nuclear Bomb (1945), Pebble Bed Nuclear Reactor (1966), Power Plant Scrubber (1971), Chernobyl (1986), Fukushima Disaster (2011).

Pictured: Diagram of a pressurized water reactor to be used on a ship.

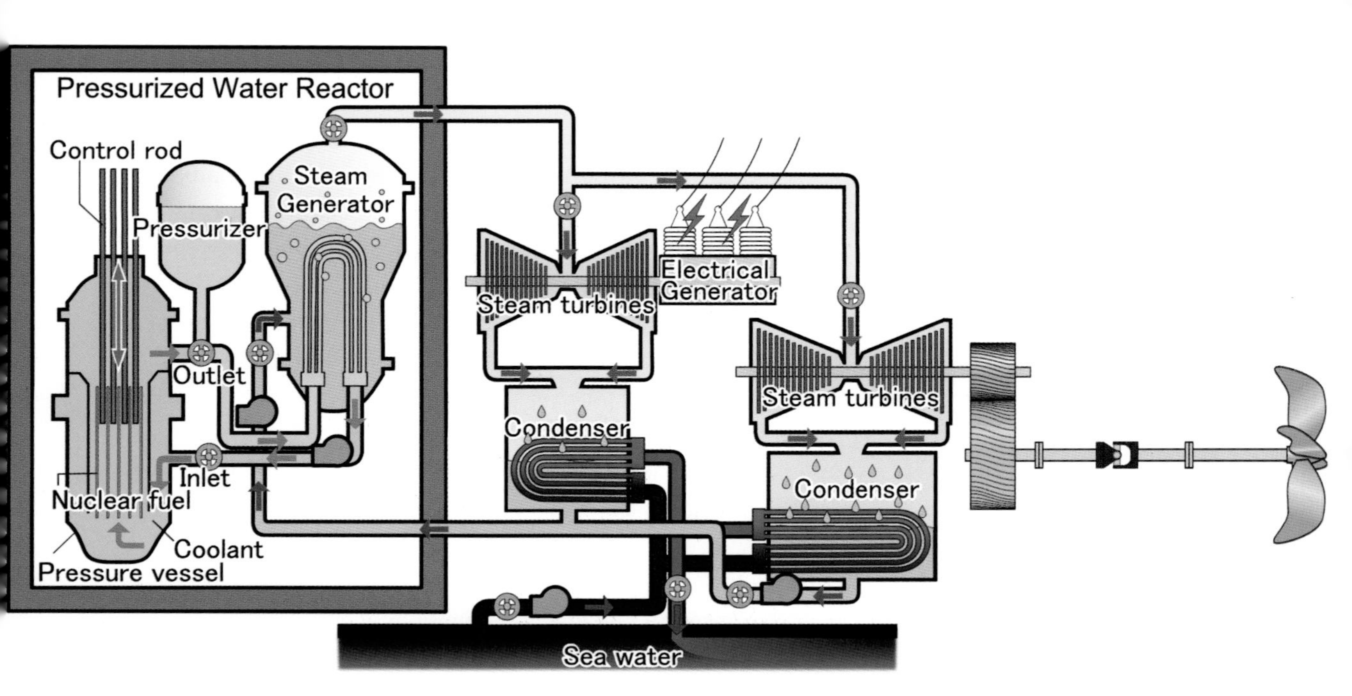
Pressurized Water Reactor
Control rod
Pressurizer
Steam Generator
Outlet
Inlet
Nuclear fuel
Coolant
Pressure vessel
Steam turbines
Electrical Generator
Condenser
Steam turbines
Condenser
Sea water

1947

AK-47

Mikhail Kalashnikov (1919–2013)

The evolution of weapons began tens of thousands of years ago with the **bow and arrow**. The AK-47 is both a horrific weapon, responsible for millions of deaths since its introduction in the Soviet Union in 1947, and an engineering marvel.

Any machine gun is a bit like a reciprocating engine. The energy generated when firing a bullet is enough to prepare the mechanism for the next shot. There is a hole drilled in the top of the barrel about halfway down its length. As the bullet passes this hole, expanding gases rush up the hole into the gas tube, blowing a piston backwards.

The piston is connected to the bolt carrier. As it slides backwards, the carrier extracts the spent shell, ejects it, pushes the hammer back, and cocks it. When pressure inside the barrel drops, a spring starts pushing the bolt carrier forward. In the process it strips the next shell off the magazine, loads it, and closes the bolt behind it. In automatic mode, the hammer immediately releases to fire the new shell, and the process repeats.

Three engineering goals are expressed in Mikhail Kalashnikov's design of the AK-47: It must be inexpensive, difficult to jam, and easy to clean. Therefore the number of parts is kept to a minimum and the receiver—the main body of the mechanism that holds everything together—is a stamped piece of steel. No fancy machining, no casting or forging. Take a piece of metal, stamp holes in it, and bend it.

To prevent jamming, the clearances between all the parts are large and there are very few moving parts. The gas tube is larger to prevent clogging. The hammer is a huge, blunt piece of metal.

Cleaning is a breeze. Push one button to release the top cover. The bolt carrier, its spring, and piston all come out easily. The trigger, hammer, bolt, chamber, firing pin, etc., are now all easily accessible for cleaning.

The engineering here is brutally elegant, designed for low cost and reliability in the worst conditions. That's why there are millions of AK-47s all over the world.

SEE ALSO Bow and Arrow (30,000 BCE), Catapult (1300), Trinity Nuclear Bomb (1945), Ivy Mike Hydrogen Bomb (1952), Cluster Munition (1965).

This example of an AK-47 was made in the Soviet Union in 1954.

1947

Transistor

Walter Brattain (1902–1987), **John Bardeen** (1908–1991)

If we were to travel back in time to 1945 or 1946, we would find a thriving electronics industry in the United States. People were buying **radios** and **television sets**. Electronic innovations like **radar** were revolutionary. And the world's first real digital computer, known an ENIAC, came into being in 1946. All of these electronic devices were powered by vacuum tubes. But vacuum tubes were a real pain. They were big (about the size of a pill bottle), hot, prone to burning out, expensive, and they used a lot of power. When ENIAC first started operation, its tubes needed to be constantly replaced. There had to be a better way.

The better way started in 1947 with the discovery of the transistor by American scientists John Bardeen and Walter Brattain at AT&T's Bell Labs. It then became reality in 1953 and 1954 when the first germanium and then silicon transistors began mass production.

A typical transistor in this era was a small three-wire device about the size of a pea. A transistor can do two different things depending on the design. It can act as an on-off switch, which is how transistors are used in computers. Or it can act as an amplifier, which is how transistors are used in radios and televisions. Transistors are small, lightweight, reliable, efficient, and (eventually) incredibly inexpensive. Engineers immediately started replacing vacuum tubes with transistors.

One of the first transistorized devices to appear in the marketplace was the transistor radio in 1954. Transistor radios were small, portable devices that could run off of a 9-volt battery and fit in a pocket. With mass production they were incredibly inexpensive. People had never seen anything like them, and billions of transistor radios were sold. The portable music craze had begun.

Also in 1954, the first transistor computer, TRADIC, came online. Compared to a vacuum-tube machine it was tiny and used only 100 watts. This meant that the computer could fly in an airplane—something that would have been impossible with vacuum tubes.

Transistors would eventually allow engineers to develop thousands of new devices. They gave engineers amazing freedom compared to the technology they replaced.

SEE ALSO Radio Station (1920), Color Television (1939), Radar (1940).

This is an early transistor used in the IBM 1401.

063

1948

Cable TV

Robert Tarlton (1914–2006), **John Walson** (1915–1993)

Cable television is usually attributed to John Walson, an owner of an electronics store. Because his store, located in Mahanoy City, Pennsylvania, was surrounded by mountains, local residents could not receive broadcasts from stations located in nearby Philadelphia. Walson's solution to this dilemma consisted of placing an antenna on top of a mountain, then using amplifiers and cables to bring the signal to the valley. In this way, the first cable systems were born.

Later, in 1950, a retailer of television sets named Robert Tarlton banded together with a group of other TV salesmen to offer broadcasts to Philadelphia-area residents for a fee. Many people wanted cable because it made the picture clearer (especially in cities, where ghosting can be a problem as signals reflect off of buildings), and it was possible to get more stations. People were willing to pay a small monthly fee for these benefits.

Once a cable system was installed in a city and enough people signed up, the system created a guaranteed audience and a guaranteed revenue stream. It therefore became possible (once regulations allowed it) for someone to create a new cable TV channel. HBO (Home Box Office) was the first such channel, created in 1972. The cable TV system already had a billing model, so HBO could easily tack its monthly fee onto the bill. HBO also created a satellite system to easily send its content to cable TV companies across the country.

WTBS—a local TV station in Atlanta that created a satellite distribution system like HBO's—was next. After that, and a few more regulatory tweaks, cable exploded both in the number of channels and the number of subscribers in the US and other countries. People wanted premium TV so badly that nearly everyone switched to cable, even though broadcast TV is free. This demonstrated that people were willing to migrate to a new system if significant benefits were offered, which lead to massive shifts such as from analog broadcasting to **HDTV**.

SEE ALSO Radio Station (1920), Color Television (1939), HDTV (1996).

Cable TV changed the way we interacted with televised entertainment.

1949

Tower Crane

Hans Liebherr (1915–1993)

Anytime a new skyscraper is being erected, we see a tower crane helping the process. But when we watch a tower crane in action, it almost looks impossible. We see this thin, leggy **truss** forming the "tower," and then another thin, leggy truss forming the arm, and then this arm appears to be able to hold a huge load that should, seemingly, cause the whole thing to collapse. How did German inventor Hans Liebherr, in conjunction with many design engineers, make a tower crane work?

One key to the whole process is the foundation, which normally is hidden. A month before the tower crane arrives, builders pour a huge reinforced concrete block. This foundation might measure 50 feet (15 meters) or more on a side, could be 10 feet (3 meters) thick, and can weigh many thousands of tons. The tower bolts to this foundation.

The vertical truss, although it looks leggy, is quite sturdy by design. It does have a maximum safe height however. It usually clips into the structure it is erecting for support as it grows taller.

The arm or jib is the same kind of truss—extremely strong despite appearances. It is counterweighted on one side to balance the tipping forces on the tower. There are usually cables employed as well, in a manner similar to a cable-stayed bridge.

With all of this in place, the tower crane is strong and stable despite appearances to the contrary. But there are definitely limits. There is a maximum load that changes as the load moves out toward the end of the jib. A crane might be able to pick up fifty tons if the load is near the tower, but only one ton if the load is out near the end of the jib.

It is that ability to slide the load horizontally as well as vertically that makes the tower crane especially useful. The sliding action makes the tower crane unique, and well suited for the work it was originally meant to do, rebuilding post-war Germany and setting the foundation for what would become the international manufacturing company, the Liebherr Group.

SEE ALSO The Great Pyramid (2550 BCE), Woolworth Building (1913), Kinsol Trestle Bridge (1920).

A tower crane enabled builders to tackle new projects.

1949

Atomic Clock

Louis Essen (1908–1997)

What is time? Scientists are not really sure. But we do know how to measure time, and engineers have been doing so with increasing precision for several centuries. We measure time with clocks.

If you think about a clock, it has two main parts: an oscillator and a counter. The oscillator is something that happens at a known frequency. So a pendulum in a **pendulum clock** swings back and forth once every second, say. That is a simple oscillator. A simple counter uses gears to move the hands on the face of a clock.

If you want a more accurate clock, you make the oscillator more and more precise. A quartz crystal oscillates more precisely than a pendulum, and much faster. You will never measure 1/1000th of a second with a pendulum, but you can with a quartz crystal.

Even faster and more precise are the oscillations of a known atom. The official definition of a second is 9,192,631,770 oscillations of a Cesium-133 atom, established in 1967. These atomic oscillations are the basis of humankind's most accurate clocks. They have an obvious name: atomic clocks. The first was built by the US National Board of Standards in 1949. English physicist Louis Essen developed the first accurate atomic clock later, in 1955.

How does an engineer make an atom oscillate precisely? And then count those oscillations? Many ways have been invented. One way is to excite a cloud or a stream of atoms with microwave energy. By determining the exact frequency of the microwaves that excite the most atoms to the right level, the engineers create a super-accurate oscillator.

The first atomic clocks were huge. But they have been shrinking and getting more reliable and precise. The newest, smallest commercial atomic clocks are about the size of a matchbook. They are called "chip scale atomic clocks."

At the time of writing, the most accurate atomic clock uses ytterbium atoms excited by a laser. Clocks like these might lose a second over the course of billions of years. Compared with pendulum clocks, which can gain or lose seconds per day, the accuracy is nearly perfect.

SEE ALSO Mechanical Pendulum Clock (1670), Big Ben (1858), Atomic Clock Radio Station (1962).

The first atomic clock, built 1949 at the US National Bureau of Standards (now National Institute of Standards and Technology) by Harold Lyons and associates.

• NATIONAL BUREAU OF STANDARDS •

1949

Integrated Circuit

Werner Jacobi (1904–1985), **Jack Kilby** (1923–2005), **Robert Noyce** (1927–1990)

Prior to the widespread deployment of integrated circuits, electronic devices like computers were made of discrete transistors. Each **transistor** came in a small can about the size of a pea, with three wires attached. These transistors would be soldered onto a printed circuit board. A computer of this era might have used several thousand transistors, soldered onto ten or more boards each the size of a piece of paper or larger. The boards of the simplest computers filled a box the size of a filing cabinet. The largest computers filled entire rooms. Computers were big, heavy, expensive, slow, and power hungry as a result.

The integrated circuit changed all of this. In 1949, German engineer Werner Jacobi proposed the first integrated circuit-like device. In 1958, electrical engineer Jack Kilby, working for Siemens, demonstrated the first working integrated circuit, and shortly after, Robert Noyce improved upon it by creating a silicon chip. Integrated-circuit technology transformed the computer and electronics industries.

The basic idea is simple and it starts with a polished silicon wafer sliced from a large silicon crystal grown in a lab. Three things then happen to the wafer: 1) parts of the wafer's surface can be selectively doped with the substances used to create transistors, 2) layers of oxide can be grown to act as insulators, and 3) metal wires can be deposited to connect transistors together. The engineering has gotten increasingly sophisticated in recent years, with many metal layers and 3D transistors making smaller, more complicated circuits possible.

Integrated circuits are one of the best examples of incremental improvement in engineering. Engineers have radically improved cost, capabilities, and performance over time and the pace of improvement has been staggering. Every two years, the number of transistors on a chip has doubled—a rate known as Moore's Law. So today's microprocessors have billions of transistors running efficiently at low voltages and high speed. We all benefit from the integrated circuits in our **smart phones**, **tablets** and home computers.

SEE ALSO ENIAC—The First Digital Computer (1946), Transistor (1947), Microprocessor (1971), Smart Phone (2007), Tablet Computer (2010).

Close-up of a chip in an integrated circuit.

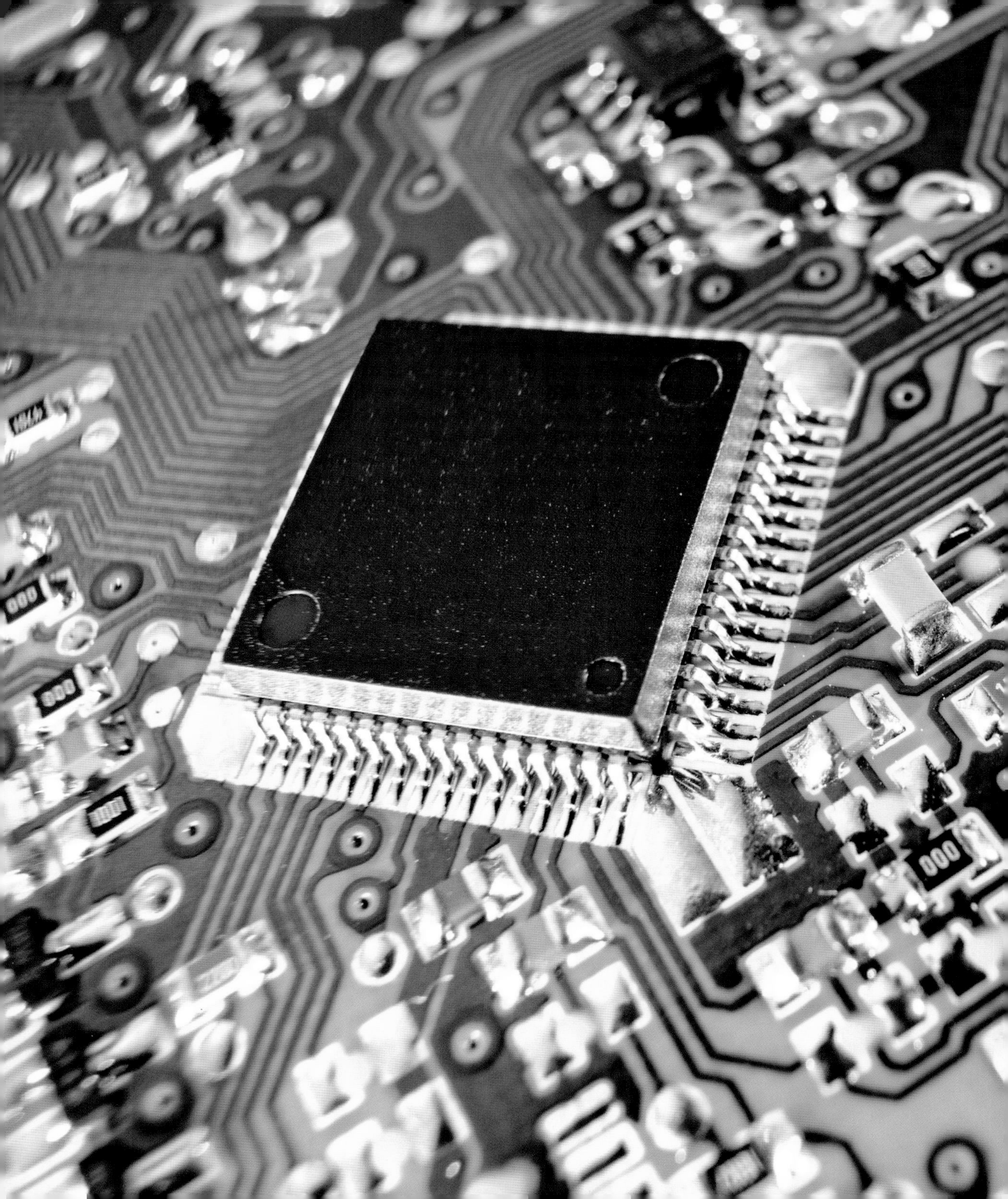

1950

Chess Computer

Alan Turing (1912–1954), **Claude Elwood Shannon** (1916–2001)

In 1950, American mathematician Claude Elwood Shannon wrote a paper about how to program a computer to play chess. In 1951, British mathematician and computer scientist Alan Turing was the first to produce a program that could complete a full game. Since then, software engineers have improved the software and computer hardware engineers have improved the hardware. In 1997, a custom computer called Deep Blue, developed by IBM, beat the best human player for the first time. Since then, humans have not had a chance because computer chess hardware and software keeps improving year after year.

How do engineers create a computer that can play chess? They do it by employing machine intelligence, which in the case of chess is very different from human intelligence. It is a brute force way to solve the chess problem.

Think of a board with a set of chess pieces on it. Engineers create a way to "score" that arrangement of pieces. The score might include the number of pieces on each side, the positions of the pieces, whether the king is well protected or not, etc. Now imagine a very simple chess program. You are playing black, the computer is playing white, and you have just made a move. The program could try moving every white piece to every possible valid position, scoring the board on each move. Then it would pick the move with the best score. This program would not play very well, but it could play chess.

What if the computer went a step further? It moves every white piece to every possible position. Then on each possible white move, it tries every black move, and scores all of those boards. The number of possible moves that the computer has to score has grown significantly, but now the computer can play better.

What if the computer looks multiple levels ahead? The number of boards the computer has to score explode with each new level. The computer gets better. When Deep Blue won in 1996, it was able to score 200 million boards per second. It had memorized all common openings and gambits. It could prune out vast numbers of moves by realizing certain paths were unproductive. Today the computational power in a laptop **smart phone** allows it to beat most people at chess using the same techniques.

SEE ALSO Microprocessor (1971), Smart Phone (2007), Watson (2011), Brain Replication (c. 2024).

Pictured: IBM's supercomputer, Deep Blue.

03
IBM
White
Black

1951

Jet Engine Testing

One thing engineers have to consider, and we out in consumer land rarely think about, is all of the edge cases. What happens to a product when it is not operating under standard conditions?

One of the most fascinating set of edge cases can be seen when testing jet engines. It is one thing for a jet engine to operate in smooth, clean air at 30,000 feet. But jet engines often operate in very different conditions, and they have to be able to handle these situations with ease. Engineers take these scenarios into account during the design process and then test rigorously to make sure the edge cases do not cause catastrophe.

Take the simplest example: rain. Rain can have a big effect on a jet engine. The engine inhales the rain directly and then pushes it straight into the combustion chamber. The engine needs to handle the water without extinguishing itself, even in hurricanes. Therefore a large jet engine might be tested with 1,000 gallons (3,800 liters) of water flowing in per minute. The same kind of thing applies to snow, hail, and ice testing. Another interesting situation is a sandstorm.

The most surprising thing that jet engines must handle is a bird strike. Birds are common, so a large jet engine might get tested by shooting 5-pound (2.2 kg) chickens into the intake. The front fan blades need to slice the chicken without blade damage, and the compressor needs to digest the chicken slices. The de Havilland Aircraft Company created the first chicken gun in 1951 for testing jets.

There are limits, however, as seen in the famous case of US Airways flight 1549 in 2009. That airplane ran into a flock of Canadian geese at approximately 3,000 feet (915 meters). These birds can have 5-foot (1.6 meter) wingspans and weigh 10 pounds (4.5 kg) or more. The two engines could not process the birds and flamed out. This resulted in the historic landing in the Hudson River, where other engineers had anticipated the problem of plane flotation. The fact that there were no injuries is a testimony to the pilot and the quality of the engineering.

SEE ALSO The Wright Brothers' Airplane (1903), Boeing 747 Jumbo Jet (1968), C-5 Super Galaxy (1968).

An F119-PW-100 jet engine is tested in a full afterburner state.

GFT99

1952

Center-Pivot Irrigation

Frank Zybach (1894–1980)

The standard way to efficiently irrigate an agricultural field is to lay horizontal pipe in parallel down the field, say at 40-foot intervals. Each horizontal pipe has vertical pipes, also at 40-foot intervals, and on top of each vertical pipe is a big sprinkler head. All of the pipes connect together back to a big pump that supplies the whole system with water.

This approach certainly works, but it is not very efficient. A square 9-acre (3.65 hectares) field takes something like 10,000 feet (3050 meters) of horizontal pipe and more than 200 sprinkler heads. Center-pivot irrigation was originally patented in 1952 by farmer Frank Zybach. In subsequent years, engineers perfected the original system to radically reduce the amount of pipe and the number of sprinkler heads needed to irrigate a field.

To get an idea of how this new system works, let's imagine a 9-acre field is represented by a 700-foot (213 meter) diameter circle. At the center of the circle is the source of water from a pump. It connects into a riser and from there to a single horizontal pipe that is 350 feet (107 meters) long. The amount of pipe has been reduced by a factor of 30 in return for a longer irrigation time for the field.

The most ingenious part of this approach is the way that the horizontal pipe moves. The pipe is supported in the air by triangular towers on wheels. The towers are spaced every 100 feet, say, with one segment of pipe between each pair of towers. The wheels move very slowly via electric or water-powered motors.

Not only does this system save a lot of pipe, it also saves a lot of water. Sprinkler heads point straight down from the elevated horizontal pipe, reducing evaporation and wind loss.

By completely reconceptualizing the way they irrigate, engineers lowered cost and improved efficiency. The next time you fly over the midwestern states, look down and notice the impact of this reconceptualization. Further innovations like subsurface **drip irrigation** have proven even more useful in recent years.

SEE ALSO Green Revolution (1961), Drip Irrigation (1964).

An early morning view of a center-pivot sprinkler system in a farm field.

1952

3D Glasses

Milton Gunzberg (1910–1991)

When humans look out at the real world, we see depth because of the binocular nature of our vision systems. One eye sees the world from one perspective, the other eye sees a slightly different perspective, and our brains do the calculations to determine how far away an object is. When we look at a photograph or a TV screen, we lose that depth information and they look flat. To bring depth to these flat images, the key is to get slightly different images into our two eyes at the same time so that the brain's binocular calculator can create the illusion of depth. Engineers have brought four technologies for doing this into the marketplace.

The first technology uses the red-cyan glasses popularized in the 1950s, pioneered by inventor Milton Gunzberg, who first offered them to audiences viewing the film *Bwana Devil* in 1952. The technology is extremely simple. Two slightly different black and white images—one colored red and one colored cyan—are projected simultaneously. The red lens lets the cyan image through to one eye and the cyan lens lets the red image through to the other.

A much better technology: polarized lenses. One lens lets through vertically polarized light and the other lets through horizontally polarized light. Two projectors play slightly different versions of the movie through polarizing filters onto the same screen. The lenses in the glasses unscramble the two images. Color 3D images are delivered inexpensively.

The third technology uses **LCD** shutter glasses. A **TV** plays the two perspectives by interleaving frames. One frame goes to the left eye. The next frame goes to the right eye. The glasses block the light going to each eye. At a rate of 120 frames per second, each eye sees 60 frames a second and the brain sees 3D.

Finally, there are goggles with two small TV screens built in. One screen projects to the left eye, the other to the right eye. The user sees a binocular image. This technology is the most expensive but, done well, it provides an immersive experience. The development of small Active Matrix OLED (**AMOLED**) screens has made this technology viable.

SEE ALSO LCD screen (1970), HDTV (1996), AMOLED Screen (2006).

Formally attired audience sporting 3D glasses during the opening night screening of the film Bwana Devil, *the first full-length color 3D motion picture.*

1952

Ivy Mike Hydrogen Bomb

Richard Garwin (b. 1928)

A nuclear bomb uses nuclear fission of a material like U-235 or plutonium to create a massive amount of explosive force. What if engineers want to create an even bigger bomb? In that case they would use hydrogen fusion rather than fission. But getting hydrogen nuclei to fuse is not an easy task. It takes gigantic heat and pressure, mimicking conditions inside the sun. One way to create these conditions: a conventional nuclear bomb.

So the Ivy Mike hydrogen bomb was engineered in the following way: a conventional explosion brings a critical mass of plutonium together to create a **nuclear explosion**, and that nuclear explosion drives enough hydrogen atoms together to create a hydrogen fusion explosion. For this to work, engineers have to overcome two big problems. First, they need a collection of the right kind of hydrogen atoms in the right place at the right time. Second, they need to hold the whole thing together long enough to initiate the fusion process, while the nuclear explosion is trying to rip the bomb apart.

The first hydrogen bomb—the Ivy Mike device, designed by American physicist Richard Garwin in 1952—used liquid deuterium in a vacuum flask. While this worked, it was not practical for a reliable bomb. The breakthrough was to use lithium deuteride—a solid that decomposes to tritium when bombarded by neutrons.

There is no real magic in the casing—it is just extremely strong steel a foot (30 cm) thick. A key insight is to understand that radiation and neutrons from the fission explosion outrace the blast wave. Directing them to the secondary fusion bomb allows it to detonate prior to the bomb's destruction. Not everything is understood because of military secrecy, but we do know that engineers figured it out based on the success of test bombs.

It is possible to chain another stage into the process in the time available, so the fusion bomb explodes and ignites another fusion bomb. This was the mechanism used in the Soviet Union's Tsar bomb—the largest explosion ever created by humans at approximately 50 megatons of TNT.

SEE ALSO Trinity Nuclear Bomb (1945), Uranium Enrichment (1945), International Thermonuclear Experimental Reactor (ITER) (1985).

Aerial view of the cloud produced during the detonation of the Ivy Mike hydrogen bomb.

1953

Automobile Airbag

Walter Linderer (Dates Unavailable), **John Hetrick** (Dates Unavailable)

You are driving your car on a backcountry road when you see a deer leaping from the woods on your left, straight for your car. You swerve right. The swerve to the right is a natural instinct. Unfortunately, directly in your new path lies a giant oak tree. You are going 40 mph (65 kph). The tree is 10 feet (3 meters) away. Simple math says you will hit it in about one-tenth of a second. Your brain will never respond in time.

Let's slow time way down here. The leading edge of the front bumper touches the tree. As your car moves forward, several engineers will be doing you a big favor. One of the engineers designed crumple zones into the sheet metal. The entire front of the car is going to collapse like an accordion to absorb some of the impact. Another engineer worked on the engine mounts. The engine is going to fall underneath the car instead of landing in your lap. A third engineer created a safety cage around the passenger compartment. It will be an amazingly strong bubble of integrity in this accident. A fourth engineer worked on the seat belt, which will automatically lock and keep you from flying through the windshield. Even the windshield is engineered to keep shattered glass from puncturing your face.

And then a sixth engineer will unleash an explosion, creating a cloud of gas that can rapidly fill a big cloth bag. Instead of your head slamming into the steering wheel, it will hit this rapidly inflating bag. This is the driver-side airbag.

An accelerometer tells a computer that it is time to deploy the airbag. The **computer** fires a squib, which ignites the explosive charge. The charge goes off, creating the high-speed cloud of gas. The airbag explodes out of the steering wheel just in time to meet your approaching face. The whole sequence takes just a few milliseconds.

The original invention of the airbag was credited independently to two men: German engineer Walter Linderer, and American industrial engineer John Hetrick, who both received patents for the device in 1953.

SEE ALSO ENIAC—The First Digital Computer (1946), Anti-Lock Brakes (1971).

A crash test dummy demonstrating the efficacy of an airbag.

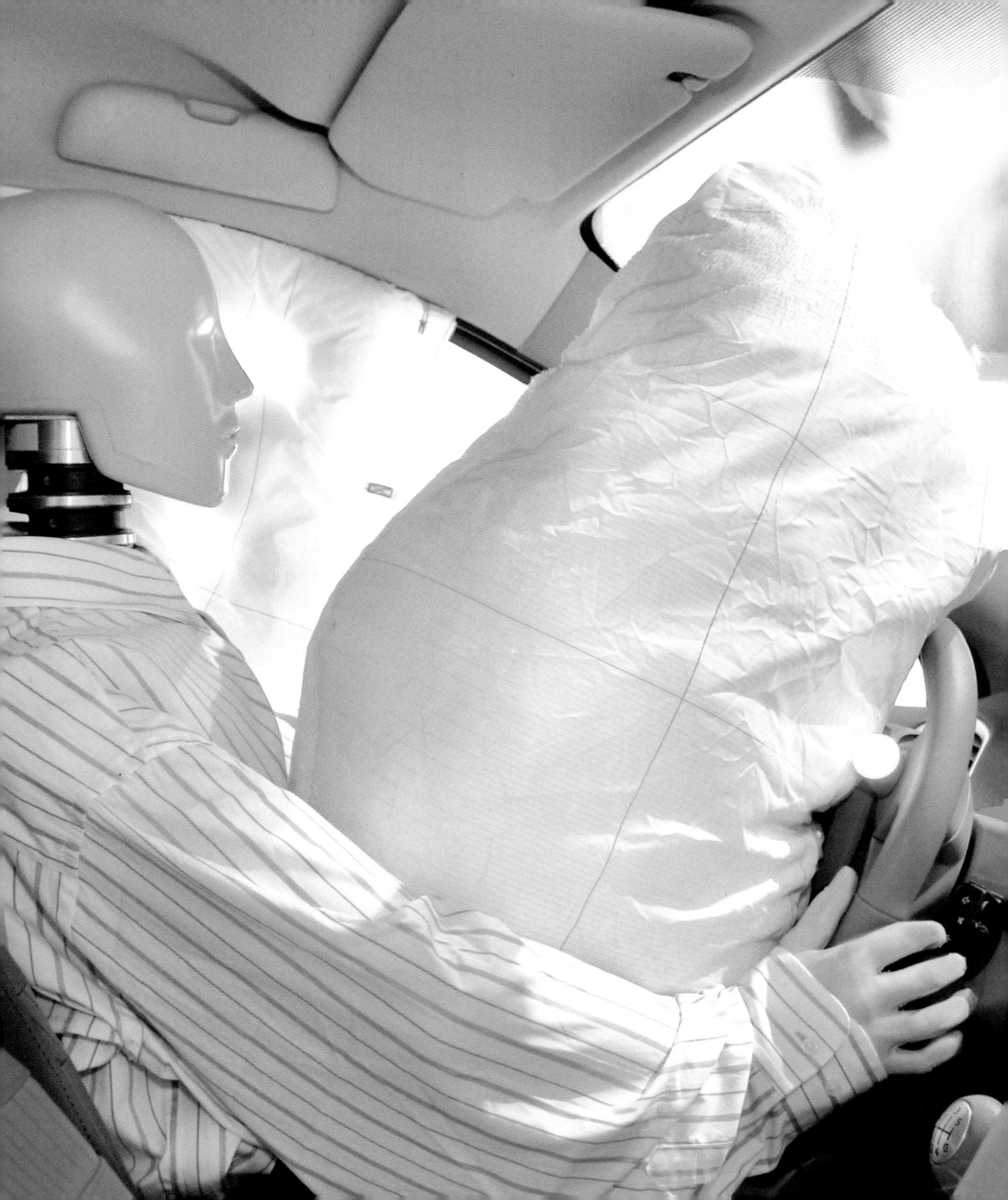

1956

Hard Disk

If we review the history of computer storage starting in the 1940s, it is quite a menagerie. Long tubes filled with mercury, giant spinning drums coated in iron oxide, tiny metal donuts handwoven into a cloth of wire . . . not to mention paper tape, punch cards, and big reels of magnetic tape. Engineers tried just about everything they could think of to store data.

All of those ideas are long obsolete. Except for one that appeared in 1956, developed by IBM for the United States Air Force: the hard disk. Engineers have had a field day making incremental improvements to hard disks for more than half a century and the results have been spectacular.

The first hard disks were as big as refrigerators and incredibly expensive. They stored just a few million characters. However, the fundamental concepts were identical to those used today. A rigid aluminum disk spins on a shaft. The coating on the disk records magnetic changes. A read/write head on an arm can move to different tracks on the disk. To keep from wearing out the head and the magnetic coating, the read/write head flies on an air cushion above the disk.

Engineers have improved everything since then. The coating on the disk allows denser and denser recording. The head gets smaller to take advantage of the density. The arm gets more precise and faster. Where engineers once stored thousands of bits per square inch of disk surface, they now store millions. But the biggest difference is cost. It once cost $15,000 to store one million bytes. Now it costs tiny fractions of a penny.

This incremental cost-lowering and improvement process is one of the hallmarks of engineering. They start with an initial idea and improve it over time. The **Wright brothers** get a plane in the air, and a few decades later, people are inexpensively flying from America to Europe at Mach 0.8. Engineers start with expensive, bulky radio phones and a few decades later we have inexpensive **smart phones** in our pockets. Initially a computer fills a room and costs millions. A few decades later we buy laptops a million times more powerful for a few hundred dollars. It is one of the best things about engineering.

SEE ALSO The Wright Brothers' Airplane (1903), ENIAC—The First Digital Computer (1946), Smart Phone (2007).

A tiny read/write head floats above the disk's surface when the disk is spinning. The lightweight aluminum arm pictured here holds the head and moves it to different tracks on the disk.

1956

TAT-1 Undersea Cable

By the 1950s, there was a powerful need to connect the United States and Europe with a reliable **telephone** line. **Radio** service did exist, but it was expensive and the quality was poor. A long wire was needed to connect the two sides of the Atlantic. By 1956, the technology to do it had become available and engineers assembled to create the TAT-1 (Transatlantic 1) undersea cable.

TAT-1 made use of several innovations. To cross the Atlantic—a distance of approximately 2,000 miles (3,200 km)—it used a pair of coaxial cables able to carry 36 simultaneous conversations. One coaxial cable carried the voices headed west to east, and the other carried those headed east to west.

When you see a coaxial cable for a TV in your home, it has a thin center conductor often made of copper, plastic insulation around it, a wire mesh shield, and then an outer protective cover. Undersea coax cables have the same kind of construction at their core, but with a more substantial center wire and shield. The TAT-1 cable was then wrapped in jute for padding, followed by an armored layer of heavy steel wires, more padding, and then an outer sheath.

Because of the distance, repeaters were spliced into the cable approximately every 40 miles (64 km). The repeaters received the signal, which had weakened over the course of the prior 40 miles, and amplified it so it could travel the next 40. TAT-1 used a line of vacuum tubes and other components sealed in Lucite to implement flexible repeaters that were spliced into the cable. Specially made high-reliability vacuum tubes with a rated life of 20 years formed the heart of the amplifiers. These tubes were burned in for 5,000 hours, inspected, and then certified for use in the cable.

Once the cable was installed, it remained in service for 22 years—a true testament to the quality of the engineering involved in its creation. Phone calls across the Atlantic were not inexpensive however. In 1956, a three-minute phone call cost $12, or roughly $100 in today's dollars.

SEE ALSO Telephone (1876), Radio Station (1920), Smart Phone (2007).

This photo depicts the laying of the world's first transatlantic telephone cable in Clarenville, Newfoundland. The project was the combined venture of the AT&T Long Lines department, American Telephone and Telegraph Company, and Canadian agencies.

1957

Frozen Pizza

Imagine that you want to make a fresh pizza in your home. For this, you do not need an engineer. You mix up some dough, spread it out flat, put some sauce on it, and sprinkle cheese over the sauce. You do this with your hands using ingredients from your pantry and refrigerator. Now imagine that you want to **mass produce** 100,000 pizzas every day because you have created a popular frozen pizza brand sold in grocery stores. For example, Celentano is often mentioned as the first brand to sell frozen pizzas in America in 1957. For this, engineers are a must. Making 100,000 pizzas in a day at a reasonable price without any glitches is a big challenge.

The kind of engineer you need in this situation is called a manufacturing engineer. To make so many pizzas cost efficiently, nearly every step in the process will be done by machine, and in huge quantities. If each pizza consists of a pound of dough, you need machines to make fifty tons of dough per day. If there are just two ounces of pepperoni on each pizza, you need to dispense 12,500 pounds of pepperoni per day. That means huge bins to hold it, conveyor belts to move it around, and a special machine to apply the pepperoni evenly on each pizza.

What if a food scientist determines that the pizza must be frozen in less than three minutes to -20°F, or else the crust will get mealy? An engineer must design a machine that can do that 100,000 times a day. It might be a giant wind tunnel in which the windchill temperature is -60°F, or a machine that dips the pizzas in liquid nitrogen to freeze them instantly.

Then, at the end of every day, the whole assembly line needs to be broken down and cleaned. This might generate 10,000 gallons of water filled with tomato sauce, dough bits, and lots of disinfectant. The city won't let you discharge this into the sewer system. So an engineer will need to design a wastewater treatment plant for the factory.

It is impossible to build a modern, automated factory without engineers, whether that factory makes cars, golf balls, or pizzas.

SEE ALSO Mass Production (1845), Modern Sewer System (1859).

An advertisement for frozen pizza.

At last! A frozen pizza with real pizzeria flavor ...party-perfect in only 8 minutes

Chef does the work. You do the enjoying. Yes, 8 minutes after you get the pizza urge, you're enjoying it. And it's pizza to perfection.

You see, only Chef Boy-Ar-Dee Frozen Pizza is made with 3 cheeses the way real pizzeria pizza is made. Tangy Romano, traditional Mozzarella, and Cheddar for that "extra cheese" flavor.

Rounding out the good flavor is a tender, crunchy crust (it's made with pure vegetable shortening) and Chef's famous slowly simmered pizza sauce.

Keep several packages in your freezer for those spur-of-the-moment parties.

Two kinds: cheese, or sausage with cheese. Both flash-frozen to keep the flavor pizzeria-fresh.

flavor-frozen...by Chef Boy-Ar-Dee®

1957

Space Satellite

Think about your typical satellite, for example a photographic satellite that takes pictures of the Earth. On the one hand it is not that complicated. It has a high-resolution digital camera attached to a telescope that acts as a lens. It has solar panels and batteries to provide electricity. It has a radio to communicate with Earth and an antenna. There's nothing really surprising. These are the same parts you would expect to find on any remote camera system—the camera itself, a power source, and a radio link.

So why does a satellite like this cost millions of dollars? It is largely because of the special considerations that go along with flying in space. Engineers have to keep the satellite functioning in an inaccessible, harsh environment. These challenges were first faced when Russian scientists launched the first space satellite, Sputnik 1, in 1957. Since then, satellites have gotten far more sophisticated.

Take for example the computer in a modern satellite. Engineers cannot use a normal computer in space. Everything must be radiation hardened at the time of manufacture to prevent cosmic rays, solar particles, and other forms of radiation from disrupting the circuits. The computer will then be triple redundant, with a voting system to detect if one of the
three fails.

The satellite has to keep itself properly oriented at all times. It usually does this with a combination of sun trackers, star trackers, and reaction wheels. The reaction wheels can speed up or slow down in order to change the orientation of the satellite. There will also be thrusters and plenty of fuel (e.g., 300 pounds or 150 kg) to last a decade or more.

The solar cells are not typical. They are radiation-hardened, high-efficiency cells. And the batteries are not standard either. Engineers have created special nickel-hydrogen **batteries** that handle tens of thousands of charge/discharge cycles and can last more than a decade.

And then there is one final thing: a massive amount of reliability testing, certification, redundancy, etc., including assembly in a clean-room environment, testing in a hard vacuum, vibration tests, and so on. There is no way to repair the satellite if something goes wrong, and it has to work for many years in space. All of this work and all of these special components make any satellite an expensive proposition.

SEE ALSO The Hubble Space Telescope (1990), Lithium Ion Battery (1991), Global Positioning System (GPS) (1994), Iridium Satellite System (1998).

Three crew members of mission STS-49 hold on to the 4.5-ton International Telecommunications Organization Satellite (INTELSAT) VI, 1992.

NASA
HUGHES

1958

Cabin Pressurization

If you look back at the high-altitude bombing runs that the Allies used during WWII, you can see an example of a workable but not very good solution. The B-17 was a typical bomber of the era. To avoid flak and fighters, B-17 bombing runs typically occurred at 25,000+ feet (7,620+ meters). The cabin was not pressurized for a variety of reasons. This meant that everyone on board had to wear oxygen masks and heavy clothing. Typical air temperatures at that altitude are -50°F (-45°C). Waist gunners had to worry about wind chill as well and wore bulky electrically heated suits.

While it would have been possible to add air compressors and the associated changes to the fuselage to piston-powered aircraft (both the Boeing 307 in 1938 and the Lockheed Constellation in 1943 used this technique), the thing that made cabin pressurization easy was the appearance of **jet engines**.

Jet engines offer two big benefits for airplanes with pressurized cabins. The first is the high-performance air compressor that is already built into every jet engine. Bleeding air off the compression stage of the engine creates plenty of compressed air for the cabin without any additional weight.

The other thing engineers had to do was to seal and strengthen the fuselage to handle the pressure. Doors, windows, and the aluminum skin all required augmentation, which added weight. The good news is that jet engines are much more efficient and powerful than piston engines, meaning that more power was available to handle the added weight.

In 1958, the Boeing 707 and Douglas DC-8 entered service and made jet travel in pressurized cabins routinely available to the general public. Pan Am launched transatlantic passenger jet service that same year—something that would have been impossible or very uncomfortable without pressurized cabins.

Cabin pressurization is a great example of one new technology giving engineers a handy side effect that makes another new technology easier. Kind of a two-for-one special, and in this case it powered a huge boom in passenger jet travel.

SEE ALSO Turbojet Engine (1937), Norden Bombsight (1939), Boeing 747 Jumbo Jet (1968).

The Boeing 707, pictured here, was one of the first airplanes to feature cabin pressurization.

707
BOEING 707

1959

Desalination

Samuel Yuster (1903–1958), **Sidney Loeb** (1917–2008), **Srinivasa Sourirajan** (b. 1923)

As the world's population grows, fresh water in some areas becomes scarce. There is too little precipitation to handle the human population that has grown in the region.

Engineers look out at the ocean and ask, "Can we desalinate this water and make it drinkable?" Engineers and scientists have come up with many different ideas for industrial desalination. One technique: distillation. When the water in salt water evaporates, the vapor rises and leaves the salt behind. Condensing the vapor creates fresh water. The most common way to do this is with fossil fuels. The process is simple and straightforward but has a big carbon footprint. Another method: nuclear reactors. They leave a much smaller footprint, and this process is common in Japan. Co-generation is also possible, where electricity generation and desalination combine.

The saltier the water gets, the more energy it takes to desalinate it. This insight led to the creation of MSF (Multistage Flash) desalination. Imagine a series of tanks, with the coldest and least salty tank at one end and the hottest and saltiest tank at the other end. Concentrated brine from the hottest tank flows through a pipe to the next tank down, and heats it via a heat exchanger. Brine from the less salty tanks flows to the saltier tanks as it concentrates. This process repeats up and down the line, trying to capture and reuse as much heat as possible and maintaining a gradient of heat and salt.

The other big technology is reverse osmosis (RO), which became viable starting in 1959, after UCLA scientist Samuel Yuster and his two students, Sidney Loeb and Srinivasa Sourirajan, were able to demonstrate its efficacy. Water at high pressure flows through an RO filter. Only fresh water makes it through. The high-pressure pumps require about as much energy as distillation, but the pumps use electricity rather than heat. A big **power plant** (either conventional or nuclear) can run RO desalination at night when electricity demand is low and then supply the grid during the day, making good use of otherwise idle capacity.

As more desalination plants are built, engineers pay more attention to the environment. For example, too much brine discharge can make the ocean near a desalination plant poisonous. As human population grows, engineers will need to be even more creative.

SEE ALSO Power Grid (1878), Light Water Reactor (1946), Green Revolution (1961), Bath County Pumped Storage (1985).

Diagram of a reverse-osmosis water desalination plant.

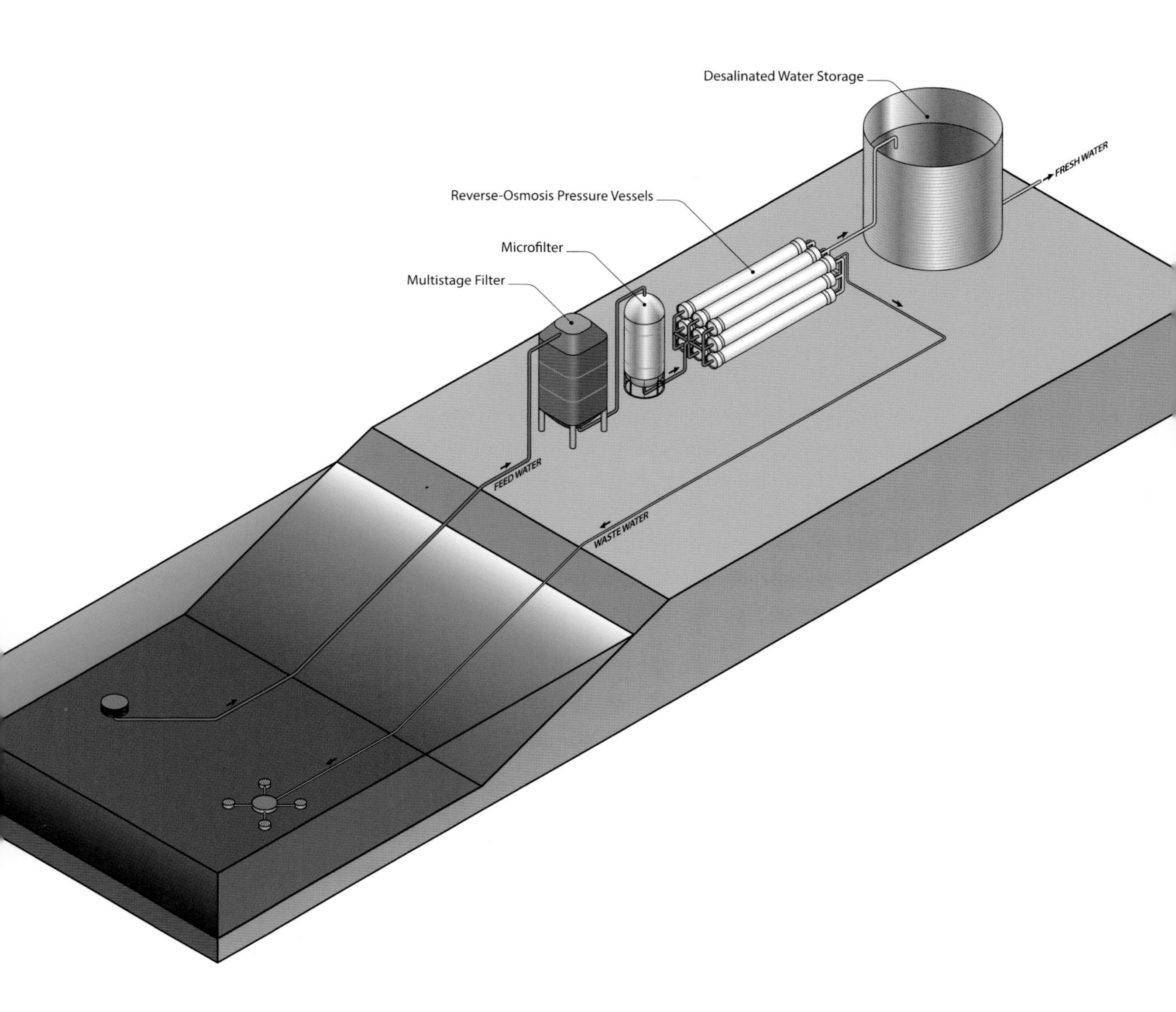
Desalinated Water Storage
FRESH WATER
Reverse-Osmosis Pressure Vessels
Microfilter
Multistage Filter
FEED WATER
WASTE WATER

1960

Cleanroom

Willis Whitfield (1919–2012)

Cleanrooms, first perfected by American physicist Willis Whitfield in 1960, have become an important part of the modern engineering landscape. They are important when manufacturing **integrated circuits**, medical devices and medicines, **satellites**, sensitive optics, and so on. The idea is to create a room that is free of dust, germs, and other impurities that could degrade the device being assembled.

The problems solved by a cleanroom are most easily understood when manufacturing an integrated circuit wafer. One tiny piece of dust on the wafer could cause hundreds of transistors to malfunction. The rooms where these wafers are handled are therefore kept incredibly clean.

Think about the air in a typical house. It is full of dust, dandruff, skin flakes, mold spores, bacteria, insect droppings, cloth fibers, etc. When the sun shines through the window and you see the particles floating in the air, the problem is obvious.

A cleanroom's primary defense against this airborne onslaught of particles has three parts. The first is extreme filtration, which removes any particles already in the air. The second is positive **pressure**, which ensures that no air from outside gets in. The third is protection from contamination inside the room. People encase themselves from head to toe in protective, lint-free suits—often called bunny suits—to keep hair, skin cells, germs, dirt, etc., from getting into the air. Everything from the paint to the tools is chosen to eliminate particles in the air.

To put this in perspective, think about typical room air you find in a house. It contains tens of billions of particles in a cubic meter (35 cubic feet) of air. We may not see the particles hanging in the air because they are so small, but we do see them when they settle out as dust on the furniture or in the furnace filter.

An ISO 1 cleanroom, on the other hand, would have at most 15 particles in a cubic meter of air. And 10 of those will be smaller than 0.2 microns. To accomplish this the filtration is extreme, the room is tightly sealed, and nothing inside the room produces particles. Engineers create the cleanest place possible.

SEE ALSO Integrated Circuit (1949), Space Satellite (1957), Cabin Pressurization (1958).

Scientists in the photolithography laboratory in the London Center for Nanotechnology cleanroom.

LAMARFLO
Acetone
AMARFLO

1961

T1 Line

In 1940, the phone company handled a long distance phone call by sending analog voice signals over pairs of copper wires, one signal per copper pair. You could think of it this way: To make a call between New York and San Francisco, a 3,000-mile-long pair of copper wired would be assembled from segments. This is one reason why long distance calls were so expensive. It took a great deal of hardware to get a voice across the country.

The phone company found new ways to transmit long distance calls. Multiple analog calls could flow though a single coaxial cable, or on a microwave signal. Now it took less hardware to get a voice across the country. Multiple voices shared the same wire, but it was still seriously expensive.

That all started to change in the 1960s. Electrical engineers working for Bell Labs in the United States created hardware that allowed voice signals to be digitized and then sent as digital data. The T1 line in 1961 could carry 1.5 million bits per second over a copper pair—enough bandwidth for 24 simultaneous uncompressed phone calls.

To do this, engineers took 24 analog phone signals and sampled each of them at a rate of 8,000 times per second, with a resolution of one byte per sample. So each eight-thousandths of a second, there were 192 bits plus one additional framing bit for the T1 line to carry. At the other end of the line, the digital signals were sent down another T1 line or turned back into analog phone calls.

The great thing about digital data is that the wire did not care if it was transmitting voices or data. A byte is a byte. So now there was a way to move data through the phone lines.

T3 lines became available as a way of aggregating T1 lines. A T3 line can handle the capacity of 28 T1 lines, or 44.7 megabits per second, or 672 phone calls.

T1 and T3 lines would eventually become extremely important to the expanding Internet. In 1987 the Internet's backbone across the country started using T1 lines. By 1991, **Internet** traffic had increased so much that these were upgraded to T3 lines.

SEE ALSO Telephone (1876), ARPANET (1969).

Pictured: T1 Internet lines.

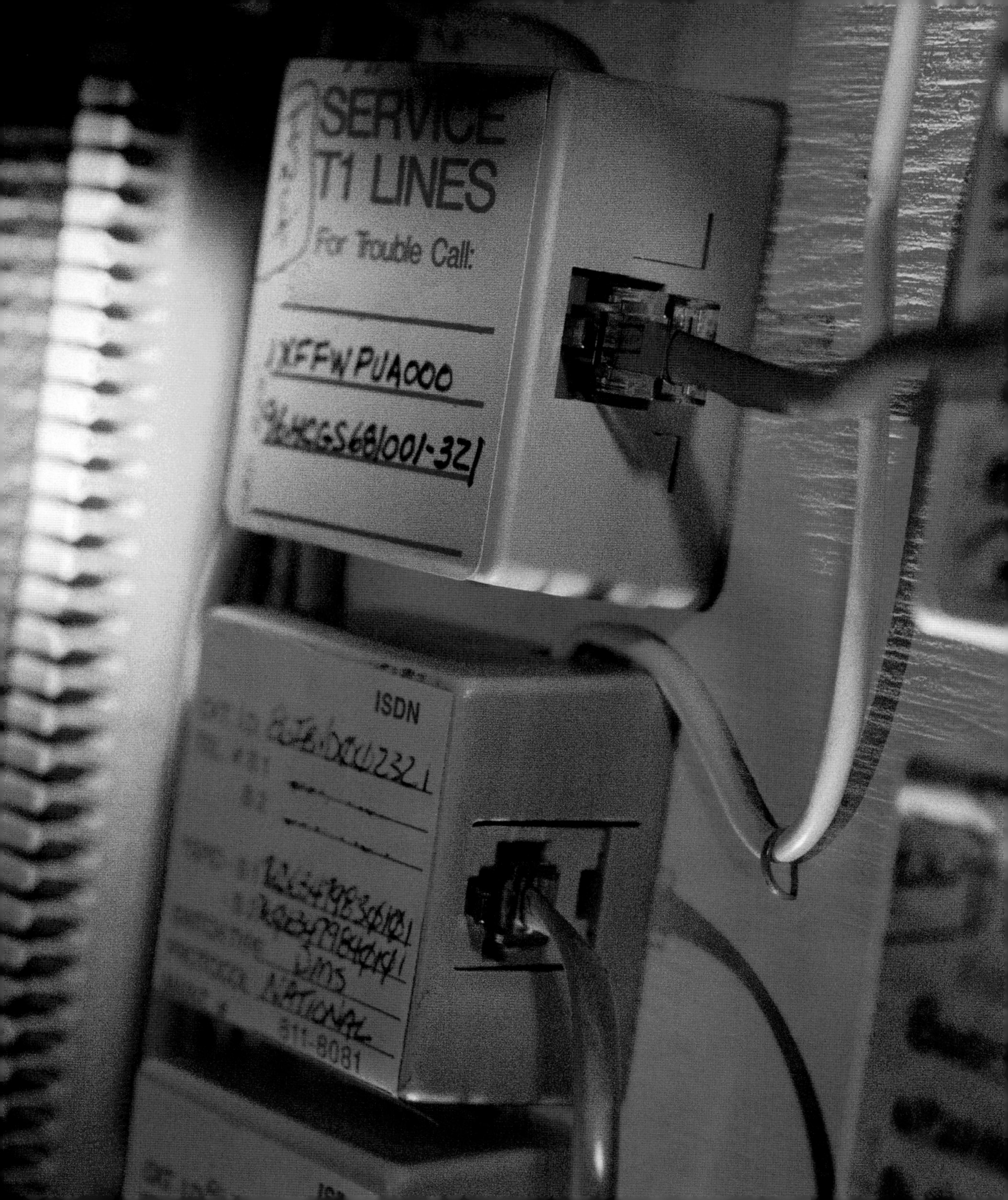

SERVICE
T1 LINES
For Trouble Call:

ISDN

1961

Green Revolution

Norman Borlaug (1914–2009)

Between 1950 and 1987, world population doubled from 2.5 billion to 5 billion people. It was a startling surge. And humans were already putting a strain on food supplies. In 1943, for example, 4 million people in India (6 percent of the population) died in a famine.

During that period of surging population, there was a problem brewing: at then-current production levels, the world had no way to produce enough food to feed everyone. The process that prevented mass starvation—the development that saved a billion or more lives—began in 1961, and was called the Green Revolution. Biologists and engineers worked together to spread advanced agricultural technologies around the world, championed by American biologist and humanitarian Norman Borlaug, who became a spokesperson for these initiatives.

A big factor in improving food production happened at the biological level, by breeding and, later, **genetically engineering** better strains of cereal grains like wheat and rice. Biologists took an engineer's problem-solving approach—they were trying to breed plants that could make use of more nitrogen, while at the same time putting the nitrogen into grain production rather than stem construction. The biologists wanted short, stocky stems so the plants would not fall over. They also wanted to reduce the time to harvest. They were able to pull this off by finding and incorporating dwarf strains and other useful characteristics to create high-yield crops.

These new strains of plants needed water and fertilizer. Engineers could respond to those needs with **irrigation** projects and new ways of increasing fertilizer production. Then they went a step further: in warm climates it is possible to put in two crops a year, but only if there is enough water to support the second crop. With the rainy season supplying the water for one crop, a country like India needed a way to store water for the second crop. So engineers built thousands of new dams in India to catch the water from monsoon rains and hold it. Now India could grow twice as much food.

The effect of these improvements was startling. World food production doubled, then doubled again. Even though population was growing, the food supply grew faster. Science and engineering together created a farming revolution.

SEE ALSO Combine Harvester (1835), Center-Pivot Irrigation (1952), Drip Irrigation (1964), Genetic Engineering (1972).

The engineering innovations that came about as a result of the Green Revolution were numerous, and included new strains of rice meant to create higher yields for famine-stricken regions.

1962

SR-71

Imagine that your job is to design the most amazing airplane ever conceived. It will fly faster than anything on earth—Mach 3+. It will also fly higher—over 80,000 feet (24,400 meters). It will be able to fly across the North American continent in just over an hour without refueling. This plane will be a celebration of aeronautical engineering.

Why would anyone need a plane like this? Two words: military surveillance. Especially during the Cold War, in 1962, the need to fly over any location on Earth without getting shot down became a military obsession. Flying at 80,000 feet at Mach 3, this plane would be untouchable. To accomplish its mission, the plane would need to hold two people, some nice cameras with good lenses, and appropriate electronics.

How would you engineer a plane for this mission? The two biggest challenges would be power and heat.

Mach 3 travel requires a lot of power to push the air out of the way. Rocket engines are one possible solution, but there are issues with the oxidizer, throttling, and in-flight refueling. Engineers chose **jet engines**, but they have a problem at Mach 3: air flows into the engine way too fast.

The solution is a movable device at the front of the engine called an inlet cone. It slows the air down to subsonic speeds before the engine inhales. In addition, bypass valves let excess air vent around the engine.

The heat at Mach 3 affects every part of the aircraft. Skin temperatures above 500°F (260°C) mean that aluminum is out of the question. Therefore, nearly the entire plane is made of **titanium** alloy. The skin and frame expand as they heat up, so the skin is loose on the ground. The plane leaks fuel before it takes off. Once up to speed, everything expands to fit tightly. Even the fuel of the SR-71 is special—JP-7 fuel has a much higher ignition temperature to handle the heat in the fuel tank.

Engineers created an amazing aircraft, and the SR-71 had an illustrious career spanning three decades.

SEE ALSO Turbojet Engine (1937), Titanium (1940).

Capable of flying more than 2,200 mph and at altitudes of over 85,000 feet, the SR-71 was an excellent platform for research and experiments in aerodynamics, propulsion, structures, thermal protection materials, atmospheric studies, and sonic boom characterization.

1962

Atomic Clock Radio Station

Engineers are known for making society more efficient. So think about this: Every year in the United States, we change our clocks forward and back for Daylight Savings Time. At least twice each year, every clock has to be set.

Now assume 100 million households in America. Assume that every household has one **clock** that needs setting. Assume it takes a minute to set the clock. Twice a year. How much time is America wasting on clock setting under those assumptions? Almost 400 human years. Is there a way to recover all that lost time?

Engineers answered the question with an ingenious solution. They envisioned clocks that can set themselves. To do it, they use **radio** waves. There is a radio station in Colorado called WWVB, activated in 1962. It has massive power: 70,000 watts. It has an enormous antenna. It transmits on a very low radio frequency of 60,000 hertz, giving it a great deal of range (especially at night). And it has an incredibly simple AM format: every second it transmits one bit, and every minute a new time code is sent. It also encodes a second phase-modulated signal as well.

With this radio system in place, a clock in your home can contain a simple AM radio receiver and a tiny **processor** to receive and decode the signal. Then, once it receives the signal, it can display the correct time. Since WWVB gets its time from a NIST-certified (National Institute of Standards and Technology) **atomic clock**, every clock using this technique has extremely accurate time.

Because of the radio station's power, antenna, and low frequency, it covers the entire continental United States. At night it can reach out toward Hawaii and South America.

The effect is this: If you buy a clock that can listen for WWVB—it will often describe itself as an "atomic clock" on the package—you can put in the batteries and one of two things will happen. Either, a minute later, it will magically set itself. Or, that night, it will set itself. It depends on how far away you are from Colorado. Engineers just saved you a few minutes of your life every year.

SEE ALSO Mechanical Pendulum Clock (1670), Radio Station (1920), Atomic Clock (1949), Microprocessor (1971).

Radio-controlled clocks are synchronized by a time signal from an atomic clock.

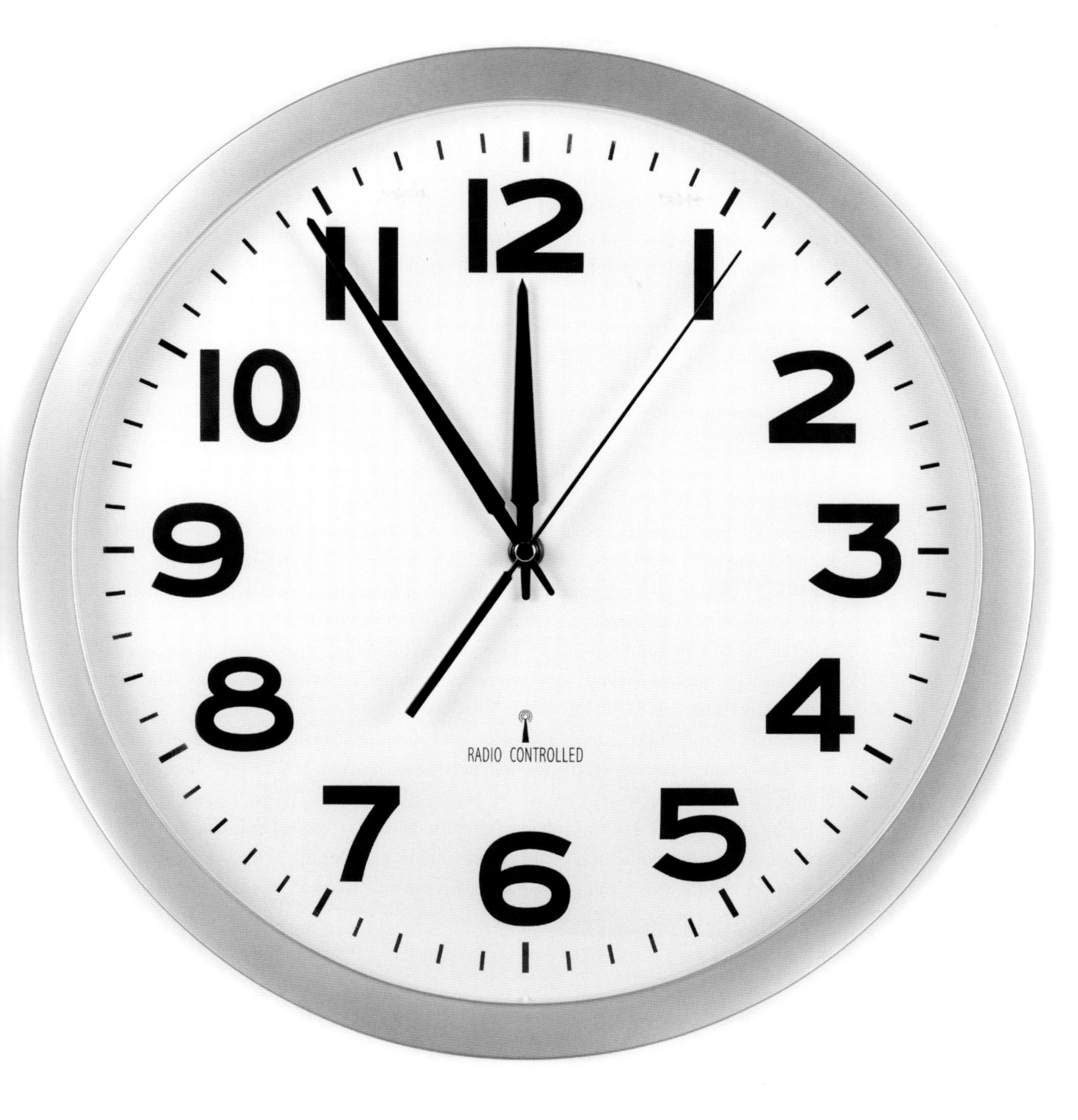

RADIO CONTROLLED

1963

Retractable Stadium Roof

David S. Miller (Dates Unavailable)

Imagine a roof engineered to cover several acres. We actually see roofs like this all the time, whenever we walk into a big-box retail store. A large superstore can cover 4 acres (1.6 hectares) or more under a single roof. If you look up at the ceiling the next time you are in such a store, you will likely see a series of steel I beams and/or **trusses** supported by evenly spaced columns across the store. It is an easy and inexpensive roof structure.

Now imagine that you want to engineer a roof of that size or bigger, and you have two additional requirements: There can be no columns in the middle, and the roof must be able to completely retract to open the floor to the sky. This is exactly the challenge that engineers face when designing stadiums with moveable roofs, first patented by American architect David S. Miller in 1963.

One common approach to a retractable roof is to use multiple panels that slide underneath one another. The four considerations from an engineering standpoint are that the panels: 1) need to be able to slide back and forth, 2) can weigh millions of pounds, 3) have to take on extra weight in some climates, like snow and ice, and 4) need to span long distances, perhaps as much as 700 feet or more. Listening to these criteria, the panels start to sound a lot like bridges, and that is how they are designed. A panel usually consists of long, deep trusses that can span the whole stadium. The roofing materials are fastened to the top of the trusses.

The two ends of these roof-bridges are on wheels similar to railroad wheels running on tracks. Electric motors set the roof panels in slow motion. It might take fifteen minutes for a roof to retract.

The advantage of a retractable-roof stadium is easy to understand. When the weather is nice, the roof can provide an open-stadium experience. In bad weather, the roof closes and the game can go on. Engineers are able to create the best of both worlds for the players and the audience.

SEE ALSO Truss Bridge (1823), Stadium TV Screen (1980).

Chase Field in Arizona, United States, features a retractable roof.

1963

Irradiated Food

Frozen pizza entered the marketplace in 1957, but there are many foods that cannot be frozen. Think about the refrigerated food section in any store. The food is packaged, yet it still needs refrigeration. The reason is because it contains bacteria. If your grocer left milk out on a shelf rather than refrigerating it, it would spoil in a few hours. At room temperature, the bacteria in the milk multiply rapidly and ruin it. The same is true of most dairy products, meats, premixed cookie doughs, and biscuits, etc.

What if engineers could develop a technology to eliminate the bacteria in packaged foods? It turns out they have—it is called canning. If you heat a can or jar of food to a sufficient temperature and then seal it, the contents are sterile. This makes the food shelf-stable, and it explains why things like spaghetti sauce and canned vegetables do not need refrigeration. The problem with canning is that the heat can change the taste and texture of the food.

What if engineers could develop a technology to eliminate the bacteria without changing the taste? This technology exists and it is called food irradiation, first approved by the US Food and Drug Administration (FDA) in 1963. Using this process, it is possible to seal fresh meat in a plastic wrapper, irradiate it to kill the bacteria, and then leave it on a shelf for a year or more. Inside its wrapper the meat is sterile, so bacteria do not spoil it. When you unwrap it, it is a piece of fresh meat.

Gamma-ray irradiation is one popular technique. Cobalt-60 is a gamma ray source and meat passes near the source to irradiate it. The process is extremely simple. The main engineering challenge is keeping people away from the radiation and keeping the radiation sources secure. Since gamma rays are high-frequency, ionizing radio waves, they pass through the meat like radio waves would, killing the bacteria while leaving nothing behind.

Irradiation technology has existed since the 1960s. NASA uses it on meat flown in space. But the technology is used on only a small number of foods. Why don't we see it commonly in grocery stores? The reason is something engineers can't do anything about: fear. When people hear the word "radiation," there is an immediate negative reaction, despite the safety and benefits of the process.

SEE ALSO Electric Refrigeration (1927), Microwave Oven (1946), Frozen Pizza (1957).

This photo shows irradiated strawberries (left) and normal strawberries (right) after several days.

1964

Top Fuel Dragster

What if you let engineers take technology to the limit? You let them evolve everything with a single goal in mind. That is what has happened in top fuel drag racing to get us to the cars we see today. Engineers want to make the fastest piston-engine vehicle possible. How far can the car evolve to make that happen?

One way is to grow the **engine's** power. Engineers start with a 500-cubic-inch (8.2 liter) V-8—the largest allowed under National Hot Rod Association (NHRA) rules. They add a **supercharger** to cram as much air as possible into each cylinder. Then they change the fuel, from gasoline to nitromethane, aka "top fuel." Nitromethane is more a liquid explosive than a fuel and was first allowed in the 1964 season. It contains quite a bit of the oxygen it needs to burn within the molecule itself (a characteristic of an explosive). The amount of fuel injected is staggering. Think of the water coming out of your showerhead in the morning. Each cylinder is receiving fuel at that kind of rate. The engine can burn 1.3 gallons (5 liters) of fuel per second.

Then grow the rear tires. They are incredibly wide to create the largest possible contact patch, and they use "wrinkle wall" technology to make the contact patch even longer at the starting line. The wing provides 8,000 pounds of downforce to keep the tires glued to the track once the car gets moving.

The power gets from the engine to the tires through a complex six-stage clutch that can provide a lot of slip at the start, and progressively less slip as the car comes up to speed. It is a direct-drive system with no transmission—clutch slip keeps the engine from bogging down. Getting the clutch parameters right is one of the most important aspects of the car's setup.

All of this technology means the engine makes 8,000 horsepower (6 megawatts), and the car has gigantic acceleration. When the light turns green, the car goes from zero to 75 mph (120 kph) in half a second, using approximately 20 feet (6 meters) of track to do it. By the end of the quarter mile the car is going over 300 mph (480 kph).

SEE ALSO Supercharger and Turbocharger (1885), Internal Combustion Engine (1908), Formula One Car (1938).

Driver Larry Dixon brings his Top Fuel Dragster down the track during a qualifying run for the Tire Kingdom NHRA Gatornationals race in Gainesville, Florida, in 2011.

1 T/F
TIRE
GOODYEAR
EAGLE
TOYOTA
AL-ANABI
Racing
Prestone

Drip Irrigation

Simcha Blass (1897–1982)

Farmers have been irrigating fields for centuries. They can let water flow into fields through irrigation channels or set up overhead sprinkler systems. But if farmers use these techniques in an arid climate, they do not work very well. There probably is not enough water available for flood irrigation, and too much water evaporates into the air with a traditional sprinkler system. **Center-pivot irrigation** would not work here.

Thus it makes sense that in Israel a new irrigation system would appear. Engineer Simcha Blass, with his son Yeshayashu and in conjunction with Kibbutz Hatzerim, patented the first surface drip irrigation emitter functional for practical use. Israel is extremely arid and fresh water is a scarce resource. The drip irrigation approach, patented in 1964, became a new standard for water-saving irrigation.

The basic idea is simple. Water drips onto the roots of each plant individually. Plastic pipes carry the water around the field, and special emitters that are immune to clogging control the rate of water delivery.

There are several important advantages embodied in this engineering approach. The main one is that it drastically reduces evaporation. The water drips directly onto the soil where it is absorbed immediately. There is no time where it is flying through hot, dry air and evaporating.

The second advantage is that the water concentrates where it is needed: in the root zone. The ground between plants, which contains many fewer roots, never receives moisture. This reduces surface evaporation and also cuts down on weeds.

Another advantage is that drip irrigation helps turn gray water and wastewater into irrigation water. If water has not been treated, it is not advisable to aerosolize it by spraying it in the air. Drip irrigation applies the water directly to the soil, or directly into the soil with subsurface emitters, so the wastewater becomes completely safe.

Drip irrigation is another great example of an engineering reconceptualization that yields multiple benefits from a different way of thinking. It has opened up far more land for farming. For example, all of the cotton grown in Israel (more than 100,000 tons per year) is drip irrigated. It would be nearly impossible to grow it otherwise.

SEE ALSO Center-Pivot Irrigation (1952), Desalination (1959), Green Revolution (1961).

Drip irrigation is a water-saving innovation that many of us use in our home gardens today.

1964

Natural Gas Tanker

Every now and again, engineers are asked to do something that initially seems impossible. But in a lot of these cases, they can look at the difficulties and knock them down one by one to find a workable solution.

Such is the case with natural gas transportation. On land the problem is solved with a pipeline. But what if we need to move natural gas over an ocean?

With oil, ocean transportation is a relatively simple process. That's because oil is a normal liquid. You can put it in a tank at atmospheric temperature and pressure, and it will sit there unattended for years.

Natural gas, unfortunately, is a gas. You can compress it, and that works to some degree. It's how engineers store natural gas in cars and buses. But compressed natural gas isn't dense enough for **shipping**. For that, natural gas needs to be liquefied.

The advantage of liquefied natural gas, or LNG, is the size reduction. Six hundred cubic meters of natural gas becomes one cubic meter of LNG. The problem with LNG is that it is a cryogenic liquid. If you have a cup of LNG sitting on your desk, it is a clear, boiling liquid at -260°F (-162°C).

To hold LNG in a ship, engineers have to design huge super-insulated tanks. The first ship to do this, the *Methane Princess* built for British Gas, appeared in 1964. The insulation often comes in the form of one foot (30 cm) or more of foam, with a metal inner liner and a strong metal shell on the outside. The largest LNG tankers of the Q-Max class hold nearly 10 million cubic feet (266,000 cubic meters) of LNG.

The super-cold temperatures cause big problems. The inner linings shrink and can shatter from cold shock, so they must be precooled. Metal pipes for loading and unloading become brittle and also shrink, so cold-resistant materials handle junctions and expansion joints. And then there is boil off. During the trip, as much as 10 percent of the LNG becomes gas. Some ships use this as fuel in the engines. Some reliquefy it and put it back in the tanks.

Even at -260°F, engineers can make it work.

SEE ALSO Oil Well (1859), Power Grid (1878), *Seawise Giant* Supertanker (1979), Container Shipping (1984).

Liquid Natural Gas tanker at port with LNG liquefaction plant in background.

L N G

1964

Bullet Train

Hideo Shima (1901–1998)

In the United States, a typical Amtrak passenger train travels at speeds below 100 mph. Meanwhile, the bullet trains, or Shinkansen, in Japan routinely run at 200 mph. Hideo Shima was the chief engineer in charge of the first bullet train, the Tokaido Shinkansen, which was launched by Tokyo National Railways in 1964. Other engineers working for the Railway Technology Research Institute also contributed to its design.

How did these engineers take train technology to speeds like that? To understand the differences, it is helpful to look at typical train track. The track bed is gravel (also known as ballast). Wooden crossties underpin the steel rails, which are held in place with railroad spikes and clips. The tracks frequently cross highways and roads at grade with crossing arms trying to keep cars off the tracks when trains go by. Two or more **diesel locomotives** pull the trains, and train cars have conical steel wheels on solid steel axles.

For high-speed trains, all of this changes. Track improvements smooth out the ride. High-speed track uses continuously welded steel rails. The rails attach to crossties upgraded from wood to concrete, and often the ballast is upgraded to concrete as well. At-grade crossings with highways and roads are eliminated to prevent interactions with cars because high-speed trains can take two to three miles to stop. All sharp turns are eliminated, favoring long smooth curves with radii up to five miles. Instead of diesel engines, high-speed train systems use overhead electric wires to supply electricity. Instead of an engine, the motors that power the train are spread out to wheels on all of the cars. This helps to improve acceleration and braking. The front car is still unique, however, because it has an aerodynamic shape to cut down on drag and also holds the operator's cabin.

When engineers combine all of these features together successfully, trains can run at 200 mph and passengers will have a smooth, comfortable ride. Engineers make the ride so smooth that passengers become unaware of the tremendous speed.

SEE ALSO Diesel Locomotive (1897), Magnetically Levitated Train (1937), Vactrains (c. 2020).

Mount Fuji in Japan, with a Shinkansen (bullet train) approaching in the foreground.

1965

Gateway Arch in St. Louis

Eero Saarinen (1910–1961), **Hannskarel Bandel** (1925–1993)

Both sides of the Gateway Arch in St. Louis, MO, were built independently, as freestanding towers. Before they were connected, each stood 630 feet (192 meters) in the air with a thin gap separating them. The topmost piece of the arch was 8 feet (2.4 meters) wide, but the sun baked the structure and there was only a 2-foot (0.6 meters) gap due to expansion. The architect who designed the arch, Eero Saarinen, and the engineers, lead by structural engineer Hannskarel Bandel, anticipated this very situation. A huge hydraulic jack sat atop the arch to create one million pounds of gap-widening force. The jack jammed the cavity open and the final piece slid perfectly into place. The alignment had a precision of less than half a millimeter.

The Gateway Arch, completed in 1965, is a world-renowned engineering achievement. No other stainless steel monument in the world is taller. And it's a timeless beauty—a perfect gleaming curve that could last for millennia and still look just as stunning.

Making the arch strong enough to withstand earthquakes and tornadoes, capacious enough to hold one hundred visitors in the observation deck, and hollow enough to house the strangest elevator in the world are just three of the major challenges engineers faced on this project. The arch starts deep underground in the form of two massive 60-foot-tall (18.3 meters) cast concrete blocks embedded into the limestone bedrock on the site. These blocks weigh 52 million pounds (24 million kg) each. On these engineered foundation blocks, the arch begins with a hollow stainless steel triangle measuring 54 feet (16.5 meters) on each side. The arch tapers smoothly so that the topmost triangle measures 17 feet (5.2 meters) on each side.

Tourists need to get to the observation deck quickly because more than a million a year want to take in the view. How do engineers create an **elevator** that is vertical at the bottom and horizontal at the top? They make the elevator cars cylindrical like the drum of a clothes dryer and let the cars rotate as they move up the arch to keep the seats horizontal. Every part of this arch is impressive.

SEE ALSO Leaning Tower of Pisa (1372), Elevator (1861), Empire State Building (1931), Burj Khalifa (2010).

Close-up view of the Gateway Arch, the symbolic gateway to the West (although there's a great deal of Midwest to go before the West is reached).

1965

Cluster Munition

There are two ways to look at cluster munitions. On the one hand, they are horrible. Like **AK-47**s, they kill people and destroy property with stunning efficiency. On the other hand, if the goal is to kill the enemy and destroy enemy property efficiently, cluster munitions are an engineering achievement.

Early cluster bombs, first deployed in Laos and Vietnam in 1965, were very simple. Imagine a bomb casing filled with a large number of small explosives similar to hand grenades. An airplane would drop the bomb, the casing would open midair to release the bomblets, and they would explode when they hit the ground. Any human being caught within the footprint of the cloud of bomblets would die in the maelstrom of shrapnel the bomblets all produced. An unfortunate side effect of the process would be the duds—bomblets that would fail to detonate for some reason on impact. They would remain dangerous for years, causing countless casualties.

Modern cluster bombs like the CBU-105, also known as a Sensor Fuzed Weapon, are completely different. When the bomb opens midair, ten BLU-108 submunitions deploy and fall slowly under individual parachutes. A radar sensor in the bottom of the BLU-108 detects the height above the ground. At the appropriate height, the BLU-108 fires a small solid rocket that spins the canister and increases its altitude. The rotational force allows the canister to throw out four spinning bomblets, each with its own sensors and intelligence to detect heat-producing vehicles on the ground. Each bomblet seeks a vehicle and then fires an explosively formed penetrator at it, destroying the vehicle. The bomblets also automatically self-destruct if they did not acquire a target, eliminating duds.

Imagine a squad of soldiers encountering a column of enemy vehicles on the move, or an encampment of enemy vehicles. One CBU-105 would selectively destroy all of the vehicles, and no duds would be left behind to harm civilians. This approach is significantly better than the simple cluster bombs that preceded the CBU-105.

SEE ALSO Trinity Nuclear Bomb (1945), AK-47 (1947).

M190 Honest John chemical warhead section containing demonstration M134 GB (Sarin) bomblets.

1966

Compound Bow

Holless Wilbur Allen (1909–1979)

The idea of an arrow powered by a bow is one of the oldest engineered technologies that humans have created. **Bows and arrows** have been in use for tens of thousands of years. Therefore it seems like engineers would have milked the technology of all of its improvements long ago. Then, in 1966, engineers came along and looked at things differently. This led to the reconceptualization known as the compound bow.

Think about how a normal bow works. You take a stick and bend it, then apply a string across the two bow tips to maintain the arch. When the archer pulls back on the string, the energy is stored by increasing the depth of the arch. The tips come closer together and also come closer to the archer. The farther back the archer pulls, the more tension there is on the string. With a strong English longbow, the draw weight can go as high as 150 pounds (670 newtons). The archer has to stand there and exert 150 pounds of effort to hold the string in the drawn position.

The key to the compound bow, patented by Holless Wilbur Allen in 1966, is to store the energy differently. There is a rigid vertical piece where the archer holds the bow, called the riser. Then there are two nearly rigid horizontal pieces called limbs. The energy will be stored by bringing the tips of the limbs closer together. The limbs become cantilever springs—the same sort of spring that powers a diving board.

The cams create a variable leverage arrangement. They allow the tension on the drawn string to change depending on the distance from the riser. Near the riser, the draw weight can be significant, but as the archer draws back further, the cams rotate to create a different lever arm. The archer feels this as a surprising reduction in the draw weight. It is much easier to hold the string in this position and aim.

This is one of those situations where engineers asked the right question: How can we reduce the draw weight without reducing the power of the bow? They found a great answer, even if it took tens of thousands of years to find it.

SEE ALSO Bow and Arrow (30,000 BCE).

Albina Loginova at women's individual compound 3rd place, 2013 FITA Archery World Cup, Paris, France.

Team
HOYT
KIA
SHIB
Platinum

1966

Parafoil

Domina Jalbert (1904–1991)

When we think of engineering, our minds often turn to hard objects: **bridges**, **airplanes**, transmissions. But it is also possible for an engineered object to be soft and flexible. The modern sport parachute is a good example—there is engineering throughout. Think about the traditional round parachute. It is a piece of fabric sewn together from a set of panels. Thin ropes attach to the edge and tie the parachute to a pair of shoulder straps. A backpack and harness attach the parachute to the skydiver's body. The skydiver basically floats straight down under the canopy, at the mercy of the wind.

The modern parafoil, first patented by Domina Jalbert in 1966, is a complete reconceptualization that has several big advantages over a round parachute. The parafoil is essentially a fabric wing with an airfoil shape. Ram air at the front of the airfoil inflates the wing and keeps it semi-rigid. Since it is an airfoil, it flies—it has forward speed and a glide ratio. The lines that attach the parafoil to the skydiver keep it aligned overhead. The back lines attach to loops that the skydiver holds in his or her hands to control the parachute. Pulling on the left toggle causes the left side of the parafoil to slow down. Ditto on the right side. Pulling down both toggles causes the parachute to slow down and flare, which is especially useful for landing.

The advantage of the parafoil is the fact that it is flying. The skydiver can steer the parachute and have much more control over the landing. Parachuting is much safer as a result. Engineers can control a number of variables when designing the parafoil: the shape, the thickness of the airfoil, the overall size, the aspect ratio, the number of cells, the fabric, etc. Some parafoils are small and thin for speed flying. Some are large for leisurely parasailing. The design depends on the goals and needs of the user.

Engineers might not have anticipated the tricks and stunts that experienced skydivers can do with parafoils. Engineers create the technology, and then enthusiasts take it to the limit.

SEE ALSO Plastic (1856), Kinsol Trestle Bridge (1920), Human-Powered Airplane (1977).

Pictured: Domina Jalbert's 1966 patent for the parafoil.

Nov. 15, 1966 D. C. JALBERT 3,285,546

MULTI-CELL WING TYPE AERIAL DEVICE

Filed Oct. 1, 1964 2 Sheets-Sheet 1

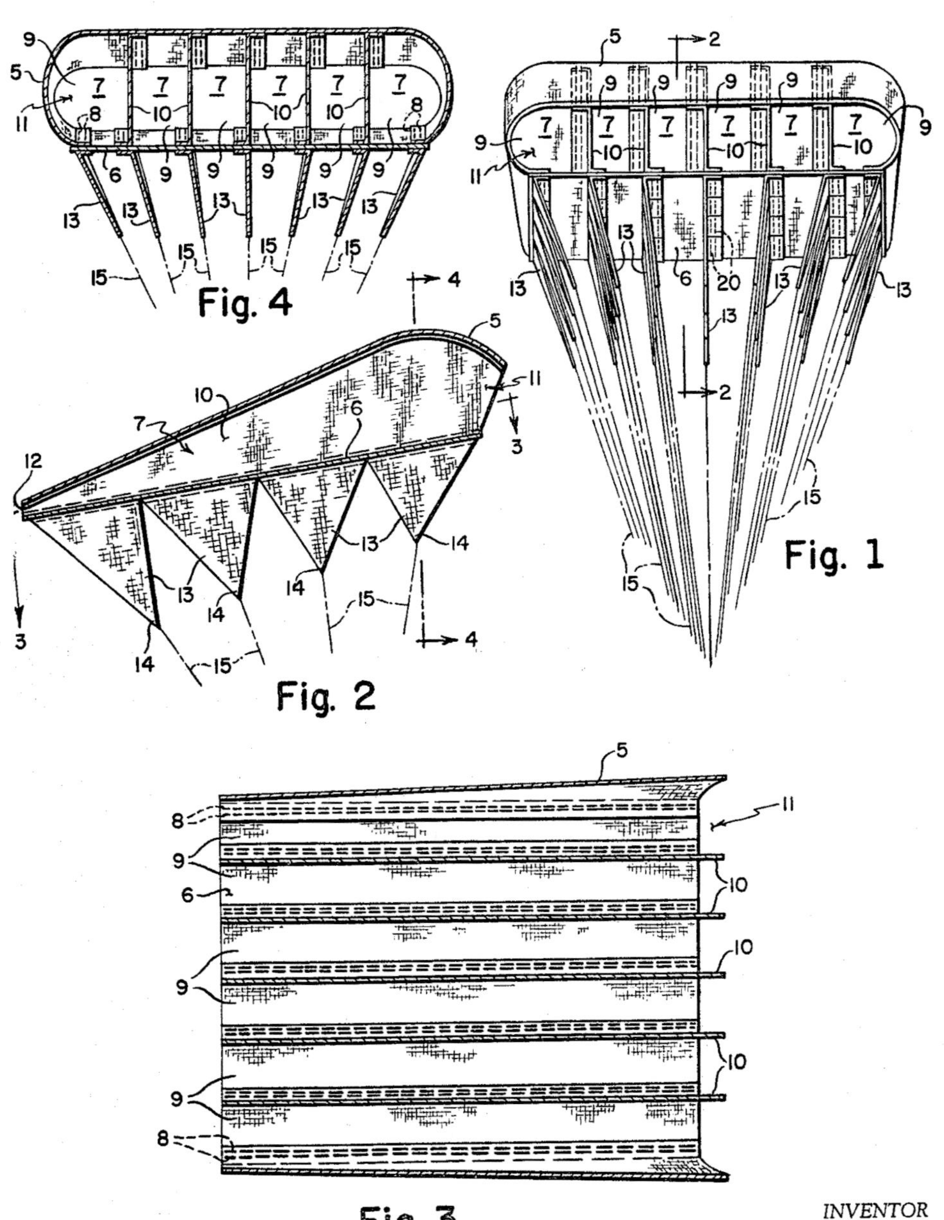

INVENTOR

DOMINA C. JALBERT

BY James N. Ogle

ATTORNEY

1966

Pebble Bed Nuclear Reactor

The pebble bed reactor is an example of a complete reconceptualization. Almost all of the major nuclear reactor systems in place around the world today use water in one way or another and have the potential for major accidents if something goes wrong with the water. The pebble bed reactor, which was first successfully demonstrated in 1966, takes a completely different approach in an attempt to make a simpler, smaller, safer reactor.

The first word in the name comes from the way the reactor packages the **nuclear** fuel. Instead of fragile fuel rods, the fuel is encased in a sphere of protective materials. Each "pebble" is the size of an orange (2.36 inches, 60 mm). Inside of the sphere are a handful of much smaller spheres. Imagine a small sphere (half a millimeter in diameter) of nuclear fuel wrapped in a barrier material and then wrapped in pyrolytic carbon. This handful of tiny spheres is then encased together to form the larger sphere.

The second word in the name comes from the way engineers house these spheres. They are piled in a tube—think of a small silo—and the collection of pebbles produces a tremendous amount of heat. The heat is extracted by blowing an inert gas through the pile of pebbles.

Pebbles can be extracted from the bottom of the pile for inspection. Old or damaged pebbles are removed, while those that pass inspection are reinserted at the top of the pile.

The inherent safety claimed for pebble bed reactors comes from the following idea: If everything else failed and all that was left was the pile of pebbles, this pile would reach some maximum temperature and then exist at that temperature indefinitely. The fuel would not melt or detonate.

The interesting thing about the pebble bed design is the ability of engineers to look at existing ways of doing things, identify the problems and weaknesses, and then creatively reconceptualize to address the problems. It is a process that engineers frequently use to leapfrog ahead. In China, there are plans to build hundreds of these reactors.

SEE ALSO Light Water Reactor (1946), CANDU Reactor (1971), Chernobyl (1986).

AVR pebble bed reactor at Forschungszentrum, Jülich, Germany.

Pebble Bed Reactor scheme

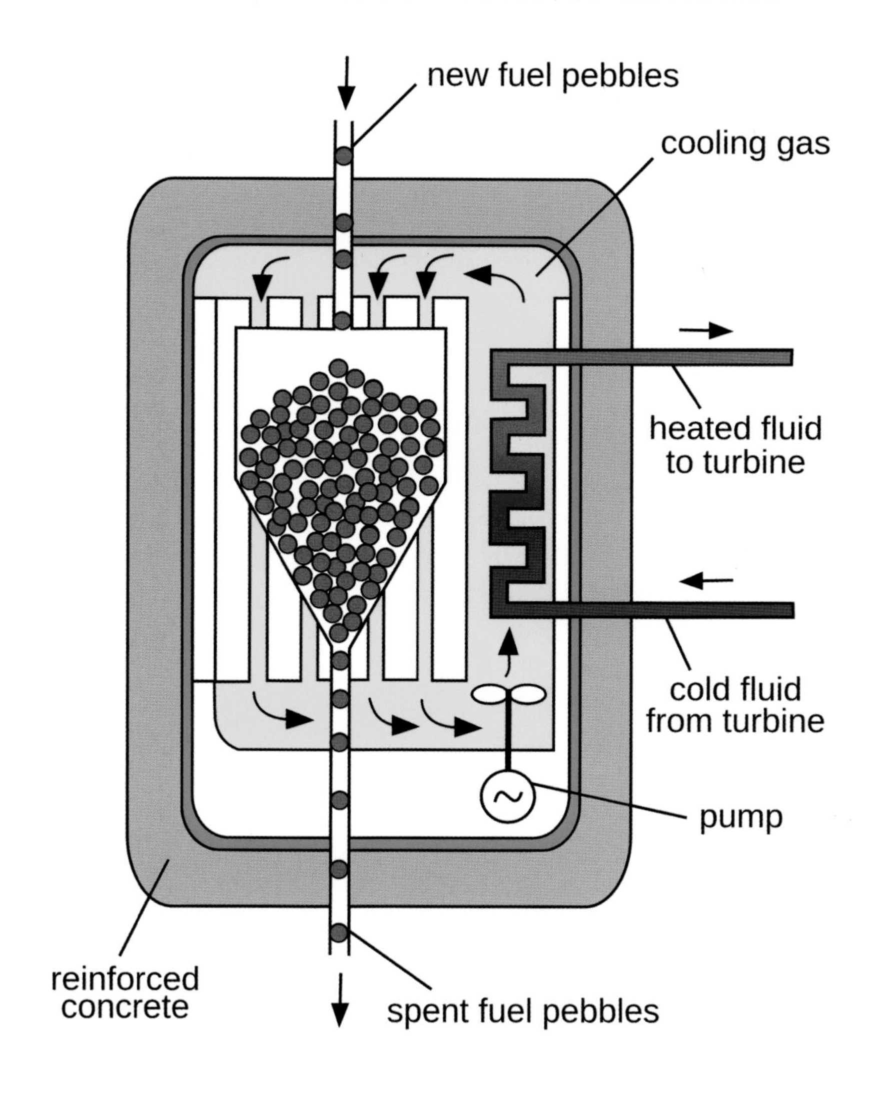

1966

Dynamic RAM

Robert Dennard (b. 1932)

Every **computer** needs RAM, or Random Access Memory. The central processing unit of the computer needs a place to store its programs and data so it can access them quickly—at the same pace that the CPU's clock is operating. For each instruction the CPU (Central Processing Unit) executes, it must fetch the instruction from RAM. The CPU also moves data to or from RAM.

Imagine you are an engineer looking at computer memory options in the late 1960s. There are two possibilities. The first is core memory, which is made by weaving tiny ferrite donuts into a wire mesh. The problems with core memory are many; it is expensive, heavy, and enormous. The second possibility is static RAM made from standard transistor circuits. It takes several transistors for each memory bit, and given the state of integrated circuits at the time, it is not possible to put much memory on a chip.

But in 1966, American electrical engineer Robert Dennard, working for IBM, tried something different in the interest of reducing the number of transistors and fitting more memory cells on a chip. He explored the idea of dynamic RAM using a capacitor to store one bit of data. When the capacitor is charged it represents a 1, discharged it represents a zero. On the surface this seems ridiculous, because capacitors leak. If you store a 1 in memory made of capacitors and do nothing, the capacitor will leak and forget the 1 in less than a tenth of a second.

But the advantage is that this approach greatly reduces the number of transistors, and therefore increases the number of memory cells on a chip. To solve the leakage problem, all of the capacitors are read periodically (for example, every few milliseconds) and rewritten, thus refilling all of the leaking capacitors containing 1s with a full charge. This approach is known as dynamic RAM (DRAM), first manifested in 1970, because it must be dynamically refreshed to keep the capacitors charged.

The dynamic RAM approach yields memory cells that are so much smaller, and therefore less expensive, than static RAM that every desktop, laptop, **tablet**, and **smart phone** today uses DRAM. It is a great example of the way engineers can reduce costs by embracing ideas that may seem initially ridiculous.

SEE ALSO ENIAC—The First Digital Computer (1946), Flash Memory (1980), Smart Phone (2007), Tablet Computer (2010).

Pictured: Dynamic SDRAM memory for a computer.

94V-0
RCE00003-01
E187565

1967

Automotive Emission Controls

In 1960, in a big US city like Los Angeles, there are several things we would be able to detect: 1) a brown layer of smog visible at the horizon, 2) a layer of lead developing in the topsoil (fallout from leaded gasoline), 3) serious particulate pollution in the air, and 4) ground-level ozone.

Public outcry, combined with the Federal Air Quality Act of 1967, led to a series of engineering advancements that cleaned up car emissions and made our cities breathable again.

The first improvement was positive crankcase ventilation, which was instituted in 1961. This is a simple system that sucks fumes out of a car's engine and burns them rather than letting them float into the air. This matters because unburned fuel and oil fumes combine with sunlight to create smog.

The next big change was the catalytic converter, combined with unleaded gasoline. The first catalytic converters in 1975 burned off carbon monoxide and unburned gasoline in the exhaust. Later models also eliminated nitrogen oxides. Engineers came up with an innovative, long-lasting ceramic honeycomb approach to bring the exhaust gases in contact with catalyst metals like platinum. Petroleum engineers had to improve **refineries** and techniques to create gasoline that did not need tetraethyl lead to artificially boost the octane rating.

Another trick that engineers devised is called exhaust gas recirculation. Exhaust contains no oxygen, so it takes up space if sent into the cylinders. Therefore it lowers the temperatures in the cylinder and reduces nitrogen oxides.

Finally there is the gas tank. When a car sits in a parking space, the gasoline in the tank evaporates. Normally these hydrocarbons would float into the air to create smog as the tank vents. Engineers developed canisters filled with carbon granules to absorb gasoline vapors, and sealed gasoline systems.

By combining all of these techniques, engineers radically reduced the lead, hydrocarbons, nitrogen oxides, and carbon monoxide coming out of each automobile on the road. The air in America has been getting cleaner ever since, even with the growing number of cars.

SEE ALSO Oil Well (1859), Wamsutta Oil Refinery (1861), Prius Hybrid Car (1997).

Innovations in automobile transport changed the world as we knew it, but also gave rise to the need for clean air standards.

1967

Apollo 1

Sometimes engineers make mistakes. Sometimes those mistakes are disastrous. The *Apollo 1* tragedy demonstrates one of those mistakes. It also shows how engineers learn from their mistakes, correct things, and move forward.

The initial design for the Apollo space capsule called for a pure oxygen atmosphere at higher than normal pressure. There were a number of good, rational reasons for this design decision. One reason is because that is the way NASA had done it on the Mercury and Gemini missions.

Having a single gas in the capsule, rather than a nitrogen/oxygen mix as on Earth, is lighter. There is only one gas to carry, meaning only one tank to hold the gas and no mixing/monitoring equipment to keep two gases in the right proportions. Also, the capsule pressure can be reduced to 5 psi (34 kPa) once in space, which is the same pressure used in a spacesuit. Low pressure is especially helpful in a spacesuit because it maximizes flexibility.

On the launch pad, however, the problem with a pure oxygen atmosphere is that it must be maintained at slightly higher than ambient pressure to keep outside gases from entering: 16 psi (110 kPa). With pure oxygen at this pressure, materials we think of as flammable become incendiary. They burn far faster and hotter. And *Apollo* was full of flammable material, particularly nylon, which is a **plastic** that burns readily. Even the flight suits were made of nylon.

During a routine ground test, the three astronauts were locked in the capsule with 16-psi oxygen. A spark started a fire. And then the fire spread with stunning speed. All three astronauts died. The last intelligible words out of the capsule were "I'm burning up!" and they came just 17 seconds after ignition.

A twenty-month review and redesign process followed the tragedy. The biggest change: a mixed nitrogen/oxygen atmosphere on the launch pad. Another was the replacement of nylon with a non-flammable material in the flight suits.

Engineers do make mistakes. The key thing is that they learn from them and make things better in the future.

SEE ALSO Plastic (1856), Saturn V Rocket (1967), Lunar Landing (1969), *Apollo 13* (1970).

Crew prepares to enter their spacecraft.

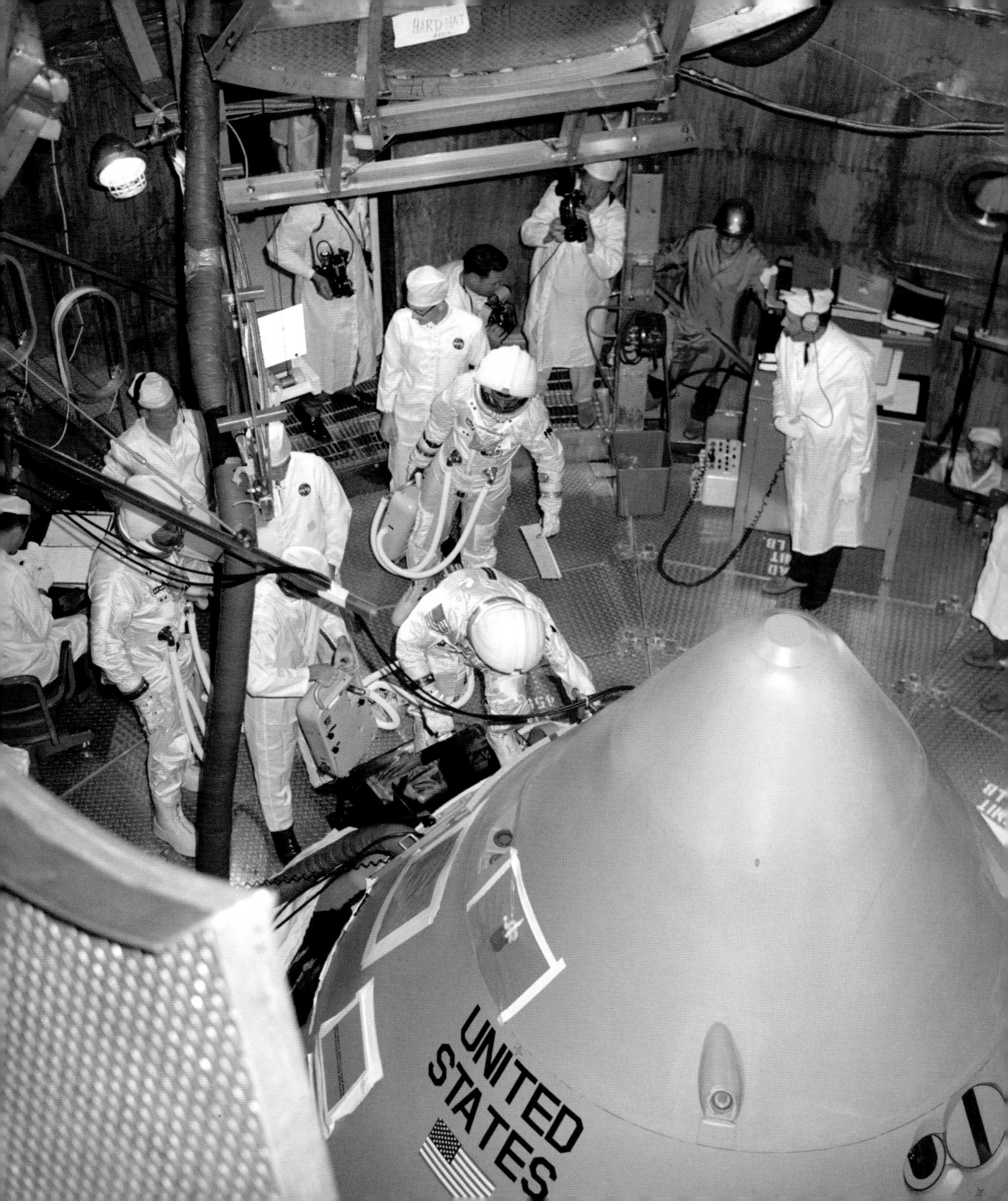
HARD HAT
UNITED
STATES

1967

Saturn V Rocket

The Saturn V rocket screams "engineering!" From its ridiculous size to its brute force power to the mission it helped accomplish, it is the most amazing rocket ever created. It holds a number of records, including its 260,000-pound (117,934 kilogram) payload capacity to low Earth orbit.

How could engineers create something this stupendous given the technology available at the time? Most engineers were still using slide rules as this rocket was conceptualized. And how did they create it so quickly? In 1957, the United States had never successfully launched anything into orbit. Yet in 1967 this colossus headed into orbit with ease.

One key was the F-1 engine. Engineers started its development for an Air Force project several years before NASA even existed—a fortunate coincidence. It meant the engine was tested and running smoothly before actually needed. The F-1 is the largest single engine ever created, with a thrust rating of 1.5 million pounds (6.8 meganewtons). Putting five F-1s together on the first stage created 7.6 million pounds of thrust. Which is a good thing because, fully fueled and loaded with payload, the whole rocket weighed about 6.5 million pounds (3,000,000 kg). To launch the rocket, each engine burned almost one million pounds (450,000 kg) of kerosene and liquid oxygen in less than 3 minutes.

Once the first stage fell away, the remainder of the Saturn V was 5 million pounds (2.3 million kg) lighter, and the second and third stages took the payload the rest of the way to orbit burning liquid hydrogen and liquid oxygen.

At the top of the third stage rested an essential component—a ring called the Instrument Unit. At almost 22 feet (7 meters) in diameter, 3 feet (1 meter) high and weighing 2 tons, this ring contained the computers, radios, monitoring equipment, radar, batteries, and other systems to control the three stages during flight and to communicate with ground control. The **microprocessor** did not exist yet, so the brain here was a custom-made, triple redundant IBM minicomputer.

Engineers created this disposable behemoth to send men to the moon, and then used it to loft Skylab as well. A true engineering wonder.

SEE ALSO Lunar Landing (1969), Space Suit (1969), Microprocessor (1971), Curiosity Rover (2012).

The Apollo 4 (Spacecraft 017/Saturn 501) space mission was launched from the Kennedy Space Center, Florida.

1968

C-5 Super Galaxy

Imagine that you are a military organization and you want to move soldiers, supplies, and equipment to a war zone at a moment's notice. You could put everything on a **ship** and send it, and that works well in terms of capacity and efficiency. But it might take the ship two weeks to make it halfway around the world. What if you need overnight shipping?

This situation is tailor-made for gigantic military transport aircraft like the C-5, which was first flight-tested in 1968. It can carry an impressive quantity of cargo at 550 mph (885 kph) and land it just about anywhere on the planet. And if it cannot land, it will instead drop it out of its rear cargo doors. In one case a C-5B dropped almost 200,000 pounds (90,000 kg) in one load—four Sheridan tanks and 73 paratroopers parachuted out the back. It demonstrates the ability to move lots of stuff quickly anywhere on the planet.

How do engineers build something this big that flies while carrying such huge loads? Think about this—the cargo area of a C-5 measures 19 feet (5.8 meters) wide by 13.5 feet (4.1 meters) high by 127 feet (38.7 meters) long. Yet it has doors that can open simultaneously at both ends. The cargo area is in essence a big tube. The wings are at the top of the tube carrying hundreds of thousands of pounds of fuel. The landing gear is at the bottom of the tube. And the cargo area can contain up to 280,000 pounds (129,000 kg).

Structural and aeronautical engineers essentially use two fuselages, one on top of the other, in the C-5. The upper fuselage is smaller and acts like a backbone for the plane. It can also hold 73 passengers. Then, in the cargo area and wings, engineers use construction techniques that add surprising strength. For example, to make the wings' skin, engineers start with massive 3,000-pound billets (1,360 kg) of aluminum and mill them down to one-fifth the weight, creating deep I-beam-like channels and a thin skin. They achieve tremendous strength and light weight simultaneously.

Perhaps the most amazing part: engineers took the C-5 from concept to first flight in just three years. It was an amazing engineering accomplishment.

SEE ALSO Square-Rigged Sailboats (1492), Boeing 747 Jumbo Jet (1968), Container Shipping (1984).

US airmen unload a C-5 Galaxy in support of exercise Patriot Hook 2012 at Naval Air Station North Island, March 16, 2012.

AIR FORCE RESERVE COMMAND

1968

Boeing 747 Jumbo Jet

Joe Sutter (b. 1921)

The 747 jumbo jet appeared to the world in 1968. Commissioned by the Boeing Airplane Company, its chief engineer, Joe Sutter, managed the design team behind the jet. The 747's immensity surprised everyone. Never before had the public seen a commercial aircraft with two floors and a staircase. The 747 also introduced the concept of a wide-body jumbo jet with two aisles and ten seats to a row. With over 600 passengers onboard in the densest seating configuration, the 747 would hold the passenger record for nearly four decades. It was two or three times bigger than other passenger planes at the time.

What happened? How could such a huge airplane pop into existence? Two engineering advancements made the 747 possible: the development of powerful high-bypass turbofan **engines**, and high-lift wings. The engines provided three big advantages: 1) much higher thrust than the low-bypass turbofans of that era, 2) much better fuel economy, and 3) noise reduction. Huge retractable flaps and slats increased wing surface area and curvature for short takeoffs and landings.

The size of the 747, combined with its high-efficiency engines, make possible a tremendous range. The 747-400 holds nearly 60,000 gallons (227,000 liters) of fuel in its wings, cargo area, and tail. Some variants of the 747 can fly over 9,000 miles (14,500 km) without refueling. That sounds like an abysmal 0.16 miles per gallon (.07 km/L). But if there are 600 passengers on board, it is 100 mpg (42 km/L) per passenger at Mach 0.85.

Where does the 747's hump come from? One goal of the 747 is to be both a passenger plane and a cargo plane. A real cargo plane needs a door at the front, with the nose of the aircraft able to rotate up for full access to the cargo area. This requirement meant that the cockpit needed to be above the cargo area rather than in the nose. Aerodynamic considerations demand that engineers build a tapered area behind the raised cockpit to smooth out airflow. That tapered area extends all the way back to the wings, and became the second floor—an engineer's creative use of the aerodynamically required space.

SEE ALSO Turbojet Engine (1937), C-5 Super Galaxy (1968).

Mock-up of 747-100 cabin, March 1967.

1969

Lunar Landing

Every part of the mission to land men on the moon involved an engineered solution: the rockets, the spacecraft, the life support systems, the **space suits**, the power supplies . . . even the food and its packaging. Nothing was left to chance.

The whole endeavor centered on the mission architecture. The goal was to land a human on the moon and bring him back safely. The time frame was limited because of fear that the Soviet Union would get there first. So the engineers designed an architecture that was both amazing and surprising.

When the Saturn V took off from its launch pad, it weighed about 6.5 million pounds (3 million kg). This was the entire self-contained package needed to fly to the moon and return. After 150 seconds the first stage was done and fell away. The package was 5 million (2.3 million kg) pounds lighter. The second stage dropped off 360 seconds later, eliminating another million pounds (454,000 kg). The third stage burned twice. The first burn put the spacecraft in orbit around the Earth. The second sent the spacecraft toward the moon, aka trans-lunar injection. When it fell away, 260,000 pounds (118,000 kg) went with it.

The lunar excursion module (LEM) came out of its fairing and docked with the command/service module (CSM). The LEM had two parts: the descent stage and the ascent stage. The LEM landed on the moon using the descent stage on July 20, 1969. The astronauts spent their time on the lunar surface. Then the ascent stage brought them back into lunar orbit to rendezvous with the CSM again. They discarded the ascent stage and the CSM brought everyone back to Earth orbit. They jettisoned the service module and the command module came back to Earth, using the heat shield and three parachutes to make a safe landing. When the command module hit the water, it weighed about 13,000 pounds (5,900 kg). It contained about 500 pounds of astronauts and 47 pounds of moon rocks.

Who could conceive of a plan so complicated, and yet so perfect for accomplishing the stated goal? And then make those many parts work flawlessly so many miles from home? Engineers.

SEE ALSO Space Suit (1969), Lunar Rover (1971).

Astronaut John W. Young on the moon during the Apollo 16 mission.

1969

ARPANET

Donald Davies (1924–2000), **Paul Baran** (1926–2011), **Lawrence Roberts** (b. 1937)

In the 1950s, there were only a few hundred computers in the world, but, by the 1960s, companies were selling thousands of computers. The minicomputer was born in 1965 with the PDP-8, produced by the Digital Equipment Corporation (DEC).

What if you wanted to use a **computer**? You needed a terminal and a dedicated communication line. To use two computers, you needed two terminals and two lines. And so people started to consider connecting computers together on networks to allow access to many computers. Electrical engineers created hardware that allowed voice signals to be digitized and then sent as digital data. The **T1 line**, invented in 1961, could carry 1.5 million bits per second—enough bandwidth for 24 phone calls. Once phone lines could carry digital data, two things could happen: computers could connect together, and different services could be implemented that take advantage of the computers and the connections between them. By organizing all of this, the Internet was born.

The first Internet-like interconnection of computers occurred when four computers connected together in 1969 under the name ARPANET, using concepts and ideas developed by American engineer Paul Baran, Welsh scientist Donald Davies, and Lawrence Roberts of the Lincoln Laboratory. Then this tiny network started growing. By 1984 the number of host computers hit 1,000. By 1987 the number was 10,000.

Two key technologies important to the early Internet were NCP (Network Control Program) and IMPs (Interface Message Processor). Together these technologies created what is known as a packet-switched network between all the computers. When a host computer wanted to send something to another computer, it would break the information down into small data packets and hand the packets of information along with a destination address to its IMP. The IMP, working in conjunction with other IMPs, would deliver the packets to the desired recipient computer. The two computers had no idea how the packets would travel over the network, nor did they care. Once they arrived, packets would be reassembled.

NCP would eventually be replaced by TCP/IP, and IMPs would be replaced by **routers**. At that point, engineers had created the Internet as we know it today.

SEE ALSO ENIAC—The First Digital Computer (1946), Fiber Optic Communication (1970), Router (1975), Domain Name Service (DNS) (1984), World Wide Web (1990).

This is a photo of the front panel of the very first Interface Message Processor (IMP). This one was used at the UCLA Boelter 3420 lab to transmit the first message on the Internet.

A CCN B
HALT INH
W.D.T.
AUTO RSTRT
MEMORY PRTCT
IMP STATUS PANEL
INTERFACE MESSAGE PROCESSOR
Developed for the Advanced Research Projects Agency by Bolt Beranek and Newman Inc.
T1 T2 T3 T4 F I A C PI ML EA DP
ON OFF PFI PFH POWER
SENSE
OP X B A P/Y M REGISTER
MASTER CLEAR STORE FETCH P P+1 OPERATION
IMP REGISTER PANEL
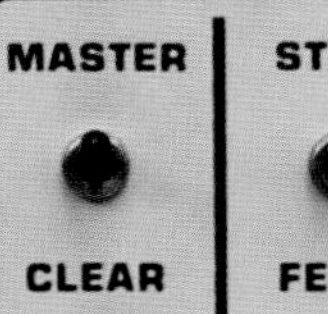

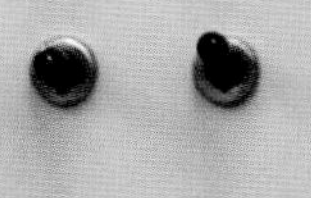

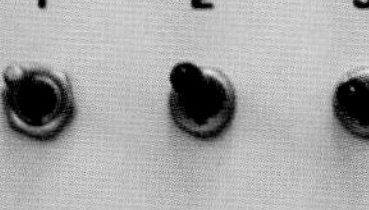

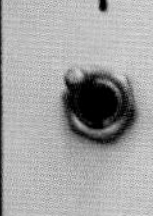

1969

Space Suit

While Soviet cosmonaut Yuri Gagarin's space suit, the SK-1, was the first to be worn in orbit, it was not meant for the challenges of a moon landing. The environment on the moon is about as harsh as it gets: a hard vacuum, intense solar radiation, and abrasive moon dust are three of the biggest challenges. Engineers working on the space suits to be worn during moon landings were faced with a unique challenge: how to produce a suit that would allow a human being to walk freely in these conditions. This isn't ordinary engineering—this is mission-critical engineering. The suit must perform flawlessly or someone will die. And the suit also has to be comfortable and flexible enough for astronauts to use it for up to eight hours one day, and then put it on and use it again the next day.

The suit itself is a remarkable, multilayer, fabric spaceship. This ship happens to be formed in the shape of a human and is designed for short-duration trips, but it is a spaceship nonetheless. The astronaut wears an undergarment laced with water-filled tubes to provide cooling. Then the suit itself starts with a sealed, rubberlike balloon made of neoprene-coated nylon. This layer holds in the pressurized atmosphere. More than a dozen layers of nylon, Mylar, Dacron, Kapton and Teflon-coated glass fibers provide protection from heat, cold, abrasion, puncture, and moon dust.

The backpack, called the PLSS or Primary Life Support System, provided the atmosphere and cooling, along with power and **radio** links. Oxygen tanks refreshed the oxygen that the astronauts consumed, and a CO_2 scrubber based on lithium hydroxide removed the carbon dioxide exhaled. A water reservoir provided for cooling.

There was also a backup system should the primary system fail. A separate oxygen tank could flow oxygen into the suit and vent it, keeping the astronaut oxygenated and cool. This system was simple and worked for about thirty minutes in an emergency.

The engineering done for the spacesuit created a highly reliable system that worked without failure on multiple missions. It is a true testament to engineering excellence.

SEE ALSO Radio Station (1920), *Apollo 1* (1967), Lunar Rover (1971), International Space Station (1998).

American astronauts were fitted with suits that could stand conditions on the surface of the moon. Yuri Gagarin's suit was made to handle flight in space, but not a lunar landing.

1970

LCD Screen

One measure of engineering success is the degree to which a technology becomes ubiquitous. By this measure, LCD screen technology certainly has succeeded. We see LCD screens in watches, calculators, clocks, **microwave ovens**, **smart phones**, **tablets**, laptops, and TVs. It is quite possible that many people today spend more time looking at an LCD than they spend looking at the real world.

If we look back to the birth of the technology, it solved a very real problem. The first digital watch in 1972, the Pulsar 1, was typical. You had to push a button to see the time because the LED display took so much power. But by 1973, engineers had perfected the twisted nematic field effect LCD, first patented by Swiss health care company Hoffmann-LaRoche in 1970. It consumed essentially zero power.

LCDs are also incredibly simple and therefore inexpensive to mass produce. An LCD display for a watch is a six-layer sandwich. At the bottom is a mirror. Next is a piece of plastic that polarizes light in one direction. Next is a piece of glass covered in a transparent conductor, usually indium tin oxide. Next is the twisted nematic liquid crystal layer. Then there's another piece of glass with the display segments or pixels etched on using a clear conductor. On the top is another piece of plastic that polarizes light, offset 90 degrees from the first.

Because of the 90-degree offset, the two polarizing filters should block all light and turn the display black. But the liquid crystal layer twists the light 90 degrees as it passes through. So the display appears white. Applying a voltage to one of the segments destroys the twisting and the segment turns black. The display uses virtually no power because no current flows—it is a field effect.

Engineers then made a color display by shrinking the pixel size, adding a backlight, and adding RGB color filters over the pixels. The backlight consumes power, but the LCD itself is extremely efficient.

It is possible that the **Active Matrix OLED** display or something similar will eventually replace LCDs but that does not detract from the massive success engineers had with the LCD for several decades. It made many of the electronic devices we cherish today possible.

SEE ALSO Microwave Oven (1946), AMOLED Screen (2006), Smart Phone (2007), Tablet Computer (2010).

Pictured: Reflective twisted nematic with liquid crystal display.

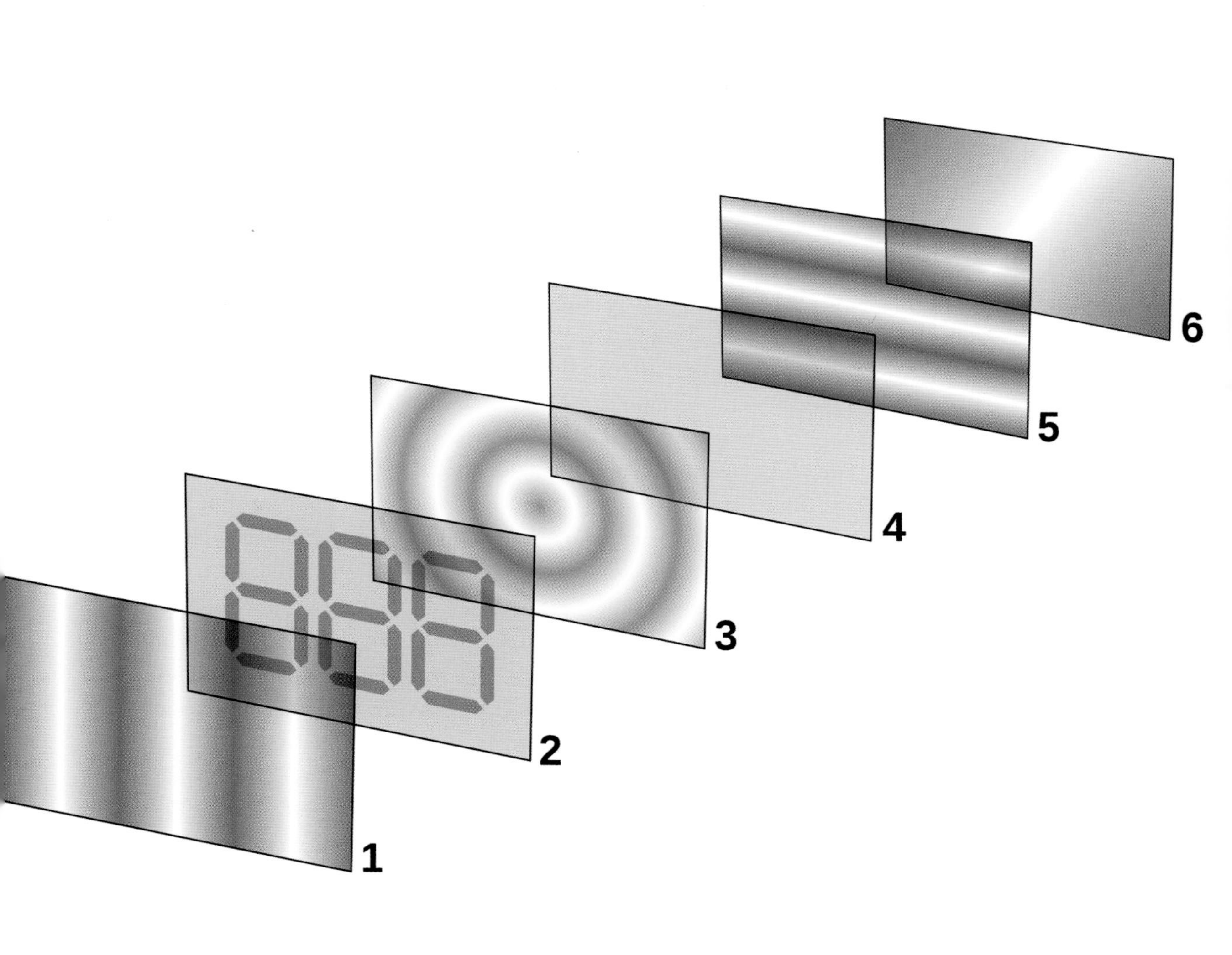

1
2
3
4
5
6

1970

Apollo 13

The successful moon missions—*Apollo 11, 12, 14, 15, 16,* and *17*—are magnificent triumphs of engineering. The mission architecture, the design process, the technology created, the execution, the pace of development, the reliability of the equipment . . . it all speaks volumes about the engineers involved in this massive undertaking.

And then there was *Apollo 13*, in 1970, which was a different sort of engineering achievement. The problem on *Apollo 13* was that an oxygen tank exploded due to an electrical problem inside the tank. The explosion eliminated the oxygen supply for the command module, meaning that the spacecraft had no oxygen for breathing or for power production with its fuel cells.

This scenario created an impressive, immediate challenge for engineers on the ground. Would they be able to use the equipment available and quickly put together a plan to bring the astronauts home safely? Amazingly, despite all the problems that appeared during the ordeal, engineers were able to use the LEM (Lunar Excursion Module) as a lifeboat and get everyone home.

The command module's only source of power was a **battery** system. So the astronauts shut down the command module and moved into the LEM to save the batteries. The LEM, unfortunately, was not designed to support three astronauts for such a long period of time, meaning that CO_2 concentration would build up to deadly levels. The astronauts were able to rig a way to use command module scrubbers in the LEM. The LEM's descent engine was also used several times to create the correct trajectory around the moon and back to Earth.

And then the final problem—the astronauts had to return to the command module and reactivate it from its shutdown state, running it on battery power through the landing. Running on battery power during reentry was normal, but the reactivation process was not. Engineers, astronauts, and flight controllers found a sequence that would allow reactivation with the available power. Then the return to Earth and splashdown proceeded normally.

It was an amazing example of on-the-fly problem solving, engineering, and system re-appropriation. And all of the astronauts returned safely. Rather than being one of the world's bigger engineering disasters, it was a triumph.

SEE ALSO *Apollo* 1 (1967), Lunar Landing (1969), Lithium Ion Battery (1991).

Deke Slayton (checked jacket) shows the adapter devised to make use of square Command Module lithium hydroxide canisters to remove excess carbon dioxide from the Apollo 13 LM cabin.

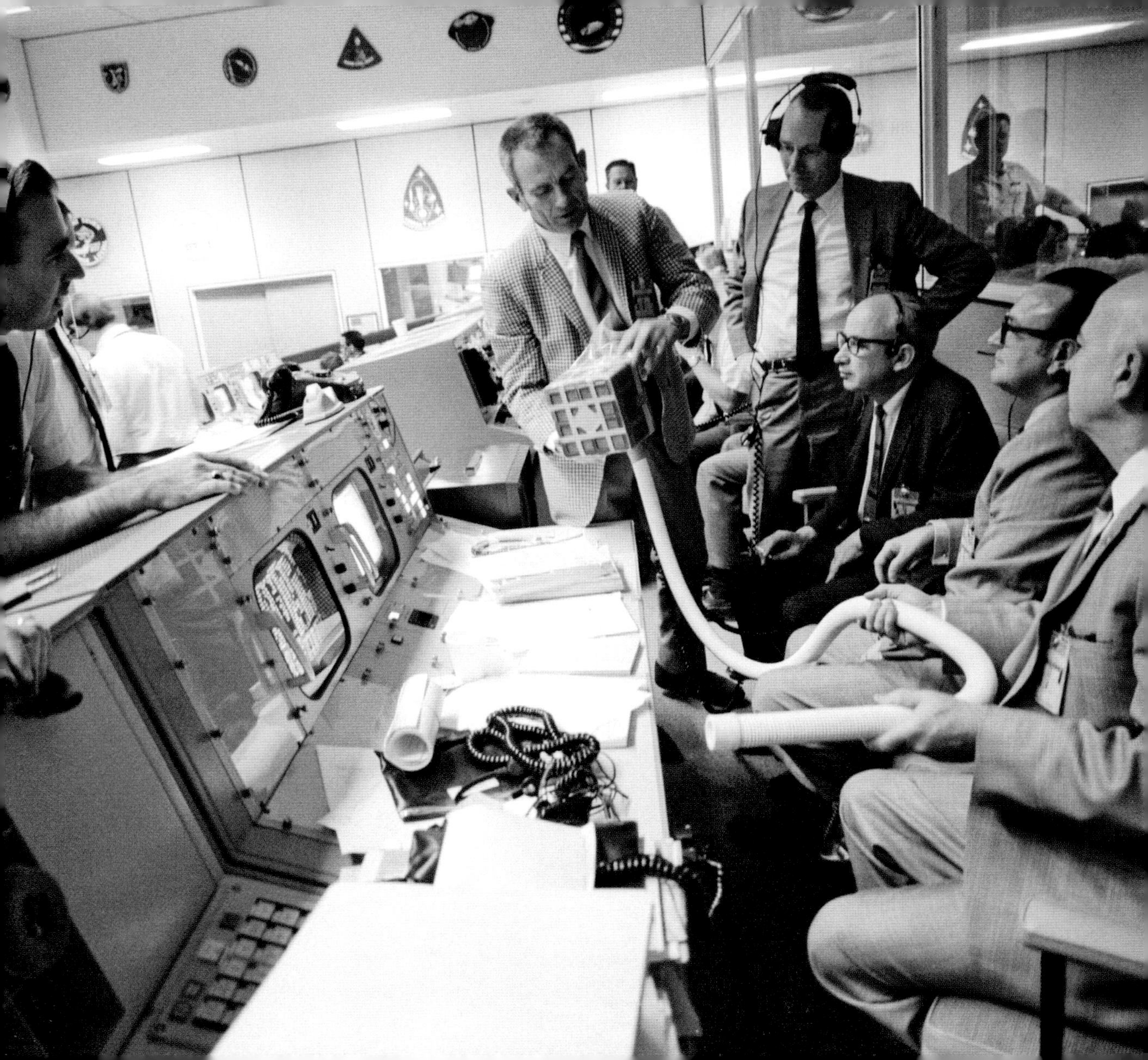

1970

Fiber Optic Communication

Robert Maurer (b. 1924), **Donald Keck** (b. 1941), **Peter Schultz** (b. 1942)

If we were to look back at the history of communications cables, say for long distance **phone** calls, we would find that the first cables were bundles of copper pairs. A cable like this is big, bulky, and heavy. The next step was coaxial cable. Many phone calls could be multiplexed onto the cable. This used less copper per call and reduced the number of repeaters (signal amplifiers) required.

But the real breakthrough came with the commercialization of optical fiber technology for data transmission. Drs. Robert Maurer, Donald Keck, and Peter Schultz, working for Corning, began to explore the possibilities of low-loss fiber. This resulted in the first optical fiber capable of maintaining laser light signals long distance in 1970. The first commercial lines arrived around 1977 and the technology has exploded since then. Nearly every bit of data (including phone calls) that travels across the country or around the world flows through a fiber optic line today.

The transition from copper to fiber has happened because fiber is better than copper in every way. Fiber is lighter, less expensive, immune to electromagnetic interference, and difficult to tap. Most importantly, a fiber optic line can transmit further before it needs a repeater and it can carry far more data on a single line.

The basic idea behind fiber optic lines is also incredibly simple. If you take a glass rod, heat it, and stretch it, you can make a thin strand of glass many miles long. If you stick a laser diode at one end of the strand and a light sensor at the other end, and then flash the laser on to represent a 1 and off to represent a 0, you can send digital data. If the glass in the rod is exceedingly clear, light can travel for 50 miles (80 km) or more before it needs a repeater.

SEE ALSO Telegraph System (1837), Telephone (1876), TAT-1 Undersea Cable (1956), ARPANET (1969).

In terms of long distance communication, fiber optic cables are superior to copper coaxial cables in just about every way.

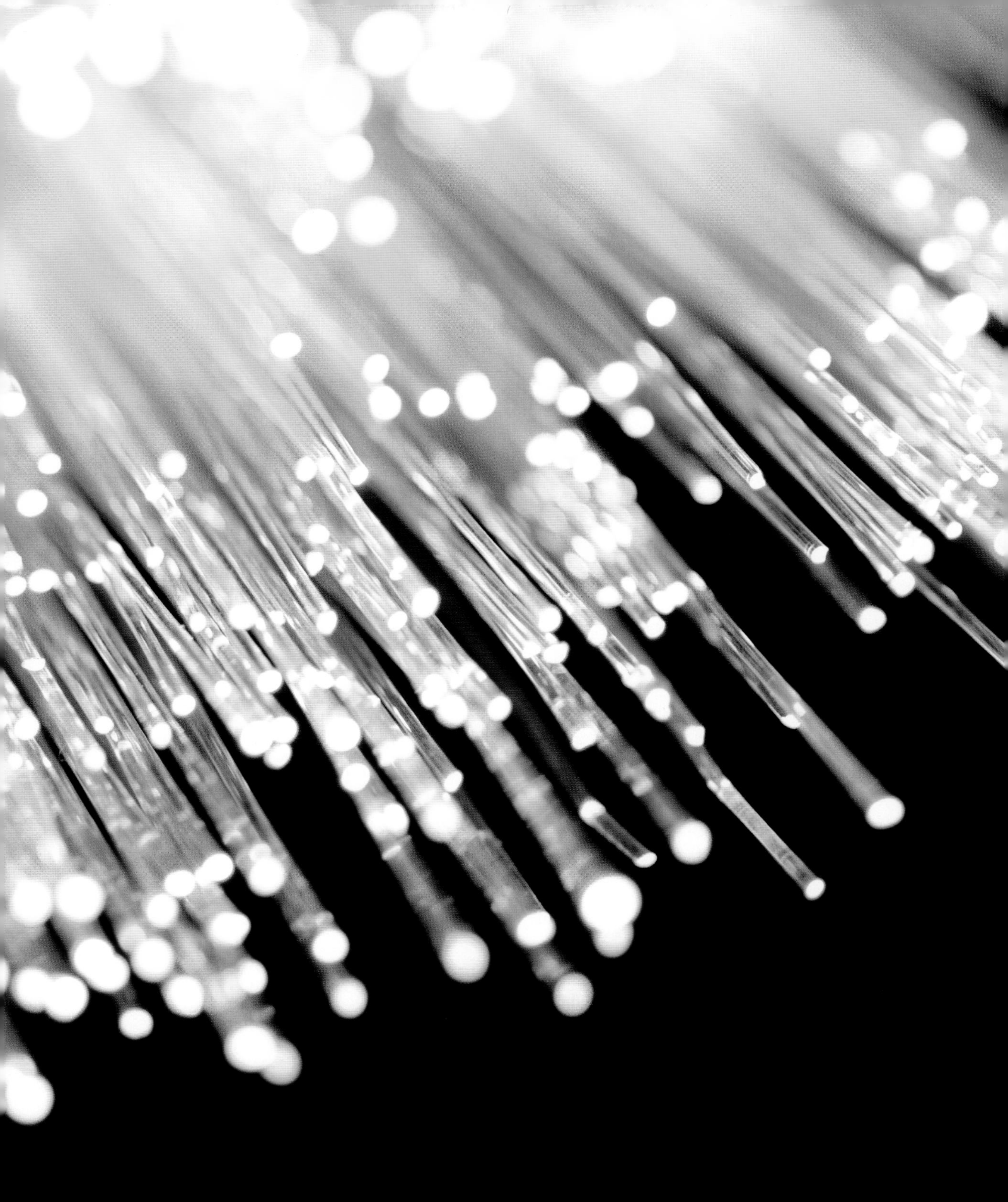

1971

Anti-Lock Brakes

When we hear screeching tires, it often means that a driver has made a panic stop and has lost control of the vehicle. The tires are sliding rather than rolling. Steering input has no effect and the car is following a ballistic path carried by its momentum.

What if we kept the tires rolling during a panic stop? The screeching noise would disappear and the driver would maintain some control. Stopping distances would also go down. Engineers went to work and the anti-lock braking system was born and popularized in the 1970s. The first autos sold with the anti-lock braking system (ABS) were commercially available beginning in 1971.

The basic idea behind anti-lock braking is very simple. Rotation sensors monitor all four wheels. When the driver presses the brake pedal, a computer checks the rotation rates. If it sees any of the wheels locking up, it pulses the brake on that wheel to keep it rolling. The driver feels a buzz in the brake pedal as the computer rapidly pulses the hydraulic pressure in the brake line to keep the wheel spinning.

Engineers seek efficiency and reuse. Once there is a computer on board that can sense wheel rotation and control braking, what else can the system do? It can, for instance, detect a tire that is going flat. Its rate of rotation will be different from the others. Another useful capability is traction control. If one tire starts spinning uncontrollably, the ABS can apply the brake on that wheel to keep it from spinning. More power gets transferred to the wheel that has grip. There is also stability control, made possible by the addition of a steering sensor and gyroscopes. If the driver turns the steering wheel but the car does not respond, the computer can apply the brakes on the inside of the turn.

Between anti-lock braking, traction control, and stability control, engineers have greatly improved the safety of cars.

The thing engineers cannot do is to keep a driver from being a knucklehead. If someone is driving too fast on an icy road, he is probably going to crash. Engineers compensate for knuckleheads with **airbags**.

SEE ALSO Automobile Airbag (1953).

Pictured: Close-up of an anti-lock braking system.

AUK 127 99 677 B

1971

Lunar Rover

The moon missions represented a pinnacle of engineering achievement and they were amazing to the billions of people who watched the spectacle. Humans had never before traveled so far, so fast, in such harsh conditions, for so long.But then on the *Apollo 15* mission in 1971, engineers took it one step further. Astronauts unpacked a spindly electric moon buggy from their lunar lander, unfolded it, and then drove up to 3 miles (5 km) away from their spacecraft. The utter audacity of it was breathtaking.

The lunar rover was surprisingly simple because engineers focused on weight and reliability. It consisted of a folding, three-part aluminum frame, four electric wheel motors (0.25 hp each) and brakes, two **batteries** and a power controller, a simple suspension system, and four-wheel electric steering. The entire vehicle weighed only 462 pounds (209 kg). One horsepower (745 watts) doesn't seem like much, but the top speed was 8 mph (13 kph), and the low moon gravity made everything lighter.

A dish antenna allowed television transmissions back to Earth. Astronauts would park, align the antenna toward Earth and then mission control could control the camera while the astronauts carried out their explorations.

What if something went wrong? Each wheel motor could be disabled individually. There were two batteries in case one failed, and two steering systems as well. But what if the vehicle failed completely? The excursions were designed so the astronauts always had enough time to walk back to the LEM before their life support backpacks ran out of consumables. But what if a life support backpack failed? There was backup oxygen in each backpack, and then a hose that could connect the astronauts together to share the cooling system in the working backpack.

Engineers had thought of every failure mode and had a way around it. So while it looked audacious to drive three miles away from their only way home, the astronauts were safe. Engineers had created good equipment, and if something did go wrong, there was a backup plan.

SEE ALSO *Apollo 1* (1967), Lunar Landing (1969), Space Suit (1969), *Apollo 13* (1970), Lithium Ion Battery (1991).

Apollo 15 *was the first mission to use the LRV. Powered by battery, the lightweight electric car greatly increased the range of mobility and productivity on the scientific traverses for astronauts.*

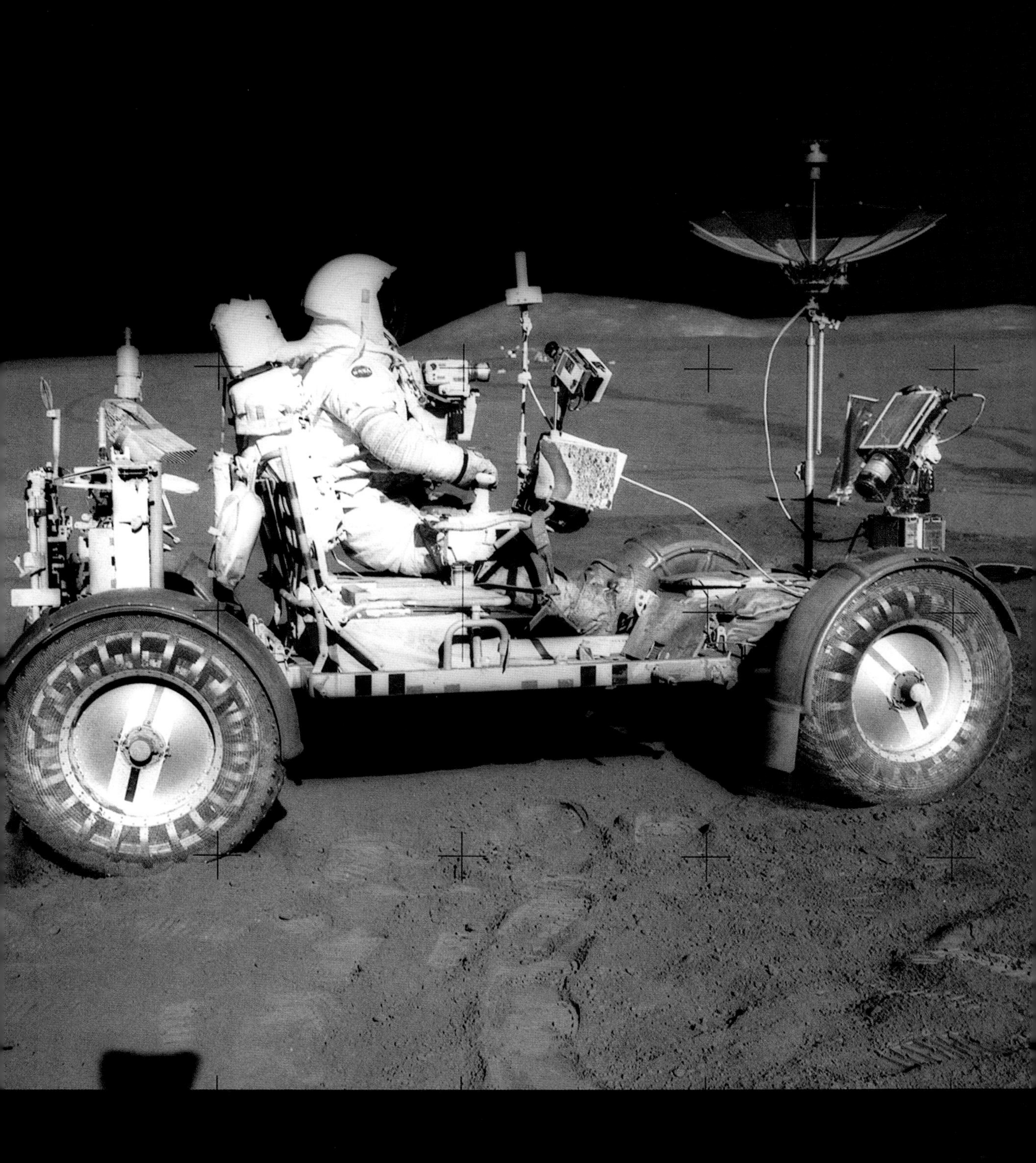

1971

Microprocessor

Marcian Hoff (b. 1937), **Federico Faggin** (b. 1941), **Stanley Mazor** (b. 1941), **Masatoshi Shima** (b. 1943)

Without microprocessors, none of the things we take for granted today could exist: calculators, digital clocks/watches, remote controls, desktop computers, laptops, **tablets**, **smart phones**, **HDTV**, digital displays and keypads in **microwaves**, DVD players, dashboards, **radios**, thermostats, printers. **Digital cameras** could not exist, nor MP3 players, ECUs, etc. Every single thing around any normal person was mechanically controlled in 1969. The **Saturn V rocket** had a computer in it, sure. It was huge and cost millions.

Then in 1971, with the Intel 4004 chip's arrival, everything started to change. This was the first microprocessor—the first time a single silicon chip held all of the circuits for a complete computer: Arithmetic Logic Unit (ALU), registers, memory addressing, instruction decoder. It was pathetic by today's standards: roughly 2,000 transistors, 10 micrometer feature size, pMOS technology, 740,000 hertz clock, 8 or 16 clock cycles per instruction, 46 instructions. It could perform less than 100,000 instructions per second on 4-bit numbers. And yet at the time it was amazing that such a thing existed.

To make it, Intel's engineers—including Marcian Hoff, Stanley Mazor, and Masatoshi Shima, and later, Federico Faggin—laid out the masks for the 2,300 transistors by hand using strips and rectangles of plastic.

Since then engineers have made so many advances. Four-bit registers moved to 64-bit. Feature size shrank from micrometers to nearly nanometers. Clock speeds from 740,000 hertz to 3,000,000,000. Transistor counts have gone from ~2,000 to billions. Not to mention all the conceptual improvements: floating point units, pipelining, multilevel cache memories, multiple cores, superscalar CPUs and hyperthreading, etc. Small microprocessors today cost pennies and use nearly zero power.

The microprocessor now touches every part of our lives. A typical car can have two dozen. A typical household many more. There will come a day when a microprocessor exceeds the power and complexity of the human brain. Engineers will have engineered their replacement.

SEE ALSO Radio Station (1920), Microwave Oven (1946), Saturn V Rocket (1967), Digital Camera (1994), HDTV (1996), Tablet Computer (2010).

Designer Federico Faggin points out the intricacies on an enlarged blueprint of the Intel 4004, which he helped design. It became the world's first microprocessor in 1971.

4004

1971

CANDU Reactor

The **light water reactors** that went into service in the United States and other countries in the 1960s had several problems. First, the uranium needed to be enriched from its natural state of 0.7 percent U-235 to 3 percent U-235 in order for the reactors to operate. Second, light water reactors have to shut down for refueling because the pressure vessel must be depressurized and opened.

What if engineers wanted to avoid or eliminate the **enrichment** step? Since uranium enrichment is a complex and expensive process, avoiding this step could save a lot of money. What if a reactor design could avoid the shutdowns? That would improve the amount of time the reactor is online making electricity and money. These were two of the core questions engineers thought about in conceiving the CANDU (CANada Deuterium Uranium) design. Also weighing in was the fact that Canada did not have the industrial capacity to make the huge pressure vessels needed for light water reactors.

A key element of the CANDU design, first introduced in 1971 at the Pickering Nuclear Generating Station in Ontario, is the use of heavy water. The water we drink is mostly light water, made up of two normal hydrogen atoms plus one oxygen atom. Heavy water has two deuterium atoms in place of the two hydrogen atoms. While a hydrogen atom (also known as protium) has one proton and one electron, deuterium has one proton, one neutron, and one electron. Although it is rare in nature, heavy water can be extracted from regular water or formed chemically. Making heavy water is expensive, but not as expensive as enriching uranium, and heavy water is not dangerous as is enriched uranium. There is no way to turn heavy water into a **bomb** as there is with enriched uranium.

With an appropriate reactor design and a supply of heavy water, it is possible to "burn" natural uranium in a reactor. The heavy water slows down neutrons without capturing them, so the low concentration of U-235 in natural uranium is enough to produce power. And the power production is more efficient (uses less uranium per unit of power) than a light water reactor.

CANDU represents an engineering reconceptualization with the goal of safer, cheaper operation.

SEE ALSO Power Grid (1878), Uranium Enrichment (1945), Trinity Nuclear Bomb (1945), Light Water Reactor (1946), Ivy Mike Hydrogen Bomb (1952).

Nuclear power station located on Lake Ontario in Pickering, Ontario, Canada, generating 20 percent of Ontario's power.

1971

CT Scan

Godfrey Hounsfield (1919–2004)

The modern CT (Computed Tomography) scan machine is an impressive piece of equipment. These machines combine the skills of several engineering disciplines to form rich, complex medical images depicting the inside of the human body, paving the way for more complex system such as the **MRI**.

Imagine the typical X-ray machine, which sits on one side of a body part—for instance, a hand. A piece of photographic film (or a digital sensor) sits on the other side. X-rays are so energetic that they can travel through the body part and hit the film. Varying tissues inside the body obstruct the X-rays to different degrees, so bones largely block X-rays while soft tissue is much less opaque. An X-ray machine can form an image of bones and some softer structures, but the image is two-dimensional and can be difficult to interpret.

Now imagine an X-ray source producing a pencil-like beam of X-rays, with a sensor six feet away that can detect the strength of the ray after passing through the body. Mount these two parts on a ring. Lay a person down with the ring around her. And spin the ring. The X-rays shoot through the body at hundreds of different angles. Now slide the ring down the body, forming slices of data as it goes.

A computer can look at the data coming off the sensor. Imagine that it receives 180 readings for all 180 degrees of the half circle. By applying complex algorithms to the readings, a computer can use the data from the sensor to reconstruct a slice of the patient's body, showing different tissues and their positions inside the body. By stacking the slices, the computer can build a detailed 3D model of the interior of the patient's body. This is a CT scan, which first became available to doctors in 1971. Invented by British electrical engineer Godfrey Hounsfield in 1967, it ultimately earned him the 1979 Nobel Prize for Physiology and Medicine.

SEE ALSO MRI (1977), Surgical Robot (1984).

Pictured: A high-definition CT scanner.

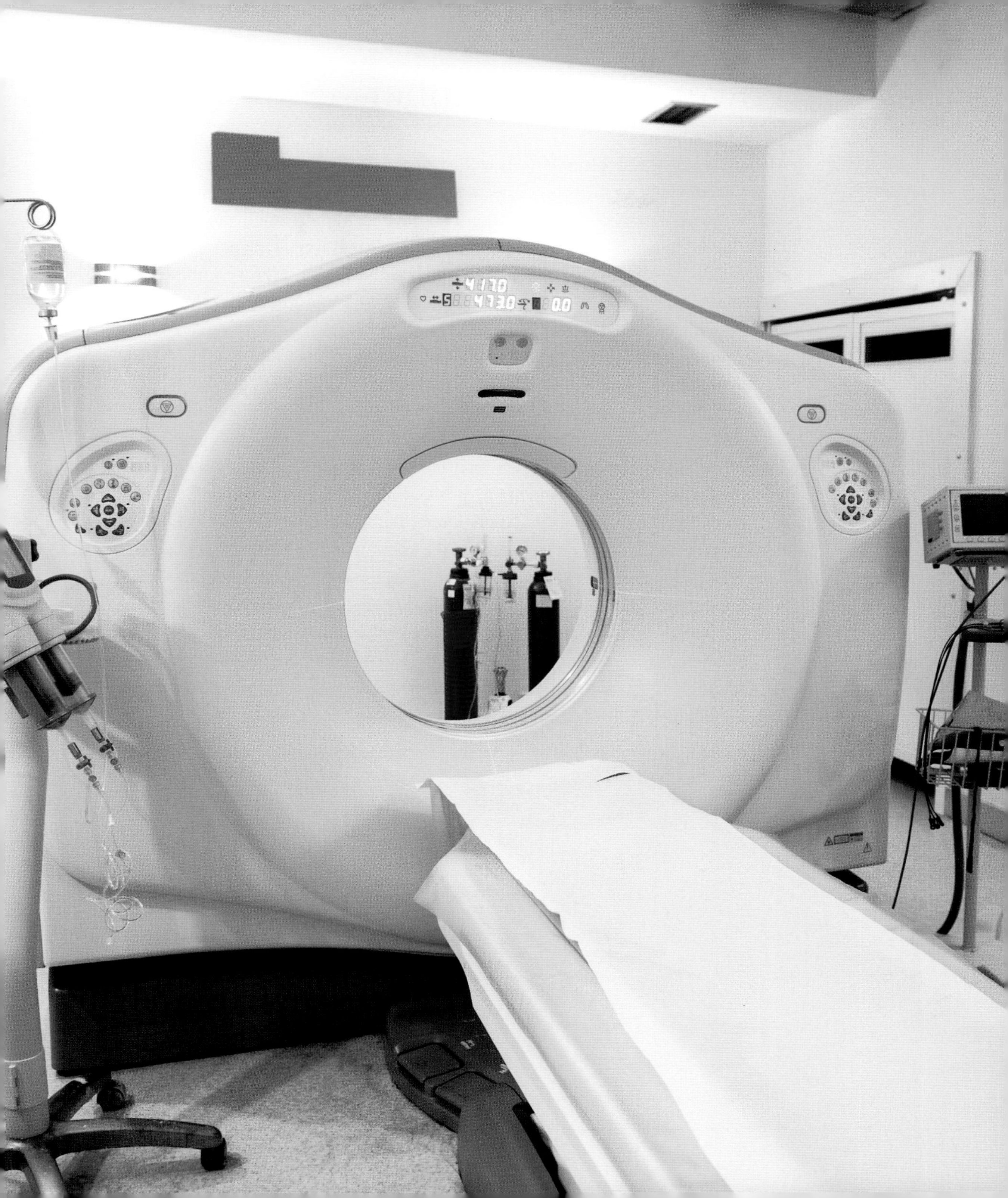

1971

Power Plant Scrubber

Modern society needs electricity. As with any product, the less expensive the electricity becomes, the better. One inexpensive way to produce electricity has been to burn coal.

In the ideal case, coal would contain nothing but carbon and hydrogen. Unfortunately, coal often contains mercury and sulfur as well. The mercury goes straight into the air and then falls as rain into lakes and rivers. Sulfur combines with oxygen to form sulfur dioxide, which, when mixed with moisture in the air, becomes sulfuric acid. This is where the term "acid rain" comes from. Sulfur from coal-fired **power plants** is the main source of acid rain.

So engineers were asked to clean the sulfur and mercury from the flue gases of power plants. They solved the problem with scrubbers—but only because the Environmental Protection Agency forced them to with sulfur dioxide laws in 1971.

To eliminate mercury, engineers spray powdered activated carbon into the exhaust stream. The mercury atoms bind to the carbon, which then allows them to be filtered out.

The sulfur removal process is more interesting. By spraying lime—$Ca(OH)_2$—mixed with water into the exhaust stream, the sulfur combines with the lime to create $CaSO_3$. By adding oxygen and water, gypsum gets formed. So engineers have taken sulfur pollution out of the air and used it to create gypsum as a product. The power plant can sell the gypsum and it is used to create things like wallboard.

What engineers have not figured out yet is the inexpensive, easy way to extract all of the carbon dioxide from the exhaust stream and store it somewhere safe. Once they do that, we will have truly clean power plants.

SEE ALSO Power Grid (1878), Light Water Reactor (1946), Carbon Sequestration (2008).

The power plant scrubber is an innovation that emerged due to concerns about phenomena such as acid rain.

Kevlar

Stephanie Kwolek (1923–2014)

Think about how you might create a bulletproof vest. The goal is easy to understand—keep a bullet from penetrating the vest and entering the body of the person being shot. But the materials available all have problems. Steel, in the thickness necessary to stop a bullet, is too heavy. Aluminum is too. Ceramic plates are useful, at about half the weight of steel, but still fairly heavy, and ceramic is expensive. The plate to cover the torso (10 inches by 12 inches or 25 cm by 30 cm) might weigh 5 to 6 pounds (2.5 kg). High-density polyethylene has been used because it is lighter than ceramic, but unfortunately even more expensive. And every one of these materials is rigid.

This is why Kevlar was so exciting when it was introduced in 1971. Stephanie Kwolek at Dupont accidentally discovered it in 1965.

By weight, Kevlar is about five times stronger than steel. It can be woven like nylon to create fabric. Fabric layers can be stacked, and the result is extremely strong. Kevlar allows for body armor that is lightweight, thin, and flexible—so light that people like police officers can wear it daily. Yet it is extremely effective. Thousands of police officers have survived shootings because of their Kevlar body armor. It is also used in helmets, where it is much lighter and stronger than the steel it replaced.

Because it is so strong and light, Kevlar has lots of other applications. It is possible to replace steel cable with Kevlar in some cases and the cables become incredibly light. Mooring lines for large **ships** are one place where Kevlar is useful. Small applications, like paracord in high-performance **parachutes**, are another example. Kevlar can replace nylon to create incredibly strong **sails**. Kevlar fabric is sometimes used in combination with glass fabric in fiberglass composites to make the material much stronger.

Why didn't Kevlar replace nylon? For one thing, Kevlar remains more expensive than nylon because it is harder to manufacture. Second, it is not as abrasion-resistant as nylon. Even so, Kevlar is an extremely useful material for engineers.

SEE ALSO Square-Rigged Sailboats (1492), Plastic (1856), Carbon Fiber (1879), Parafoil (1966), Container Shipping (1984).

This bulletproof vest is fabricated from Kevlar.

1972

Genetic Engineering

Paul Berg (b. 1926)

When we think about engineering, we generally think about creating a new object: a new building, a new device, a new mechanism. Genetic engineering is a different type of endeavor. Here we are taking an existing system that is quite complicated and that we do not completely understand—a genome—and we are tinkering with it. Genetic engineers add new genes to a genome to create new behaviors.

The predecessor to genetic engineering is selective breeding. Breeders select desired traits. With selective breeding we have created all the various breeds of dogs.

But genetic engineering, which appeared in 1972, when American biochemist Paul Berg created the first recombinant DNA molecules, is something altogether different: Engineers are injecting new genes into genomes in ways nature could never accomplish. For example, Berg combined two viruses. Other more recent applications have included a jellyfish gene that produces a green fluorescent protein added to a fish or a mouse, creating fluorescent mice. A gene that makes a plant immune to an herbicide gets added to soybean plants so that the herbicide won't kill them.

In one of the most bizarre examples, genetic engineers took genes for producing spider silk and added them to goats. The proteins of spider silk appear in the milk of a female goat. The goal was to extract the proteins to create super strong, highly elastic materials.

There are different ways to inject the genes of one organism into another, and the gene gun is one popular tool. The technique is so simple, it is amazing that it works. The gene to be injected is added, in liquid form, to tiny particles of tungsten or gold. The particles are shot out of a gun, **shotgun** style, at a petri dish full of target cells. Some of the cells get punctured, but not killed, and they pick up the new gene.

By injecting genes from one organism into another, genetic engineers are able to create new organisms. One of the most beneficial examples is human insulin produced by *E. coli* bacteria. Developed in the 1980s, genetically engineered insulin is used by millions of people today.

SEE ALSO AK-47 (1947), Green Revolution (1961).

Glofish, the first GMO designed as a pet, was first sold in the United States in December 2003.

1973

World Trade Center

Minoru Yamasaki (1912–1986)

The twin towers of the World Trade Center in New York City became two of the most famous buildings in the world because of what happened to them in 2001. But prior to their destruction, they were two of the most famous buildings in the world because of their engineering and their iconic architecture. They truly were works of art.

The lead architect on the project was Minoru Yamasaki, assisted by other architects and a structural engineering firm. The structural engineering of the towers was impressive for several reasons. The buildings had a dense core to hold the primary steel support structure along with all utilities and services. Four dozen steel box columns ran from the foundation to the roof in this core, which measured 135 feet (41 meters) by 85 feet (26 meters). Stairs, elevators, restrooms, HVAC, plumbing, electrical, and phone services all ran in the core. The outer wall of the building consisted of lighter weight steel columns (60 on each side) running from the ground to the roof. Steel **trusses** then extended 65 feet (20 meters) from the core to the outer wall without any interior columns. The trusses supported poured reinforced **concrete** floors.

The elegance of the design came in its beautiful repeatability. In addition, the floor space between the core and the outer wall was completely open—no supporting columns on the interior at all. Each floor was almost a perfect acre in size, with three-quarters of an acre open, usable, and infinitely configurable space on every floor. The design also meant that the building was quite light for its volume—perhaps one-quarter the weight per volume of an older building like the **Empire State Building**. This lower weight meant lower cost.

Most skyscrapers have a number of setbacks as they rise. This happened because of zoning ordinances designed to avoid a canyon-like effect in New York City. The setbacks kept skyscrapers from blocking too much wind and sunlight from neighboring areas. The Twin Towers were able to rise without setbacks because there was so much open space around them—another truly unique feature.

SEE ALSO Truss Bridge (1823), Woolworth Building (1913), Empire State Building (1931).

The original World Trade Towers were surprisingly light for their volume.

1975

Router

Virginia Strazisar (Dates Unavailable)

It is not uncommon for engineers to experience a paradigm shift or a reconceptualization—a change in the conventional way of thinking that offers significant advantages. The switch from analog to digital in music is one example of a major reconceptualization. The switch to **HDTV** was a tectonic paradigm shift.

The fundamental architecture proposed for the Internet was another major reconceptualization. The existing paradigm for communications networks involved circuit switching. To make a call from New York to San Francisco, the **phone** system would create a complete, defined circuit for the call across that 3,000-mile distance.

The architecture of the Internet uses packet switching instead. When you speak a sentence in Rome, your sentence is broken into hundreds of packets. Those packets are sent to the Internet, and each packet could conceivably take a different path to get to Paris. Each packet has a destination address, called an IP address, that specifies where it wants to go. The IP address identifies a specific machine in San Francisco. When the packets arrive at the destination, they are ordered, combined back together, and delivered to the person on the other end of the line.

Each packet needs to find a route from New York to San Francisco across a set of network segments. Routers—developed in 1975 by Virginia Strasizer in conjunction with a team involved in a DARPA-lead initiative at BBN Technologies—are the machines that pick the routes, and they do that dynamically. A router is a computer that is connected to a number of network segments. The router accepts packets, looks at where they are heading (the IP address), consults an internal table to decide the best route at that moment, and sends each packet down a network segment along to the next router. A packet might pass through ten or more routers to get from New York to San Francisco.

Routers come in many sizes. You probably have a small one in your house. Big routers, called core routers, sit on major **Internet** backbone segments and can process millions of packets per second.

SEE ALSO Telephone (1876), ENIAC—The First Digital Computer (1946), ARPANET (1969), HDTV (1996).

Pictured: An early example of a router.

WPS
RESET
WiFi

1976

The Concorde

The mid-1960s was aviation's nirvana. The **SR-71** was flying at Mach 3. Commercial jet service was routine. Humans would soon **land on the moon**. And the next big thing was on the horizon: the SST, or supersonic transport. There was a widespread idea that soon, very soon, everyone would be traveling at supersonic speeds. Instead of six hours to cross the US, it would take three.

So engineers began designing the future, and Europe had a huge lead, which is why the Concorde is a European aircraft. But even with the lead, engineers had a lot of problems to solve. Supersonic flight has a number of difficulties not experienced below the speed of sound.

Problem #1: Take off and landing. Supersonic flight favors short wingspans to cut drag. But short wings are terrible for takeoff and landing. Solution: extreme delta wings which, at a steep angle, have good lift at low speeds. Then the Concorde's articulated nose lets pilots see the runway.

Problem #2: Thrust. To fly across the ocean at Mach 2, the engines need supercruise ability. And they need to be compact to cut drag. Solution: Pure **turbojet engines** with no bypass fans. Mechanical intake flaps slow down the supersonic air entering the engine. Afterburners give the engines extra power for takeoff and the jump to supersonic speeds.

Problem #3: Heat. At Mach 2, passing air heats the plane's surface. The nose reaches 260°F (127°C). The plane's interior is warm to the touch. Flying at 60,000 feet (18,300 meters) lessens the effect. Special aluminum alloys handled the heat, but also set maximum speeds because of the metal's temperature limitations.

One problem engineers could not solve was the economics. Drag considerations limit the width of the aircraft, allowing fewer passengers. The extra speed demands extra power. The Concorde had 31,000 gallons (117,350 liters) of fuel on board for just 110 people in the cabin. A fully loaded 747 might use one-tenth the fuel per passenger mile, offering much cheaper travel. Engineers can do many things, but they can't change the laws of physics and their economic effects. Cheap supersonic travel awaits a reconceptualization.

SEE ALSO Hall-Héroult Process (1889), SR-71 (1962), Lunar Landing (1969).

The Concorde landing at the Amsterdam Airport.

British

1976

CN Tower

Imagine that your city needs a tower from which to transmit **radio** and **TV** signals. Engineers could design and erect a boring, vertical metal **truss** with guy wires. But what if you want to make a statement and create a phenomenal tourist attraction? In that case, you would ask for the CN Tower in Toronto, Canada, completed in 1976.

Structural engineering was overseen by a firm called NCK. Construction began by exposing bedrock and anchoring the tower to it with steel and concrete. With this foundation prepared, an innovative concrete process known as slipforming created the tower itself. Engineers use slipforming to create a variety of **concrete** structures such as silos, cooling towers, and bridge posts. The basic idea is to create concrete forms that surround a small ring of the vertical structure. Quick-curing concrete flows into these forms and sets. Then the forms jack up past this new concrete so that the next pour can occur. Workers stand on platforms around this ring to add rebar, direct the concrete, and vibrate the concrete. The advantage of slipforming is that the structure is essentially one continuous piece of concrete that is extremely strong.

In the case of the CN Tower, the structure has a hexagonal core for utilities, stairs, and elevators, and then three concrete buttresses that start at the bottom and taper away toward the top of the tower. The concrete structure stops at 1,500 feet (457 meters) and from there a 315-foot (96-meter) metal tower begins. A large **helicopter** lifted segments of the metal tower up one at a time for assembly. Various radio, TV, and microwave antennas attach to this metal structure. The total height is 1,815 feet (552 meters), making it the tallest freestanding structure in the world at the time of construction.

Why make it so tall? One reason was bragging rights, but there was also a practical purpose for the height. With all of the skyscrapers in the area, the tall tower allows interference-free broadcasting for the entire city. Adding the Lookout and Skypod turned the tower into a popular tourist attraction, with more than two million visitors per year.

SEE ALSO Concrete (1400 AAA), Truss Bridge (1823), Radio Station (1920), Color Television (1939), Helicopter (1944).

Canada's CN Tower makes a visual impact, but also allows interference-free broadcasting.

1976

VHS Videotape

If you go back to the beginning of television, everything was done live. The reason is because there was no way to store a **TV** signal. If you think about it, a TV signal contains quite a bit of information—far more than the audio signal for a radio station. The TV signal contains the audio for the show, plus all the information to construct 30 complete pictures per second.

The obvious solution would be to store the video signal on tape. But how? At the time, **sound recording** was made possible by recording the audio signal linearly down the length of the tape. To record a TV signal in the same way would require the tape to be running past the record/playback head at a ferocious rate. It simply was not practical.

The engineering breakthrough that made video recording possible is the helical scan approach. The record/playback heads are placed on a spinning cylinder that is canted at a slight angle. The tape path wraps the tape halfway around this spinning cylinder. As the tape moves slowly past, the head is spinning at 1,800 rpm. Each frame of the show is written as a diagonal stripe on the tape. The audio channel is written linearly at the bottom of the tape. With this approach, engineers had a way to record video on tape at a reasonable tape speed.

What happened next was surprising: the video format wars. Two groups released two variations on the helical scan model into the marketplace simultaneously. One format was Betamax by Sony, the other was VHS by JVC. They fought for dominance in the marketplace, and consumers voted with their dollars. It appears that two factors caused VHS to win: lower cost for the hardware, and longer recording time. The first home device capable of playing VHS tapes was introduced to the marketplace in 1976.

For the first time, engineers had given consumers control over their TV experience. People could rent prerecorded tapes, or tape TV shows and play them later. These two features were incredibly attractive to consumers, and once the format war was won, VCR sales took off like a rocket. Engineers had created a device nearly as popular as the TV.

SEE ALSO Tape Recording (1935), Color Television (1939), HDTV (1996).

Pictured: The interior of a VHS tape.

Human-Powered Airplane

Dr. Paul B. McCready (1925–2007)

The human-powered airplane seems like such a simple, obvious idea, but in fact represents a huge engineering challenge. The problem lies in using a human as a motor.

On a bicycle, on level ground, the human motor works pretty well because the weight of the motor has minimal effects. A human being who is in fantastic shape (like a *Tour de France* athlete) can produce half a horsepower (375 watts) over several hours. With half a horsepower, a bicycle can move along at a good clip.

In an airplane, the weight of the motor does matter. It not only moves the vehicle forward, it keeps the vehicle in the air. Gravity is constantly pulling the airplane to the ground, and the motor must supply the power that opposes gravity. In this situation, a motor that weights ~160 pounds and produces half a horsepower becomes a huge problem. The power-to-weight ratio is horrible.

So the human-powered airplane must be as light as possible. Yet it needs big wings because it is flying so slowly with the power available.

The Kremer Prize, started in 1959 by industrialist Henry Kremer, offered £50,000 to pioneers in human-powered flight. The Gossamer Condor, designed by American aeronautical engineer Paul McCready, was the first human-powered airplane to win the Kremer Prize in 1977. To win, it flew a figure eight around two pylons half a mile (0.8 km) apart.

The engineering is amazing. Even though the plane has an unbelievable 96-foot wingspan (29 meters), a huge propeller, and a surprising front-mount canard to control it, it weighs only 70 pounds (32 kg). To do this, engineers combined thin aluminum tubes with super-lightweight foam ribs, a Mylar skin, and thin wire. The "chain" is made of steel cable and plastic.

Even more amazing was the MIT Daedalus HPA, which in 1988 flew 71 miles (115 km) from Crete to Santorini in just under four hours. Made of **carbon fiber**, foam, Mylar, and **Kevlar**, it looked like a traditional airplane. With a wingspan of 111 feet (34 meters), it is an engineering plus athletic achievement of the highest order.

SEE ALSO Carbon Fiber (1879), Hall-Héroult Process (1889), Kevlar (1971).

The Daedalus 88 human-powered airplane, with Glenn Tremml piloting, is seen here on its last flight for the NASA Dryden Flight Research Center, Edwards, California.

MIT
NASA
ΔΑΙΔΑΛΟΣ
United Technologies
DAEDALUS

1977

Tuned Mass Damper

Sometimes engineers go to great lengths to solve problems, and the solutions are very cool. Then everything they do is hidden and no one gets to see it in action. The tuned mass damper found in many tall skyscrapers like **the World Trade Center** is an example of this phenomenon.

One way to think of a skyscraper is to imagine it as a long, flexible stick stuck in the ground. When the wind blows, the top of the stick will sway. The same thing happens with a tall skyscraper—the top of the building can sway in a strong wind. The wind is blowing against the side of the building and the building bends in that direction. Then the wind changes speed or direction and the top of the building shifts. Or the wind sets up a vibration in the building. No matter how stiff engineers make the building, this will happen to some degree. But the motion, especially on the top floors, can be uncomfortable. So how to lessen or eliminate it?

This is where the tuned mass damper comes in. The John Hancock building in Boston was the first building to use a TMD in 1977. It needed it because the top floors of the 60-story building swayed so much that they caused motion sickness. The tuned mass damper is a huge weight (typically hundreds of tons) that is free to move (imagine a pendulum) but is tied into the building's frame with springs and/or hydraulic cylinders. Now when the top of the building tries to sway, the weight will try to stand still (an object at rest tries to remain at rest). With its springs and shocks, the weight's stationary desire counteracts the sway, reducing it significantly.

One building where you can actually see the TMD is the Taipei 101 tower in Taiwan. A tourist caught the TMD in action during an earthquake and posted it to YouTube. Considering that the damper weighs nearly 1.5 million pounds (660,000 kg), seeing it move so freely is rather amazing.

Water also works. In one building, 100,000 gallons (380,000 liters) of water sloshes through baffles to counteract sway.

SEE ALSO Woolworth Building (1913), World Trade Center (1973), Burj Khalifa (2010).

Tuned mass damper on display in Taipei.

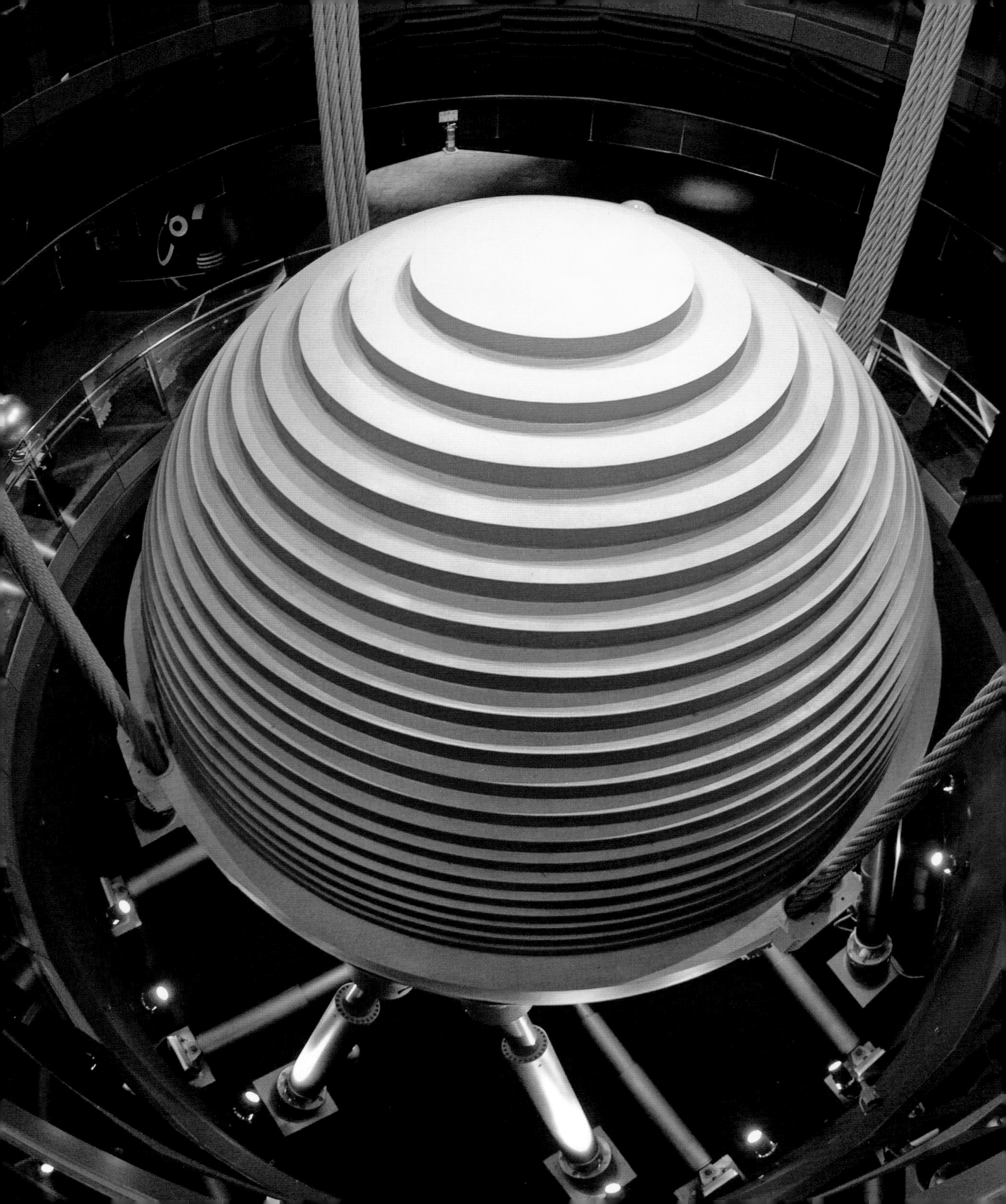

1977

Voyager Spacecraft

Think about the greatest engineering achievements of humankind over the years. Many of them are gigantic, or exceptionally powerful. But the two Voyager spacecraft are different: they are impressive pieces of scientific equipment that have been operating on their own for decades. The most surprising stat is this: The *Voyager 1* spacecraft, launched September 5, 1977, is the first human-made object to leave our solar system. And we can still communicate with it.

How does an engineer even begin to think about the design of a complex object that must last for decades without any chance for repair or refueling, and that will need to communicate with Earth even when it is more than ten billion miles away?

One thing to think about is power. Without power for the electronics, **computers**, **radios**, and heaters, the spacecraft is dead. Solar power is out because the sun is a tiny pinpoint at that distance. The engineers went nuclear, with a device called a Radioisotope Thermoelectric Generator. Inside the RTG are plutonium-238 oxide spheres that naturally produce lots of heat as they decay (with a half-life of 87 years). Thermocouples arranged in the casing surrounding the spheres convert the heat directly to electricity. Initial output of the RTG was about 480 watts. By 2025, decay of the plutonium will cause the spacecraft to run out of power.

Another consideration is communication. The spacecraft itself has a 23-watt radio paired with a directional dish antenna 12 feet (3.7 meters) in diameter. The signal is incredibly faint once it reaches Earth. But on Earth, the receiving antenna is 100 feet (30.5 meters) in diameter. By using frequencies rarely used by humans, the weak signal makes it through.

With power and a radio the spacecraft is able to communicate. The computer gathers data from the eleven instruments and does the communicating. How can a computer run so long? Step one is to engineer the system using radiation-hardened parts. Step two is to engineer redundant systems. There are actually three separate computers on board, and two copies of each one. Step three is for the software engineers to write the code very carefully. In addition, engineers have reprogrammed the Voyagers many times during the mission.

SEE ALSO Radio Station (1920), ENIAC—The First Digital Computer (1946), Space Shuttle Orbiter (1981).

An artist's concept of the Voyager spacecraft.

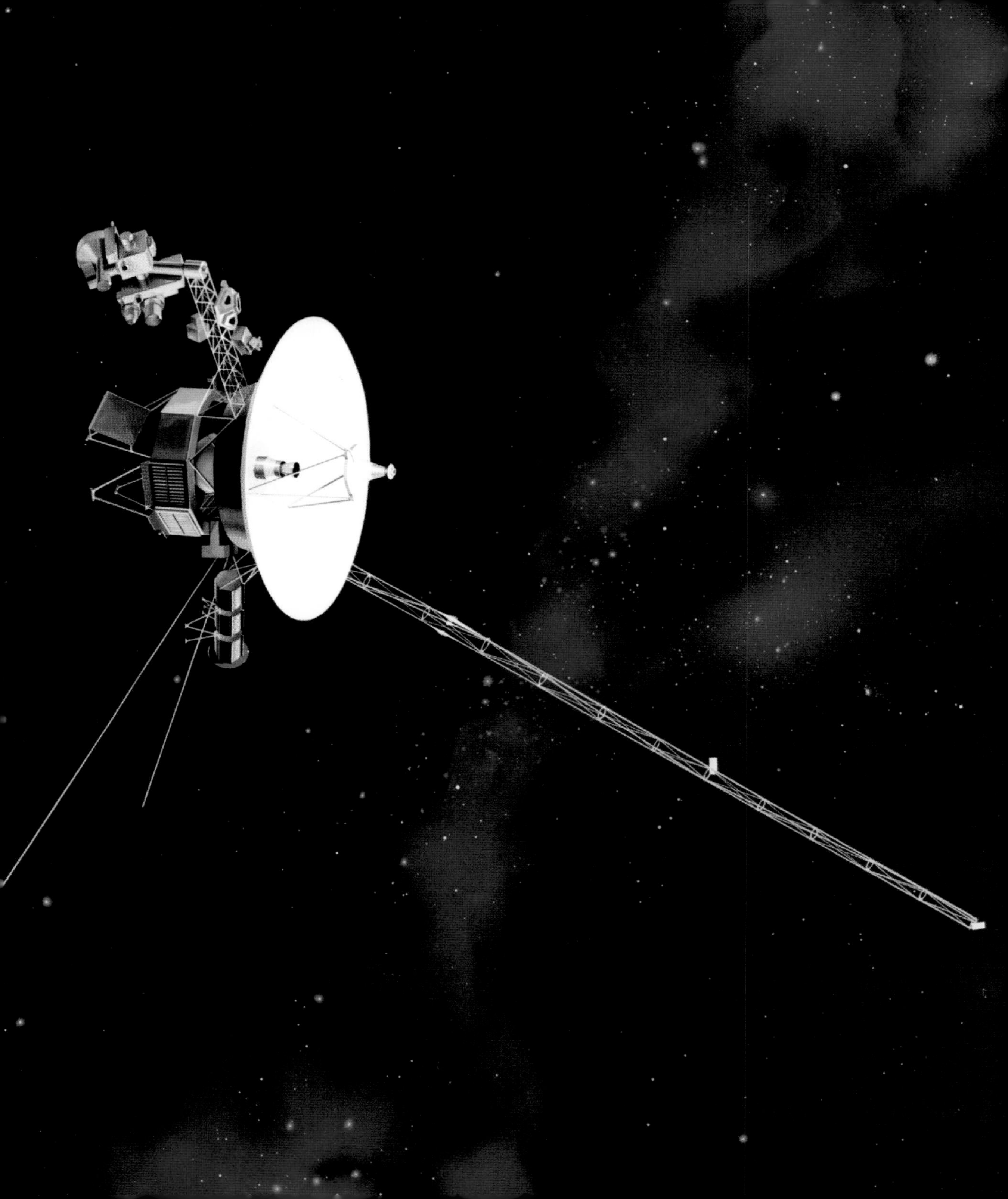

1977

Trans-Alaska Pipeline

Imagine that you have discovered a major new source of crude oil—twice as big as anything else in North America. The oil is plentiful (eventually reaching two million barrels a day), but there is one minor problem—the **well** is in the middle of nowhere far above the Arctic Circle.

For engineers, the question becomes, "What is the safest and most efficient way to get the oil from the well to its market?" Engineers settled on a hybrid approach: An 800-mile -long pipeline (1,300 km) moves the oil overland across a north-to-south traverse of Alaska, and then **supertankers** carry the oil from the port in Valdez, Alaska, to refineries.

A pipeline is just a big pipe, right? First, it's not quite that simple. Second, the unique engineering problems of Alaska came in several forms: 1) Keeping the permafrost near the pipeline from melting, 2) Dealing with earthquake activity, and 3) Keeping the oil flowing in such cold temperatures.

Oil flow is handled with eleven massive pumping stations (able to pump in excess of 60 million gallons of oil per day). The pumps keep the oil warm, pressurized, and flowing through the 4-foot-diameter (1.22 meters) pipeline.

Burying a pipeline full of hot oil naked underground would melt the permafrost, creating erosion and wildlife problems. When the pipeline must be buried, it is done using foam and gravel linings along with chillers. That's expensive. Therefore, the vast majority of the pipeline is above ground. The pipe rests on horizontal beams. The beams attach to vertical posts. The posts contain heat pipes that dissipate heat into the air.

The pipeline, commissioned in 1977 by the Alyeska, can move side to side on the horizontal beams. That ability to move, combined with a serpentine path for the pipeline, lets the pipe shift during expansion and contraction. It also offers earthquake protection.

Along the Denali Fault, the pipeline rests on beams approximately 50 feet (15 meters) long. In 2002, a magnitude 7.9 earthquake proved that the system worked when the ground along the fault shifted by 14 feet (4.26 meters).

Engineers solved 800 miles of problems to create a workable solution.

SEE ALSO Oil Well (1859), *Seawise Giant* Supertanker (1979), Earthquake-Safe Buildings (2009).

The trans-Alaska oil pipeline during a July snow shower.

562

1977

MRI

Raymond Vahan Damadian (b. 1936)

Think about how useful it is for a doctor to be able to see inside the human body without cutting it open. A doctor can detect things like broken bones, tumors, internal bleeding, etc.

Flat X-ray images provided the first window inside. They are good for looking at bone problems. **CT scanners** take X-rays to the next level and create a 3D view. The problem, however, is that X-rays create a cancer risk because they use ionizing radiation.

Then American medical practitioner Raymond Vahan Damadian figured out a new way to look inside the body in 1977. If the body is placed in an impressively strong magnetic field, it is possible to locate a specific point inside the body (imagine a millimeter-sized cube) and get a reading off the hydrogen atoms it contains. Differences in hydrogen atom signals identify different types of tissue. By probing millions of these tiny cubes inside the body, a computer can construct a very accurate 3D image.

Once the science got figured out, it was time for engineers to make MRI scanners that work as precisely, reliably, inexpensively, and safely as possible. One big problem is the magnetic field that the MRI machine requires. The field needs to be uniform, with a strength of 2 teslas or more. Usually a set of ring-shaped electromagnets creates this field, with the patient lying inside the center of the rings. For these magnets to be reasonable in terms of size and power consumption, they are cooled with liquid helium to make them superconducting. The use of liquid helium creates its own engineering problems.

One goal of engineering is improvement over time. With MRI machines, upgrades include the speed of the machine and the size of the cube that the machine can query. Higher-temperature superconductors would eliminate the need for liquid helium. There has also been progress in creating more open MRI machines to help people with claustrophobia. One day, home MRI machines? Maybe not, but engineers can dream.

SEE ALSO CT Scan (1971), Neodymium Magnet (1982).

MRI scan of a human spine.

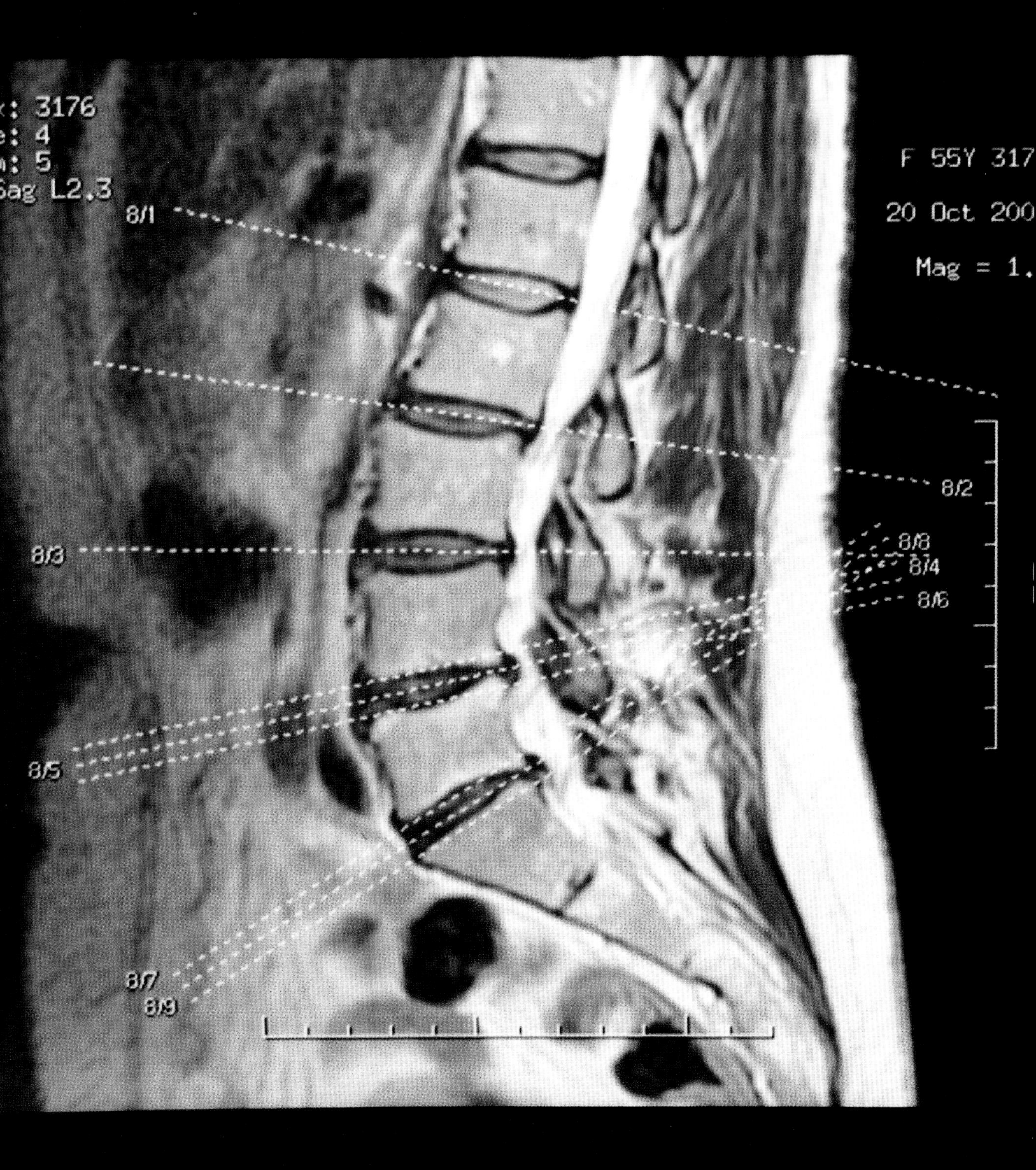
x: 3176
e: 4
n: 5
Sag L2.3
F 55Y 317
20 Oct 200
Mag = 1.
8/1
8/2
8/3
8/8
8/4
8/6
8/5
8/7
8/9

1978

Nitrous Oxide Engine

Say you are an engineer looking for a way to increase the performance of an existing **engine** without adding much weight. One way to do it is to increase the amount of oxygen in the cylinder. A **turbocharger** or supercharger is one way to do this, but these are complicated devices that are heavy, and they require engine power to do their thing.

Since air is 80 percent nitrogen and only 20 percent oxygen, one way to increase engine power without increasing engine size would be to increase the concentration of oxygen. With more oxygen in the cylinder, the engine can burn more fuel and increase its power. One way to increase the oxygen concentration is with a high-pressure oxygen tank, like you might find on a welding rig. The problem with these tanks is that they are extremely heavy and they do not hold that much oxygen.

Is there another option? Nitrous oxide systems are one engineered alternative. Nitrous oxide is N_2O, so it contains more oxygen than atmospheric air. Even better, N_2O compresses into a liquid, allowing dense storage. One liter of liquid N_2O expands by 400 times when it becomes a gas. And even better, as the N_2O liquid turns to gas, it does so at -88°C (-126°F). This chills the intake air flowing into the engine, which makes it denser and therefore adds even more oxygen to the cylinder.

The effect is impressive. An engine can improve its horsepower rating by 20 percent or more with a very simple hardware setup. Basically you simply spray the N_2O into the intake manifold in a controlled way to get the benefits, and increase the amount of fuel accordingly.

For these reasons, N_2O systems, produced by a company of the same name since 1978, are common at race tracks to boost performance. For the same reasons, N_2O systems are illegal for street cars. And on the track there is some need for care in the application of N_2O. Too much of a good thing can cause an engine to explode if it is not strong enough to handle the effects of the N_2O.

SEE ALSO Supercharger and Turbocharger (1885), Internal Combustion Engine (1908).

Drag race engine with blower and nitrous oxide injection.

NOS
NITROUS OXIDE SYSTEMS
Holley

1978

Bagger 288

When digging for coal there are two kinds of mines: underground and surface. When working a surface mine, the mindset is pretty simple: scoop the coal up and ship it to a customer by **rail** or barge.

What if engineers want to create the biggest, fastest digging machine possible? That would be the Bagger 288, also known as a bucket wheel excavator, built by the German company Krupp beginning in 1978.

This machine is so immense that it nearly defies imagination. The digging end of it is a wheel 70 feet (21 meters) in diameter with 20 excavator scoops lining the wheel. To dig, this wheel spins. The buckets on the forward end of the wheel scoop up coal. When a scoop rotates to the top of the wheel, the coal falls out the open back of the scoop onto a conveyor. Since each scoop holds something like 15 cubic meters of material, and a scoop dumps its load every few seconds, the machine can easily move 100,000 cubic meters or more of coal in a day.

The machine has three parts. The top part consists of the bucket wheel, its motor, part of the conveyor, the operator's cab, and the crane-like superstructure that supports it. The superstructure can move the bucket wheel up and down, and it sits on a turntable so it can rotate left and right. This structure is 310 feet (95 meters) tall and resembles a cable-stayed bridge like the **Golden Gate Bridge**. The second part is the rear conveyor belt and the truss structure that supports it. Together with the bucket-wheel extension, this creates a vehicle over 700 feet (210 meters) long. The third part is a set of 12 very wide caterpillar treads so the machine can move around.

This machine needs somewhat more than 20,000 horsepower (16 megawatts) to operate. It gets that power from electric motors, so it uses the world's biggest extension cord to operate.

Could engineers make something bigger? That is difficult to imagine, because if they did it could no longer move. And the nature of the mine is that the machine has to move forward as it digs. This is about as big as moving land machines can get.

SEE ALSO Transcontinental Railroad (1869), Golden Gate Bridge (1937), Light Water Reactor (1946), *Seawise Giant* Supertanker (1979).

Bucket-wheel excavators 288 and 258 in Garzweiler surface mine.

RWE
288
KRUPP
SIEMENS

1979

Seawise Giant Supertanker

The modern world has an incredible appetite for oil and **refinery** products. Something approaching 100 million barrels of oil get consumed every day on planet earth. How to move all that oil around? When oil has to cross an ocean, then a supertanker is the ticket. The largest supertankers carry more than three million barrels of oil and do it with incredible efficiency.

A big supertanker is an immense, engineered structure. Fully laden, the largest ships can weigh 600,000 tons and measure 1,475 feet (450 meters) in length. The longest ship ever built was the *Seawise Giant*, constructed by Japanese company Sumitomo Heavy Industries in 1979. Designing something this big and this heavy falls into a branch of engineering known as naval architecture. Engineers are creating an object the size of the **Empire State Building**, but this object moves across the water, endures storms and waves, may run into obstacles, etc. Safety is a special concern because crude oil is so toxic.

The dominant architecture used in supertankers today relies on two huge internal steel beams running the length of the ship. These beams give the ship its strength. They create three areas (center, port, and starboard) for tanks that hold the oil. Rather than one long tank running the length of the ship, there are dozens of smaller tanks that minimize sloshing. The tanks also make a double hull possible. The tank itself represents one hull, while the outer skin of the ship is another, with significant space between the two. If the ship runs into something, the hope is that only the outer hull takes damage, with the spacing acting as a buffer that protects the tanks.

The largest **two-stroke diesel engines** in the world power the largest supertankers. To give you a sense of the size, imagine one cylinder of the engine with the piston at bottom dead center. If you stand on top of the piston, the cylinder wall rises more than 8 feet (2.4 meters) high and the piston diameter is more than 3 feet (1 meter). An engine has 14 of these cylinders and weighs 2,300 tons, producing over 100,000 hp. The engine alone is an engineering marvel, riding inside another engineering marvel.

SEE ALSO Wamsutta Oil Refinery (1861), Two-Stroke Diesel Engine (1893), Empire State Building (1931), Container Shipping (1984).

Aerial view of crude oil tanker and storage tanks in the port of Coruña, Galicia, Spain.

1980

Flash Memory

Fujio Masuoka (b. 1943)

People using computers in the 1990s had a problem. The **RAM** (Random Access Memory) chips used in computers are volatile, meaning that they forget everything when the power turns off. And they are expensive. To solve these two problems, all computers had **hard disks**—spinning platters that store data magnetically. A hard disk can remember data for years even if powered off, and has huge capacity compared to RAM.

But hard disks are mechanical devices, meaning they are slow and big and unhappy when dropped. What the world needed was a type of electronic memory that could remember things when the power went off.

To get there, engineers went through a progression. They started with ROM, or Read-Only Memory, developed by Fujio Masuoka while working for Toshiba in 1980. Chips would be manufactured with bit patterns permanently encoded on them. Next came PROM, or Programmable Read-Only Memory. A user could burn a bit pattern onto the chip one time. Then came EPROM, or Eraseable Programmable Read-Only Memory. A user could burn a bit pattern onto the chip, and then expose the chip to ultraviolet light to erase the whole thing and reprogram it. That was followed by EEPROM, or Electrically Eraseable Programmable Read-Only Memory. The whole chip could be erased with an electronic signal. And then, in 1980, an engineer took it one step further so that individual blocks or cells could be erased electronically. This idea became flash memory, although it took more than a decade to bring it to market.

Two of the first applications of flash memory were the newly invented MP3 players and **digital cameras**. The small size, low power requirements, and ruggedness of flash memory were perfect for these portable devices. Small USB memory sticks were next. Smart phones and tablets use flash memory exclusively. And as capacities have increased, flash memory now competes with traditional hard disks in laptops and desktop computers because, even though it is more expensive, it is faster.

Engineers have tried many other systems to create permanent storage that is purely electronic. Flash memory is the first one to survive in the marketplace.

SEE ALSO Hard Disk (1956), Dynamic RAM (1966), RFID Tag (1983), Digital Camera (1994).

The advent of flash memory greatly increased the capacity of, for example, digital cameras.

CF
SanDisk
Code 8477

1980

M1 Tank

The idea behind a tank is pretty simple: let's take a cannon and put it in a car so that people can drive it around. As soon as you do that, however, the other side decides they too need cannon-cars. The cannon-cars start shooting at each other.

In order to protect the people inside the cannon-car, they need some shielding. So engineers encase them in a hull of thick steel. That makes for an extremely heavy vehicle, so engineers give it a big engine and put it on tracks instead of wheels. The modern tank is born.

The M1 tank, which Chrysler Defense built for the United States Army in 1980, takes this progression about as far as engineers can with today's technology. The outer armor is no longer steel—instead it is a composite, sometimes called chobham, formed from ceramic, steel, **plastic**, and **Kevlar**. The goal of the armor is to stop both kinetic penetrators and molten metal penetrators. It may be augmented with an additional layer of reactive armor or slat armor, which causes molten metal penetrators to discharge prematurely.

The engine is a 1,500-hp (1,100 kw) gas **turbine** along with a 500-gallon (1,900 liters) fuel tank, giving the tank approximately a 250-mile (400 km) range. This sounds like poor fuel economy, but when you consider that a fully loaded M1 tank weighs somewhere around 140,000 pounds (63,500 kg), it makes more sense.

The cannon—the whole reason for the tank's existence—is the centerpiece. It is 4.7 inches (120 mm) in diameter and can shoot kinetic penetrators, molten metal penetrators, and anti-personnel rounds. The tank also has a .50-caliber machine gun and two other smaller machine guns.

To protect the four-man crew, the tank includes an NBC system to filter out nuclear, biological, and chemical threats. The crew compartment is over-pressurized like a **cleanroom**. The interior of the tank is lined with Kevlar mats to protect the crew, and a fire suppression system.

Engineers have created what is considered to be a state-of-the-art tank in the M1. It is the best cannon-car that money can buy.

SEE ALSO Plastic (1856), Steam Turbine (1890), Cleanroom (1960), Kevlar (1971).

Marines with Alpha Company, 1st Tank Battalion, put their M1A1 Abrams main battle tank into a defensive position, May 11, 2014.

A22

1980

Stadium TV Screen

If you have been to a stadium or an indoor coliseum recently, chances are that it had a giant TV screen so that visitors can watch replays and get a better view of live action on the field. These giant displays are a great example of a device that had to wait for the right technology to come along.

Think about this from an engineering perspective. If you want to create a giant outdoor screen, how are you going to do it? The standard solution for a large-format display is a projector, but you cannot use a projector in daylight. You could maybe create a black-and-white screen with incandescent bulbs, but there are problems: the amount of power required and heat generated by the bulbs, the slight delay turning on the bulb as the filament heats, the high burnout rate for the bulbs, etc.

There was a period of time when manufacturers made small CRT devices that could be used to make a display. But these were not as bright as might be desired, they were bulky, and they had a short life as well. The displays were simply too expensive.

Then along came inexpensive LEDs (light-emitting diodes). They are small, bright, efficient, long lasting. Red LEDs appeared first, then green, and, after a long delay, blue LEDs finally appeared at a reasonable price. By that point, the price of the **microprocessors** to control the LEDs had also fallen. Everything engineers needed was ready, and large **TV** screens started to become common. Engineers could package a "pixel" by combining, say, three red, three green, and three blue LEDs in a single unit. Ten by ten of these pixels could combine to form a square-foot module. The modules could tile together to form a screen.

In 1980, the Diamond Vision Board, manufactured by Sony, allowed fans to see the game on the big screen at Dodger Stadium. Today you can find TVs in stadiums that are bigger than a basketball court. The reason is because engineers finally had a reasonably priced, reliable RGB light source they could use to make the screen.

SEE ALSO Color Television (1939), Microprocessor (1971).

A woman wins a house on the big screen as Giants' Barry Zito warms up before a game, AT&T Stadium, San Francisco, California.

DIAMOND
OF CALIFORNIA
Culinary Walnuts
DIAMOND
PlayStation
IT ONLY DOES
EVERYTHING.
PS3

SAFEWAY
Ingredients for life.

Coors LIGHT

PG&E
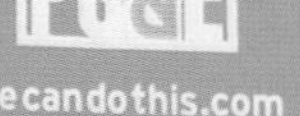
wecandothis.com

GIANTS
D.R.HORTON
America's Builder
GIANTS
VISA

YAHOO!
KETTLE
19

1981

Bigfoot Monster Truck

Bob Chandler (b. 1941)

Sometimes the art of engineering can take some amazing turns. One great example is the monster truck. The progression starts innocently enough. A pickup truck owner gets some new rims that allow for bigger tires. But if the tires are too big, they start rubbing against the body. The simple solution to this is a body-lift kit—a set of spacers that separates the truck body from the ladder frame. Or suspension-lift kits that lift both the frame and the body.

The first monster truck was Bigfoot. Bigfoot's owner, Bob Chandler, took 48-inch (1.2 meter) wheels and tires off of farm equipment and added them to his pickup truck. These tires demanded bigger, more complex axles. Turning became an issue, so hydraulically controlled steering for both the front and rear tires appeared. All of this added weight, requiring bigger engines. Then the tires got even bigger, to 66 inches (1.7 meters) because, why not?

Monster trucks completely abandon the original pickup truck that started the whole craze. Tubular steel frames resembling **trusses** hold together all the parts, and then fiberglass body panels bolt on to give the appearance of a pickup.

Looking at it from the outside, the engineering evolution happened at an amazing pace, fueled by a combination of envy, one-upmanship, fan accolades, sponsor dollars, and the creation of a monster truck racing circuit that gave these vehicles huge visibility. The whole trend started when Bigfoot crushed several cars in a farm field in 1981. The surprisingly large public response led to car-crushing stadium shows and monster truck races drawing tens of thousands of fans to every event.

It's a great example of how rapidly engineering can move when a lot of money and public interest is involved. Think about how quickly airplanes evolved once the **Wright brothers** unlocked the basic principles. Or how fast space technology developed from the first satellite in orbit in 1957 to men on the moon in 1969. Given enough incentive and funding, engineers can make amazing things happen.

SEE ALSO Truss Bridge (1823), The Wright Brothers' Airplane (1903), Lunar Landing (1969).

A monster truck jumping over cars in a show in Uppsala, Sweden.

Media Markt
Media Markt
THOR
STARTA
Kendall
MOTOR OIL
ssonsmaskiner.se

1981

Space Shuttle Orbiter

The US space shuttle orbiter, first launched in 1981, looks like something from outer space. It has a strange shape. It has rocket engines sticking out the back. The entire midsection has huge doors that cover an immense cargo bay. And the whole thing is covered in very strange ceramic tiles.

Yet at its core, the space shuttle is an **airplane**. It is made of aluminum using the same techniques that we see in an ordinary **passenger jet**. Then there are a number of additions that let this airplane orbit the earth in the vacuum of space. It is a great example of engineers taking a list of special requirements and bringing them into reality.

The first thing that makes the orbiter unique is a crew compartment that doubles as a space capsule. It seals tight and has a life support system that replenishes oxygen, removes CO_2 and humidity, and maintains the correct pressure.

The second big point of uniqueness is the need to handle reentry heat. A normal aluminum airplane would quickly melt and disintegrate during reentry because temperatures on parts of the orbiter go as high as 1,600°F (3,000°C). So the orbiter has different types of thermal protection for different temperature areas. The nose of the orbiter hits the highest temperatures, so it has a cover made of reinforced carbon-carbon. The underside is the next hottest, so it uses one type of silica tile. Other parts of the orbiter use other types of tiles, or flexible insulation blankets.

The third thing is the rocket engines. The three big ones on the back are used strictly to provide thrust during launch, and are placed there so they can be more easily reused. Smaller engines provide thrust to speed up or slow down the orbiter and also to change its orientation in space.

Other unique systems on the orbiter include a five-way redundant **computer** system, a space **toilet**, the massive cargo bay and cargo bay doors for bringing large payloads into orbit, a radiator system built into the cargo doors to dissipate heat in the vacuum, and a robot arm used to manipulate things in the cargo bay.

SEE ALSO Hall-Héroult Process (1889), The Wright Brothers' Airplane (1903), ENIAC—The First Digital Computer (1946), The Concorde (1976), Low-Flow Toilet (1992).

The launch of the STS-1, the first in a series of shuttle vehicles planned for the Space Transportation System, uses reusable launch and return components.

USA

1981

V-22 Osprey

The V-22 Osprey is one of those projects where you wonder if engineers bit off a little more than they could chew.

It sounded simple enough when the United States Department of Defense commissioned it from Bell Helicopters and Boeing in 1981: Create an aircraft that can act like a helicopter for takeoffs and landings, and then convert into a normal turboprop airplane. Being able to land like a helicopter means you don't need a runway. Turboprops use less fuel than **helicopters** in straight and level flight, and have a higher top speed than helicopters. It's the best of both worlds.

Then engineers went to design the plane. One problem: engine failure. In a normal twin-engine airplane, the failure of one engine is not a big problem. But on the V-22 in helicopter mode, one engine failure would be catastrophic. So the V-22 needs a driveshaft running through the wing to connect the two engines. But not just a straight shaft. It has to curve over the fuselage and deal with flexing wings, so it has fourteen segments.

The Marines added another requirement—the wings need to rotate to a position parallel to the fuselage, and propeller blades need to fold so the plane takes up less space in storage. That creates lots of complexity. Then mechanical engineers had to add the powerful hydraulic mechanisms to rotate the engines and props during flight. Because this rotation process is so important, the hydraulic system is triple redundant and runs at very high pressure.

Keep in mind that wing tips are normally light. In this plane they are massively heavy. So the wings need to be extra beefy. When you add up the weight of the drive shaft and transmissions, the folding wings and props, the beefy wings and the rotating engines, this plane is carrying a lot of extra baggage. To shed weight, light **carbon composites** are used in almost half of the aircraft. All of that extra complexity plus the carbon makes the V-22 expensive. Several crashes and design reviews meant the airplane took more than two decades to reach operational status.

Is it worth it? There are a lot of critics. The plane is super-expensive. But engineers did accomplish the goal.

SEE ALSO Carbon Fiber (1879), Helicopter (1944).

A sailor with USS Kearsarge *salutes MV-22 Osprey pilots as they take off.*

05
LSE
LSE

1982

Artificial Heart

Robert Jarvik (b. 1946)

In healthy people, the heart beats without pause for an entire lifetime, pumping 5 million gallons (19 million liters) of blood or more during 70 or 80 years of operation. But when something goes wrong and a heart needs to be replaced, the scarcity of natural replacement hearts creates a big problem. Thus engineers set about trying to design and build artificial mechanical hearts. However, nature's pump is very difficult to duplicate.

There were four problems that doctors and engineers had to solve to make an artificial heart work: 1) Finding materials with the right chemistry and properties so that they did not cause an immune reaction or internal clotting in the patient, 2) Finding a pumping mechanism that did not damage blood cells, 3) Finding a way to power the device, 4) Making the device small enough to fit inside the chest cavity.

The Jarvik-7 heart, designed by American scientist Robert Jarvik and his team in 1982, was the first device to meet these requirements reliably. It features two ventricles like a natural heart. Its materials avoided rejection and were smooth and seamless enough to prevent clotting. The pumping mechanism used a balloon-like diaphragm in each ventricle that, as they inflated, pushed blood through the one-way valve without damaging blood cells. The only compromise was the air compressor, which remained outside the body and transmitted air pulses to the heart with hoses running through the abdominal wall. The basic design was successful and has since been improved as the Syncardia heart, used in over 1,000 patients. One patient lived almost four years with the heart before receiving a transplant. The Abiocor heart uses a different approach that allows **batteries** and an inductive charging system to be implanted completely inside the body. It also has a diaphragm arrangement but fluid flow rather than air fills the diaphragms. The fluid flow comes from a small electric motor embedded inside the heart.

Two different engineering approaches: One goes for complete embedding. But if something goes wrong, it probably means death. In the other, much of the system is outside the body for easy access and repair, but tubes pass into the body from outside.

SEE ALSO Defibrillator (1899), Heart-Lung Machine (1926), Lithium Ion Battery (1991).

The CardioWest TAH-t, pictured, is the first and only FDA, Health Canada, and CE-approved temporary Total Artificial Heart in the world.

1982

Neodymium Magnet

There are times when a new innovation is extremely useful to engineers. It can alter many existing products in different ways and make new products possible. Neodymium magnets represent a great example of the phenomenon. The innovation here was the development of the $Nd_2Fe_{14}B$ alloy by General Motors and Sumitomo Special Metals in 1982, and then the manufacturing processes that turns the alloy into useful magnets. The technology that preceded neodymium magnets created samarium cobalt magnets, which are much more expensive. The inexpensive magnet type that neodymium magnets replace are normal ferrite (Fe_2O_3) magnets and alnico magnets from the 1930s.

So neodymium magnets appear, and engineers suddenly have a new inexpensive type of magnet that is far more powerful than the existing types. How much more powerful? Compare the "stickiness" of a refrigerator magnet that uses neodymium to ferrite. Neodymium magnets can sometimes be so powerful that they are difficult to remove from the refrigerator. This strength means that engineers can make things that use magnets lighter and smaller.

One of the first places to receive the magnets were the head actuators on **hard disks**. Today there is a linear motor in all hard disks that moves the arm holding the read/write heads across the disk platters. With neodymium magnets, this motor is small, compact, and very fast. Then there are the iconic earbuds that Apple popularized. Headphones that small and powerful would not be possible without neodymium magnets. We see remote-controlled electric **helicopters**, airplanes, and **quadrotors** at the toy store. These all use small, lightweight electric motors. They would not be possible without neodymium magnets. Electric bicycles and even some **electric cars** use hub motors—powerful, compact electric motors that can fit in a hub. These are made possible because of neodymium magnets. We see neodymium magnets in all sorts of new places. Toys snap together with these magnets. The covers on **tablet computers** use them and so do power plugs on laptops. We see lids and closures that use them. Most **MRI** machines use superconducting electromagnets cooled with liquid helium. Neodymium magnets make it possible to consider a permanent magnet solution.

SEE ALSO Helicopter (1944), MRI (1977), Prius Hybrid Car (1997), Quadrotor (2008), Tablet Computer (2010).

Pictured: Nickel-plated rare earth magnets.

1983

RFID Tag

Charles Walton (1921–2011)

Imagine that you want to take an inanimate object—say a box of cereal—and you want it to be able to say something about itself. One thing you can do is put a label on it. The label contains words that tell a human being what is inside. If you want a machine to be able to read the label, you can add a bar code or QR (Quick Response) code. But a bar code still depends on someone or something lining it up with the barcode scanner. What if you want the inanimate object to be able to say something about itself to a machine without a visual scan and without alignment worries?

The current engineering solution to this problem is the active or passive RFID (Radio-frequency identification) tag, with passive tags being more common because they are much less expensive. The first patent for the modern RFID tag was issued in 1983 to electrical engineer Charles Walton.

The technology is impressive and clever. A passive RFID tag is a small computer chip and **radio**. But computers and radios need power to operate. So there is also a power-gathering antenna that can create electricity from induction in a strong magnetic field. When the tag is placed in such a field, it powers up. The simplest tags, which first appeared in 1983, transmit a unique hard-coded ID number that is 96 bits long. More complicated tags can send much more data. The RFID tag in a US passport contains all personal information plus an encoded photo.

Because engineers have pushed down the price of tags and readers, we now see RFID tags in many places: credit cards, library books, shoplifting prevention systems, toll collection, employee IDs, etc. When you get your pet "chipped," that is an RFID tag encased in glass injected under the pet's skin.

One of the early dreams for RFID is instant checkout in a grocery store. Rather than standing in a checkout line, you walk through a gate with your cart and your total price gets instantly calculated. Engineers may yet lower tag prices to make this dream a reality.

SEE ALSO Radio Station (1920), NCSU BookBot (2013).

Close-up of an RFID tag showing the chip and antennae.

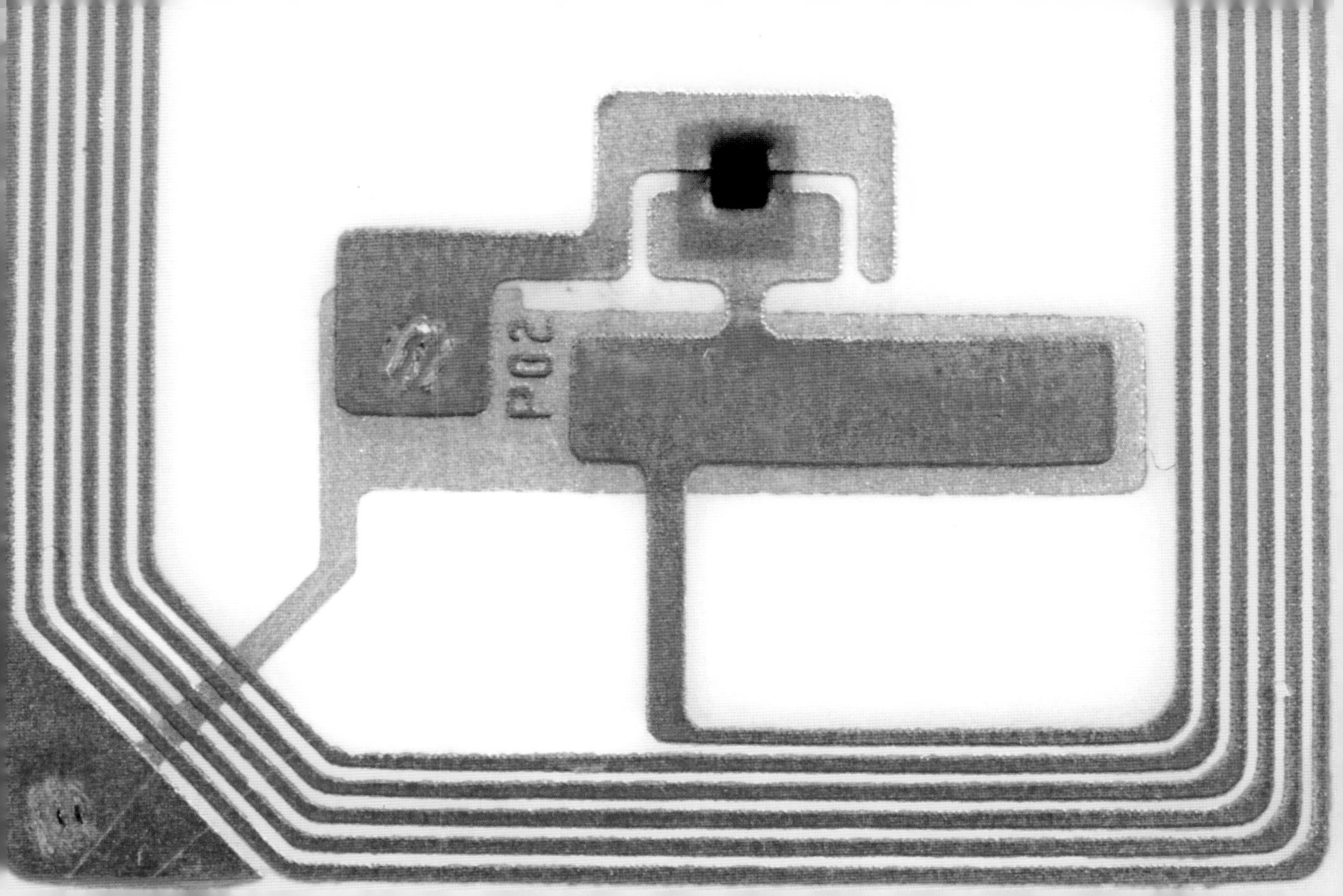
P02

1983

F-117 Stealth Fighter

Radar came into being during WWII and creates an all-seeing eye. Is it possible to make a military airplane that is invisible to this eye?

What are the options? Engineers can make an airplane smaller to cut the radar cross-section. Flying very close to the ground renders ground radar useless in most cases. By combining these two techniques, engineers created cruise missiles. But what about larger planes that need to fly high?

What happened next was a two-part process. To understand the first part, get into the mindset of evading radar. Think of an airplane as a mirror-coated cigar-shaped cylinder. Now shoot a laser at it. Because of the shape, some portion of the laser light reflects back to the source of the **laser** beam. Radar uses a beam of radio waves instead of a laser, but the concept is nearly identical. The plane's tendency to reflect the radar beam back to the source creates the vulnerability.

What if, instead, the plane was shaped like a mirror-coated box—nothing but flat surfaces. Now when the beam hits, it reflects away from the source. This is the essence of stealth technology—radar energy hits the airplane but reflects away from the receiving antenna.

Once engineers had this insight, there was another problem—how do you make an airplane with nothing but flat surfaces? And how do you control such an airplane? The answers produced one of the strangest-looking airplanes to ever fly—the F-117 stealth fighter, launched by the United States Air Force in 1983.

The F-117 does not look very aerodynamic because we expect aerodynamic shapes to consist of smooth curves. It is so boxy and angular that it doesn't look like it should fly at all. However, a combination of computer control and angular aerodynamics produces an airframe virtually invisible to radar that can actually fly.

To create stealth technology, engineers needed an insight, then a reconceptualiztion of what an airplane could look like, and then **computer** intervention between pilot and control surfaces to make it all work.

SEE ALSO Radar (1940), ENIAC—The First Digital Computer (1946).

The B-2 stealth bomber, pictured, completed its first flight at Edwards Air Force Base, CA, July 17, 1989.

1983

Mobile Phone

Martin Cooper (b. 1928)

The mobile phone is one of the most audacious engineering achievements of the twentieth century, ranking right up there with the **Hoover Dam** and the **Golden Gate Bridge**. In 1980, a device called a radiophone existed, used by only a few thousand people. It received transmissions from one big base station antenna in the center of a city, and required a rather massive **radio** system, which was installed in a user's car. Its strength was perhaps 25 watts transmitting up to 30 miles. There were a handful of radiophone lines for the entire city. A caller picked one of the frequencies to make her call. That is what preceded mobile phones.

A group of engineers at Motorola, lead by electrical engineer Martin Cooper, decided to completely reconceptualize the technology. They foresaw a world where everyone would eventually have a phone in their pocket. However, to accomplish this, they needed to make a number of changes to the system. First, there would be hundreds of towers spreading across an entire city every few miles. Then they would get the Federal Communications Commission to allocate nearly 2,000 separate radio frequencies—enough for each tower to talk to dozens of phones simultaneously. The advantage? The phone in your pocket would only have to transmit about two miles, which drastically reduces the power of the phone's transmitter, and therefore its size and **battery** consumption.

Then they added another layer. If you are traveling in your car, a computer in your phone would coordinate with two towers—the one fading in the distance and the one you are approaching ahead. The phone and the towers would perform a seamless automatic handoff from tower to tower as you drove through the city.

Each tower costs over $1 million and a big city needs hundreds of them. The only way this vision would work financially is if millions of people started to use mobile phones. Luckily, the engineers were right. Several companies started building towers in 1983 and the public response was incredible. Everyone wanted a mobile phone, and prices fell rapidly. The rest is history.

SEE ALSO Hoover Dam (1936), Golden Gate Bridge (1937), Lithium Ion Battery (1991), Smart Phone (2007).

Michael Douglas (as Gordon Gekko) demonstrating an early model of mobile phone.

1983

Gimli Glider

Imagine that you are the pilot flying a state-of-the-art **Boeing 767**. You are cruising at 41,000 feet (12,500 meters) over Canada. It should be a routine flight, but suddenly you start hearing warning chimes and there are lights flashing on the instrument panel. Multiple parts of the **airplane** have started to complain all at once about low fuel pressure. The plane has run out of fuel midflight.

Both **jet engines** shut down. The 767 is now a glider.

The thing to understand is that a 767 is a hydraulically powered plane. The steering yoke and pedals drive hydraulic cylinders that move the control surfaces like the rudder and the ailerons. Without engines to pump the hydraulic fluid, there is a big problem.

This incident, which is known as the Gimli Glider because of the airport where the plane ended up landing in 1983, shows one of the true hallmarks of good engineering. Engineers not only solve problems; they can anticipate them and design for fail-safe redundancy.

In the case of this 767, there was already redundancy. Both engines have the ability to pump hydraulic fluid. So if one engine fails, the other one can handle the load. But both engines are dead. Now what? Another level of redundancy: the electric backup hydraulic pump in case both engine pumps fail. However, it depends on an electrical generator working on one of the engines. With both engines dead there is no electricity.

A small battery backup system provides minimal electrical power in a situation like this. It kicks in to power a few instruments and the radio. But it has nowhere near the power to run a hydraulic pump. But these batteries can open a small door on the underside of the fuselage. And out pops something amazing—the ram air turbine (RAT). Engineers designed it to handle exactly this situation—a dual engine failure. It uses wind passing by the fuselage to spin a turbine that pressurizes the hydraulic system. The plane landed without any casualties because of the RAT.

It is a great example of engineers anticipating unexpected problems in order to save lives.

SEE ALSO The Wright Brothers' Airplane (1903), Turbojet Engine (1937), Boeing 747 Jumbo Jet (1968).

The Gimli Glider is shown here.

AIR CANADA

1983

Ethernet

Chuck Thacker (b. 1943), **Butler Lampson** (b. 1943), **Robert Metcalfe** (b. 1946), **David Boggs** (b. 1950)

To connect multiple **computers** that need to communicate with each other, what you need is a LAN, or Local Area Network. But how will the computers connect together? This was actually an open question for engineers, and they tried out a number of network topologies in the marketplace: various star networks, ring networks, and bus networks. The technology that won out is called Ethernet. It was first developed by engineers at Xerox PARC—including Chuck Thacker, Butler Lampson, Robert Metcalfe, and David Boggs—and patented in 1975, then standardized in 1983. In the 1980s it used coaxial cable to implement an inexpensive bus topology.

A bus topology means that every computer communicates on the same single wire. One question in this arrangement is "who gets to talk now?" In the Ethernet approach, a computer that wants to send a packet first looks at the bus to see if it is clear. If it is, it starts sending its packet. It listens as it sends, and if it sees that its packet is getting corrupted, it knows that a collision occurred—two computers started sending their packets at the same time. So it stops sending and waits a random amount of time before trying again. Why random? Because if both computers immediately start resending, they would simply collide again.

It sounds like collisions would happen all the time, especially if a lot of computers share the same bus. But the system works surprisingly well even though there is no central control.

Because Ethernet cards and cable were inexpensive and easy to install, and also available from many companies, Ethernet took off. Today just about every LAN uses Ethernet and most laptops come with an Ethernet port because the hardware is so cheap.

The original Ethernet specification ran at ten megabits per second. One gigabit per second is now common at the consumer level.

Ethernet is another one of those technologies that we take for granted because engineers make it work so well. Every computer that connects into a LAN with a wire uses it.

SEE ALSO ENIAC—The First Digital Computer (1946), ARPANET (1969).

Blue Ethernet network plug connected to the back of the desktop PC.

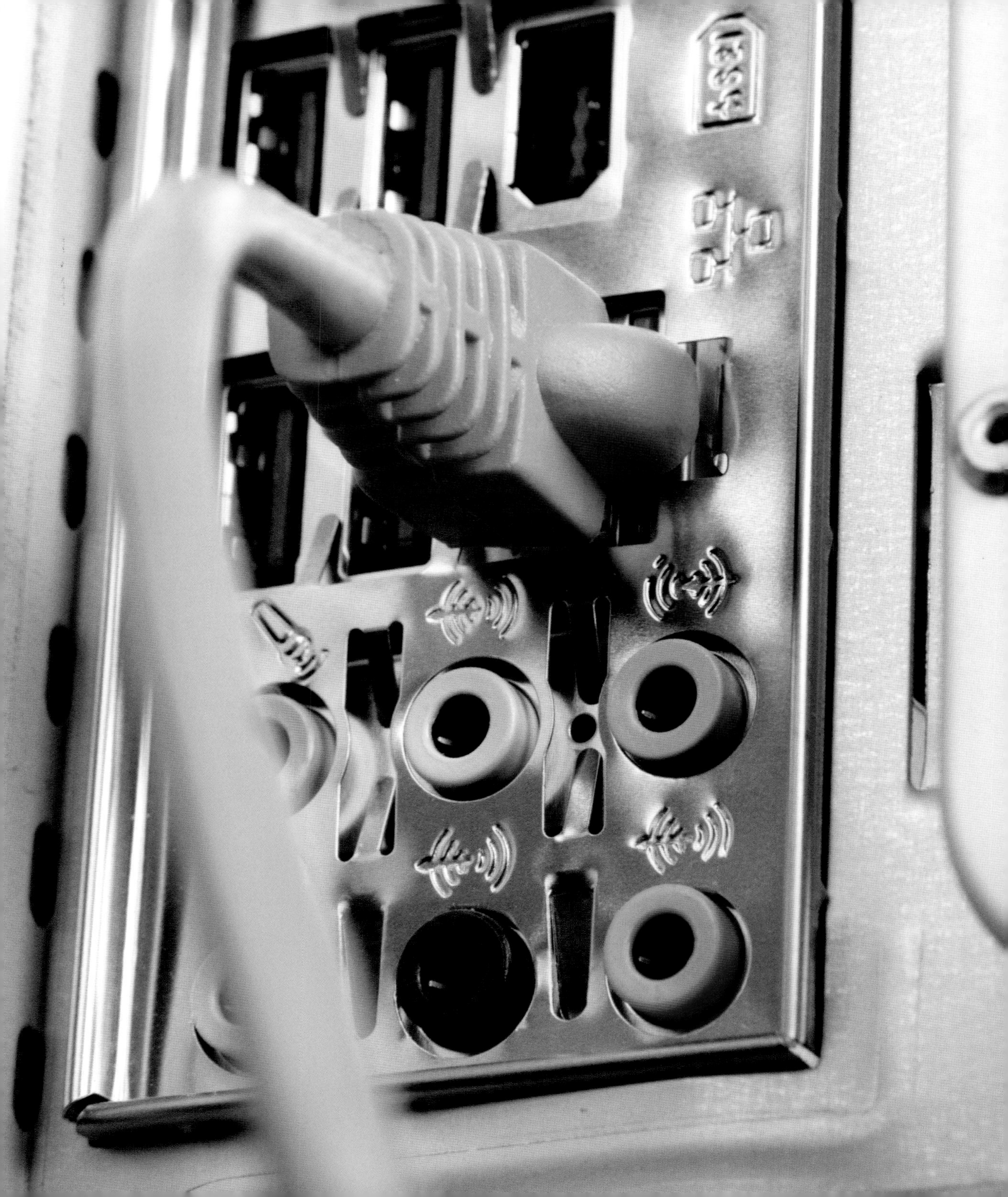

1984

3D Printer

Chuck Hull (b. 1939)

The advantage of 3D (three-dimensional) printing is that just about anyone can design a part on a computer with 3D modeling software and then print that part immediately. There are no molds to make, no high-pressure injection processes—it is immediate and inexpensive.

The rise of 3D printing has created a new development process called rapid prototyping. Engineers can design and print the parts for a product in just a day or two, fit everything together, and see how it feels. Then they can make rapid modifications to arrive at a final design in far less time than it once took. 3D printing has also given rise to the idea of instant manufacturing and home manufacturing, in which a printer can make a simple product on demand. There is even a 3D printer with all of its plastic component parts printed on a 3D printer.

3D printing was invented by American engineer Chuck Hull in 1984 with a process called stereolithography. A laser cures a light-sensitive liquid plastic. The **laser** draws one layer on a platform in a vat of liquid plastic. The platform lowers a notch (e.g., a quarter of a millimeter) and the laser draws the next layer, and so on until the object is complete. This process works exceptionally well but the machines and the liquid plastic are expensive.

3D printing really took off with the development of inexpensive 3D printers in the $1,000 price range. At that point small businesses and individuals could afford them. These printers use extrusion—ABS **plastic** (the kind of plastic used in LEGO bricks) comes into the extrusion head as a thick thread, where it is melted and deposited layer by layer to build up the product. Small pieces print in a few minutes, larger pieces take a few hours.

3D printing has unleashed an explosion of creativity. Millions of people who never could have made plastic parts because of the expense and time involved, including engineers, inventors, hobbyists, and students, can now make an exciting array of objects. Any time there is a release and expansion of creativity like this, it is great for the engineering process.

SEE ALSO Plastic (1856), Laser (1917).

3D printer manufacturing a plastic piece as designed on the computer screen.

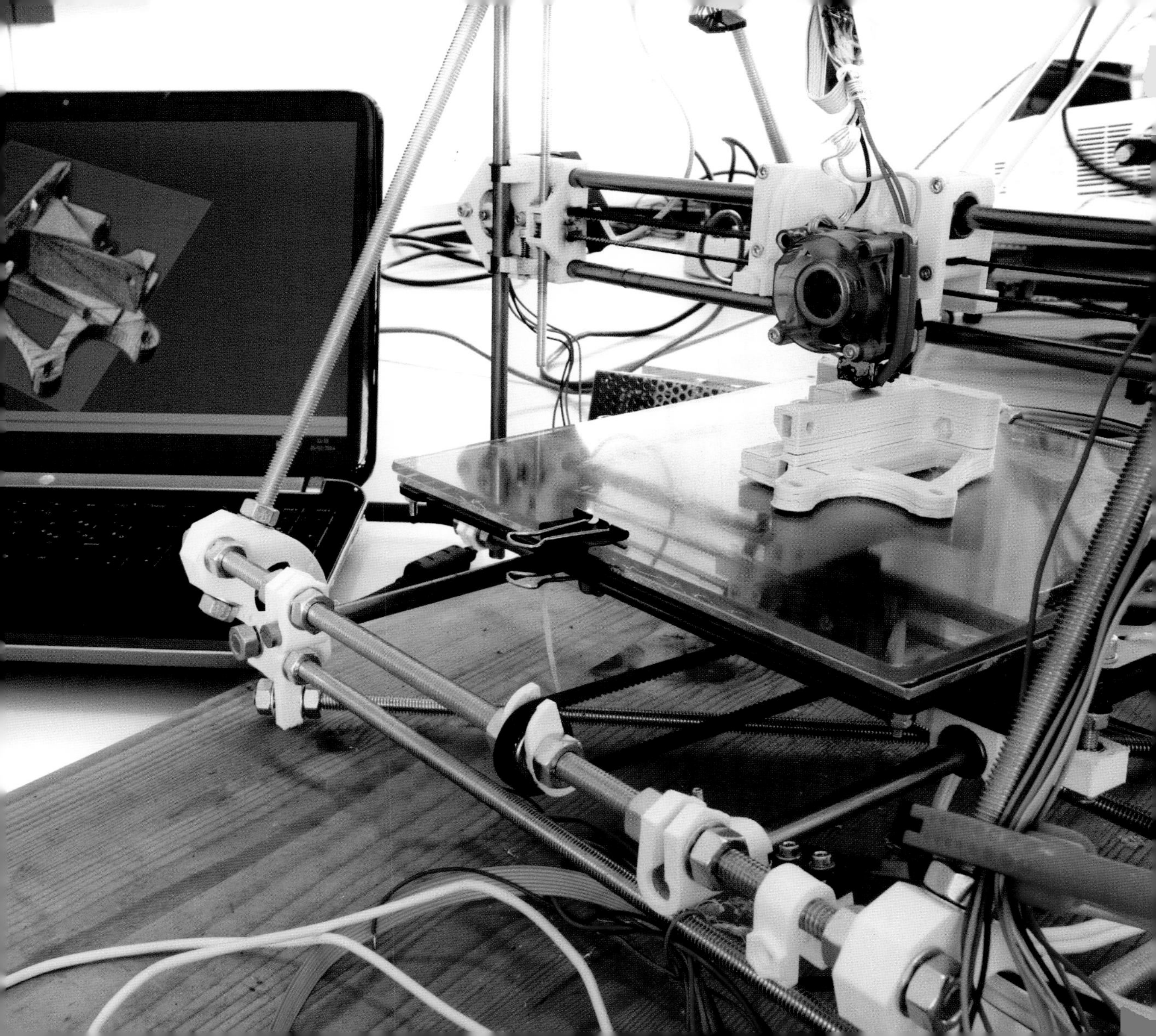

1984

Domain Name Service (DNS)

Jon Postel (1943–1998), **Paul Mockapetris** (b. 1948)

The early Internet consisted of a collection of host machines, all interconnected, with each one uniquely identified by its numerical IP (Internet Protocol) address. On the early Internet there were not many machines, so using these 12-digit numbers was okay.

But as the number of machines grew, you can imagine how cumbersome it became. People are not good at memorizing lots of random numbers. The solution would come to be known as the Domain Name Service, or DNS, developed by Jon Postel and Paul Mockapetris at the University of California, Irvine, in 1984. It is fascinating because DNS is so important to the Internet, yet so reliable and invisible that hardly anyone knows about it.

DNS works like this. You go to your **web browser** and type in a URL like http://www.marshallbrain.com. The browser hands the URL to a piece of code called the resolving name server. The RNS extracts the top-level domain name, in this case "com." There is a set of root servers that your computer knows about (because it holds a list of their IP addresses that gets updated periodically by your ISP). RNS goes to a root server and says, "Give me the IP address for a 'com' name server." It gets back that IP address. Then it goes to that com name server and says, "give me the IP address for the name server for 'marshallBrain.com'." Then it connects to that name server. It asks it for the IP address of the machine for www.marshallbrain.com. The RNS then hands that IP address to the browser, which then sends a request for the home page of www.marshallbrain.com to that IP address.

Think about all of the new web pages you look at in a day. Multiply that by the billions of people who use the Internet each day. Now consider that if the DNS system failed, the Internet would instantly become largely useless to most of us. Yet it never fails. And even though it handles billions of requests every day, the system works because it is distributed and superefficient at what it does. It represents one of the best kinds of engineering—completely invisible engineering.

SEE ALSO World Wide Web (1990).

A diagram of Domain Name Service.

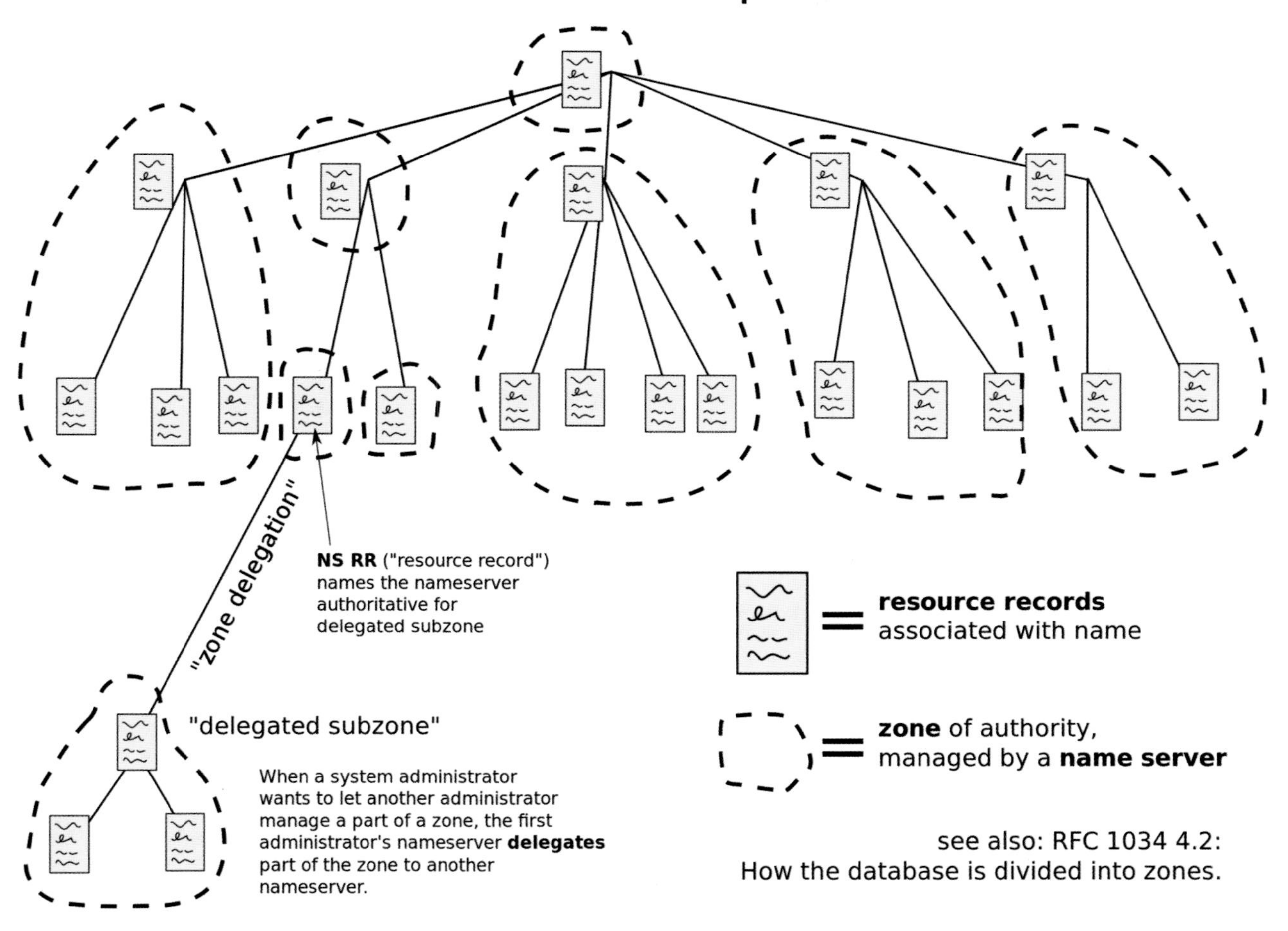
Domain Name Space
"zone delegation"
NS RR ("resource record") names the nameserver authoritative for delegated subzone
"delegated subzone"
When a system administrator wants to let another administrator manage a part of a zone, the first administrator's nameserver delegates part of the zone to another nameserver.
= resource records associated with name
= zone of authority, managed by a name server
see also: RFC 1034 4.2: How the database is divided into zones.

1984

Surgical Robot

Geof Auchinleck (Dates Unavailable), **Dr. Brian Day** (b. 1947), **James McEwan** (Dates Unavailable)

Laparoscopic surgery is a complete reconceptualization of the surgical process that has many important benefits for the patient. But it is also a pain in the neck for the surgeon, who has given up his or her visual clarity and a lot of natural dexterity in return for these patient benefits. Is there any way for engineers to relieve some of the strain on the surgeon?

This is where the world's first surgical robot, Heartthrob, comes in. Invented by a team lead by Dr. James McEwen, an engineering physics grad student, Geof Auchinleck, and physician Dr. Brian Day, the idea was to make a surgeon's job much easier.

First, the robot provides a much better vision system. Two binocularly spaced high-definition cameras provide images from the surgical site, and the doctor controls these cameras. The surgeon rests her head in a cradle to look at the two screens, which creates a 3D view of the space. This step alone reduces fatigue and neck strain while providing a significantly better view of what is going on.

With her hands, the surgeon controls three manipulator arms. One arm is typically positioned and clamped to hold something stationary. Then the surgeon directly manipulates the other two with intricate hand controls providing seven degrees of freedom in the end effector. Because the surgeon is manipulating robot arms rather than working directly with tools (like scalpels and suture needles), the intermediary computer can do things like filtering out hand tremor and guarding against accidents.

The most interesting part of the whole procedure is that the surgeon's station and the robotic arms are completely separate. Typically the surgeon is sitting just a few feet away from the patient, but with the separation, it is possible to imagine scenarios where the surgeon is many miles away from the patient. With a good communication link, remote surgery becomes a possibility. A doctor in a safe location could perform battlefield surgery, or a surgeon in India could perform surgery on a patient in another country.

SEE ALSO Laparoscopic Surgery (1910), Robot (1921).

Surgeon (lower left) performing minimally invasive surgery (MIS) on a patient's heart using a daVinci Si remotely-controlled robot surgeon (center right).

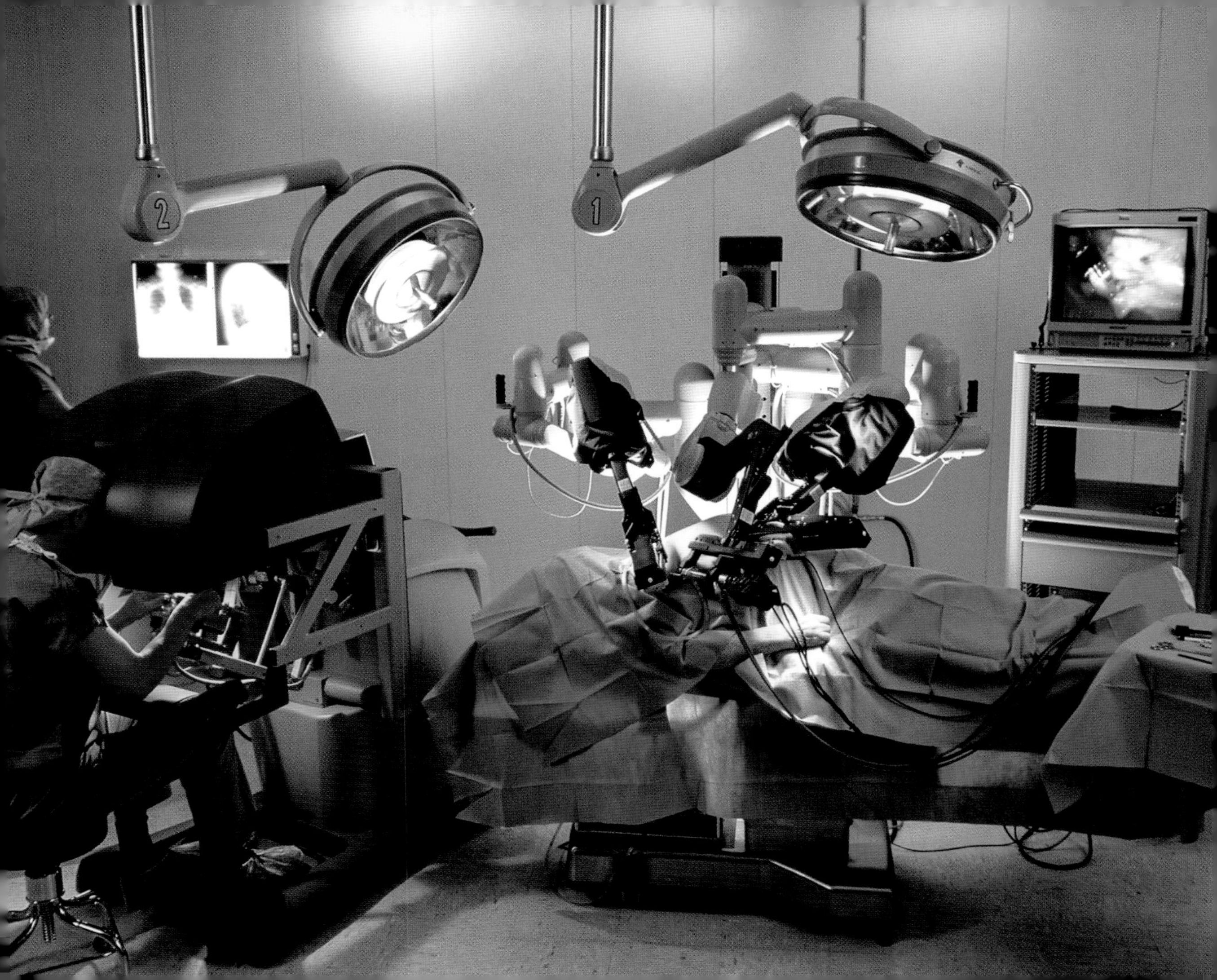
2
1

1984

Container Shipping

Think about shipping back in the WWII era. Any consumer item being shipped over the ocean had to be loaded into a ship's cargo hold on pallets using cranes. A group of people known as longshoremen—often 20 or more for a large ship—would oversee and facilitate this process. To load a ship they would move the cargo on board, pack it tightly in the hold, and secure it. Unloading reversed the process.

Containerized shipping, introduced in 1984, reconceptualized the entire international shipping process. In containerized shipping, large, reusable, standardized steel containers are loaded and sealed by the shipper. Sealed containers make their way to the dock by truck or train. At the dock, large crane systems quickly load the containers onto specially designed container ships. Because the containers move freely between ships, trains, and trucks, they are called intermodal containers.

Modern container ships are massive—some of the largest ships in the world—holding thousands of containers per trip. A typical ship might have a **two-stroke diesel engine** capable of 100,000+ hp (80 megawatts). At full power, the ship burns 3,600 gallons/hour (13,600 liters/hour), but at an economical cruise speed it is running at approximately half power. This means that high-efficiency ships might emit 3 grams of CO_2 per ton of cargo per kilometer, making it a green way to move cargo.

The containers themselves are engineering works of art. A typical container is 102 inches (259 cm) high, 96 inches (244 cm) wide, and 40 feet (12 meters) long. It is made of corrugated steel with reinforced corners and corner posts. The corners have holes in them to allow cranes to quickly attach, and connectors to interlock the containers on the ship. A typical container this size weighs about 9,000 pounds (4,000 kg) empty and can take on almost 58,000 pounds (26,000 kg) of cargo.

The introduction of intermodal containerized shipping revolutionized the shipping industry by lowering costs and dramatically increasing the speed of loading and unloading.

SEE ALSO Two-Stroke Diesel Engine (1893).

The cargo ship Inception full of containers in port on July 30, 2012 in Istanbul.

YANG MING
YANG MING
YANG MING
YANG MING
YANG MING
YANG MING
seaco
seaco
ANG MING
38
34
YM INCEPTION
MONROVIA

1984

Itaipu Dam

There are many reasons for engineers to build a dam, but one of the most compelling is for electricity. A big hydroelectric dam can produce a surprising amount of power and it is some of the cleanest and most dependable electricity possible. It is also a very easy form of electrical production to control. Simply adjust a sluice gate and a generator produces more or less power.

The Itaipu Dam, which opened on the border between Brazil and Paraguay in 1984, is the largest hydroelectric facility in the world in terms of power production—it produces nearly 100 terawatt-hours of electricity per year. For comparison, the average American home uses approximately 11,000 kilowatt-hours per year, which means this dam could power about nine million American homes. Since average home power consumption in Brazil is about one sixth of that in America, Itaipu Dam can power over 50 million homes.

How is it possible to harness so much electricity? It starts with a really big river. The Parana River is the seventh largest river in the world. At one time, it also boasted the biggest waterfall in terms of water volume, but this waterfall is now submerged in the lake behind the dam. On average, something like 4.5 million gallons (17,000 cubic meters) per second of water discharges into the river's delta.

That's a lot of water, and at that point of the dam most of the river's water is flowing through twenty giant water turbines connected to generators. Each generator is capable of producing 700 megawatts, which is the equivalent of a good-sized city power plant. In other words, this single dam produces the power of twenty traditional **power plants**. Half of the power is Paraguay's, and half is Brazil's.

The dam itself is enormous—over a mile long and 640 feet (196 meters) high, made of **concrete**. The lake behind the dam is also enormous—520 square miles (1,350 square km). The lake serves as a source of drinking water as well as an important flood-control mechanism. For these reasons Itaipu was named one of the seven wonders of the modern world by the American Society of Civil Engineers.

SEE ALSO Light Water Reactor (1946), Three Gorges Dam (2008).

Aerial view of a pumped storage hydro power station in Germany.

1985

Bath County Pumped Storage

If you are an engineer who is designing a **power grid**, there is one thing you would love to have: a big battery that can store power. The most obvious way to use the battery is to handle peak demand periods during the day. Another reason a big battery would be nice is to make use of idle time. At 3:00 a.m., when power demand is low, the **power plant** could be charging the battery.

Large chemical **batteries** are currently far too expensive. The technology that is used instead is called pumped storage. The basic idea is incredibly simple—connect two lakes at different elevations with pipes. At night, when power demand is low, pump water uphill to the higher lake. During the day when the power is needed, flip the pumps so they act like generators and let the water fall from upper lake to lower lake.

The actual implementation is more complex than that, mainly because of the scale. The Bath County Pumped Storage facility in Virginia, which opened in 1985, is a good example. Engineers built two dams to create two large lakes. The elevation difference between the lakes is about one quarter mile (0.4 km). The top lake holds about 12 billion gallons (45 billion liters). There are six pumps/turbines that are fed by pipes 18 feet (5 meters) in diameter. Each pump/turbine can produce 500 megawatts with a water flow rate of 37,500 gallons per second. That means that at the full flow rate, producing three gigawatts of power from all six turbines, it would take about 14 hours to drain the upper lake dry. Then that water can be pushed uphill that night to recharge the battery.

Pumped storage is one of the best "big battery" technologies available for power grids today. But it requires the ability to build two large lakes at two different elevations that are close together, so it does not work everywhere. In the places where the geography and water sources allow it, it is a great option.

SEE ALSO Power Grid (1878), Light Water Reactor (1946), Lithium Ion Battery (1991).

A panoramic view of the pump storage hydrostation Hohenwarte along the River Saale near Saalfeld, Germany, July 11, 2003.

RWE

1985

International Thermonuclear Experimental Reactor (ITER)

For decades we have heard that fusion power will save the world. All that engineers have to do is find a reasonable, inexpensive, controlled way to fuse hydrogen atoms together to create helium atoms, in the same way the sun does it. This fusion process would create a clean form of energy that could power the world without the risk of nuclear accidents, oil spills, and greenhouse gases.

Engineers have found one way to create hydrogen fusion on Earth: the **thermonuclear bomb**. The big problem is that it is not controlled. It also requires a **conventional nuclear bomb** to start the fusion process, which can be messy.

With the International Thermonuclear Experimental Reactor, engineers are trying to work out the construction and operation of a practical fusion reactor. Work began in 1985.

The main cavity of the ITER reactor is shaped like a donut (torus). The basic idea is that a cloud of hydrogen gas will convert to plasma while a powerful magnetic field confines it to a thin, dense ring in the middle of the donut. By running a current through the plasma, it will heat to incredibly high temperatures (100+ million degrees K). If the ring is dense enough and hot enough, the hydrogen nuclei will fuse to form helium atoms. The best kinds of hydrogen atoms to use are deuterium (one proton and one neutron in the nucleus) and tritium (one proton and two neutrons). They produce the most energy, primarily in the form of high-energy neutrons that fly out during the fusion process. Trapping the energy of these flying neutrons and converting it to heat is how the reactor will produce electricity.

Although the principles are straightforward, engineers working on ITER face a host of problems. The giant torus needs to maintain a nearly complete vacuum. The torus walls need to withstand the immense heat. Magnets this powerful are difficult to create and require helium cooling. The flying neutrons may destroy the reactor over time. And so on. It is not clear that ITER will work. For the engineers involved in the project, that uncertainty may add to the fun.

SEE ALSO Trinity Nuclear Bomb (1945), Light Water Reactor (1946), Ivy Mike Hydrogen Bomb (1952), Large Hadron Collider (1998).

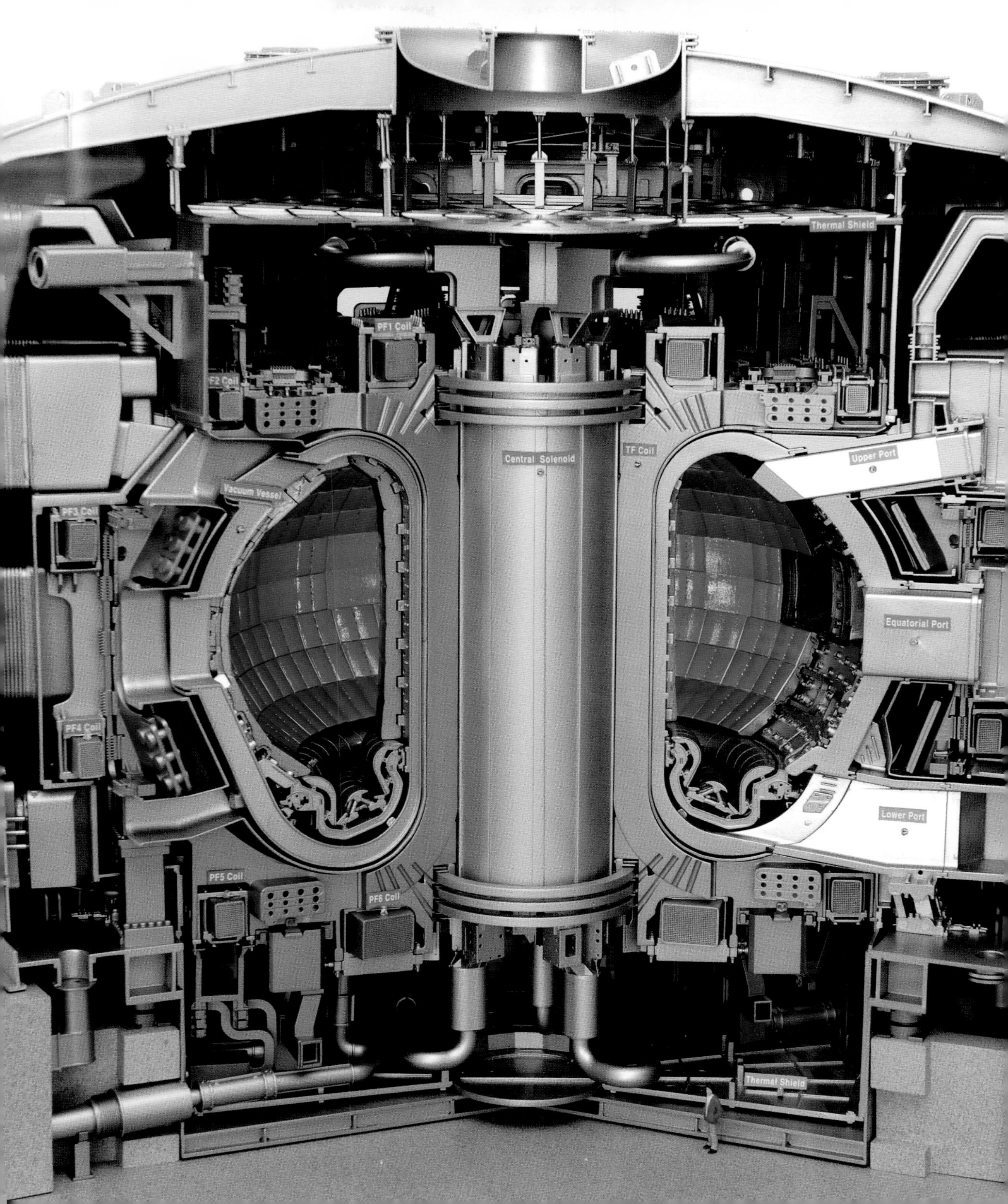

Thermal Shield
PF1 Coil
F2 Coil
Central Solenoid
TF Coil
Upper Port
Vacuum Vessel
PF3 Coil
Equatorial Port
PF4 Coil
Lower Port
PF5 Coil
PF6 Coil
Thermal Shield

1985

Virtual Reality

Jaron Lanier (b. 1960)

Engineers and the engineering mindset have been helping people cope with the real world for centuries. But virtual reality promises to let us experience artificial worlds as if they were real.

The technology needed to create a virtual reality experience has four parts. First, you need a headset that can display binocular **three-dimensional** images into your eyes. Second, you need a way to track head movement accurately and responsively. Third, you need a piece of software that can create the artificial world that you will be experiencing. Fourth, you need a way to move around in that world—it could be as simple as a video game controller or as complex as an omnidirectional treadmill.

When virtual reality started in 1985, all of these components were fairly crude. Computer scientist Jaron Lanier had popularized the term and his company, VPL Research, was working on producing virtual reality products for commercial use. What has been happening since is a set of big engineering improvements in Lanier's earlier inventions. The headsets have better resolution, a wider field of view, and a peripheral view. This means that the images fill your entire visual field and it makes the experience more immersive. Head tracking accuracy and speed have also improved. So when you move your head, the display responds instantly and exactly. This again improves immersion. And software engineers in the gaming industry have used ever-increasing CPU and GPU power to create hyperrealistic artificial worlds to feed into the headset at a full 60 frames per second. And the cost of everything has come down significantly.

Putting all of these advancements together, the virtual reality experience is improving all the time. Gaming will be transformed, but there are also things people will be able to do virtually that would have previously been impossible. Have you ever wondered why the visit to a mall, museum, show room, or tourist attraction feels so much richer than a video of the same thing? Soon you'll be able to get an immersive visual experience or move your head to see what you want to see. New virtual reality systems will fix that and will make virtual experiences seem nearly identical to the real thing.

SEE ALSO 3D Glasses (1953), *Doom* Engine (1993), *Toy Story* Animated Movie (1995).

Astronauts use virtual reality hardware to rehearse some of their duties on the upcoming mission to the International Space Station.

VRLab
MotionStar

1986

Chernobyl

The Chernobyl explosion is one of the most spectacular, apocalyptic engineering failures of all time, made possible by a combination of poor design choices and incorrect operation. It demonstrates how badly things can go wrong when engineers make mistakes.

There were actually three mistakes that worked together to cause the Chernobyl explosion. The first mistake was the way engineers used water in the reactor. They needed water to form steam, because steam is the medium that extracts the heat energy from the reactor so it can become electricity through a **steam turbine**. The problem is that liquid water absorbs neutrons much better than steam does. So if operators cool down the reactor, its core contains mostly water. If operators then heat up the reactor improperly and water flashes to steam, a power surge can occur. The quick conversion of water to steam causes a quick increase in neutrons—a positive feedback loop.

The second mistake occurred in the design of the control rods. A control rod is supposed to absorb neutrons. Unfortunately, the tips of the Chernobyl control rods were made of graphite. Therefore, as the control rods went into the reactor, the tips displaced water and led to another type of power surge.

And third, the Chernobyl reactor did not have a containment building, so when the explosion occurred there was nothing to contain the contamination.

The accident unfolded in this way: On April 26, 1986, operators improperly cooled down the core. When they started to heat it back up, water flashed to steam, creating a power surge. The control rods were inserted, with the graphite tips creating a catastrophic power surge. Fuel rods burst, jamming the control rods. A steam explosion blew open the core, letting in oxygen and starting a fire that pumped nuclear material into the air. A second explosion, possibly a small nuclear explosion that occurred as enough melting fuel consolidated, compounded the release of nuclear material.

Millions of acres of land received dangerous levels of fallout, and most of Europe received some amount of fallout from the explosion. The design decisions of a small number of engineers and the operating errors of a small number of operators affected millions of people.

SEE ALSO Power Grid (1878), Light Water Reactor (1946), Pebble Bed Nuclear Reactor (1966), CANDU Reactor (1971), Fukushima Disaster (2011).

The Chernobyl disaster was due to human error and issues relating to construction.

1986

Apache Helicopter

We tend to think of **helicopters** as fairly fragile. Generally speaking, helicopters are light, thin, and spindly.

But what if engineers want to create the deadliest helicopter in the world—one that can deliver amazing firepower and at the same time provide a level of protection for the crew? That is what engineers did with the Apache helicopter, released in 1986.

One key thing that distinguishes the Apache is its weaponry. It carries three weapon systems: Hellfire antitank missiles (up to 16), Hydra rockets (up to 76), and a 30mm chain gun shooting explosive rounds (up to 1,200 rounds at 600 rounds per minute).

One unique feature is a system that slaves the chain gun to the gunner's helmet. Sensors watch the movements of the gunner's head. The gun follows precisely. When the gunner turns to look at a threat, the gun follows and a squeeze of the trigger takes out the target. Another is the Longbow **radar** system, mounted in a dome over the main rotor. It can track up to 128 targets simultaneously.

The Apache's other distinguishing feature is survivability. It is meant to fly on the battlefield, often close to the ground, to support ground troops. That means the Apache is frequently a target. **Titanium**, **Kevlar**, and bulletproof glass protect the pilot and gunner. There is also similar protection for the engines and fuel tanks. And the helicopter blades can handle bullets up to 23mm in diameter.

If the helicopter does need to crash, this process has been thought through as well. The first part to hit the ground is the landing gear. It absorbs the initial blow of the impact. Then the area under the pilot and copilot crumples and collapses. This approach greatly improves the survival rate of the people inside.

A helicopter that carries all of this weaponry and protection is going to be heavy—almost 18,000 pounds (8,200 kg) fully loaded. To get that much mass off the ground, two turboshaft engines together producing nearly 3 megawatts (4,000 hp) provide the power.

Engineers have made the Apache helicopter so powerful and useful on the battlefield that the Army owns over 1,000 of them.

SEE ALSO Radar (1940), Titanium (1940), Helicopter (1944), Kevlar (1971).

A US Army AH-64 Apache attack helicopter prepares to depart Bagram Air Field, Afghanistan, 2012.

17E

1990

World Wide Web

Robert Cailliau (b. 1947), **Tim Berners-Lee** (b. 1955)

In the mid-1980s, the Internet existed, and people were using it. The number of host computers connected to the Internet in 1987 was about 10,000. However, nearly every person using the Internet at that time was affiliated with the universities, companies, and research organizations that provided the host computers. The public had no access.

At this time, people were using a variety of **Internet** tools to move information around. E-mail and FTP (File Transfer Protocol) were two of the most common. A person could upload a file to a FTP server and then send e-mail to people telling them that they could download the file. People could connect to computers remotely using Telnet. It all worked, but the Internet was a bit technical and cumbersome.

Then, in 1990, everything began to change, when British computer scientist Tim Berners-Lee and Belgian computer scientist Robert Cailliau, proposed a "hypertext project" for the "WorldWideWeb." The World Wide Web was born, and it made the Internet incredibly easy to use as an information tool. On the one hand it was so simple, but on the other hand it was so incredibly powerful. As a result, the web has changed so many things, including the way goods are bought and sold, the way news and information are delivered, the way we educate people, the way people communicate. In addition, it utterly leveled the playing field. Suddenly, anyone could publish information to millions of people.

There were four core ideas that had to be engineered simultaneously for the web to work: 1) the web server, which holds web pages for people to access, 2) the web browser, which can gather and assemble web pages from servers so people can view them, 3) the web markup language, called HTML, which allows people to create web pages, and 4) the web protocol, named HTTP, which allows for communication between server and browser. Once a web server existed with a web page in HTML on it, and someone had a web browser, the web was born. And then it spread like wildfire because engineers made accessing the Internet trivially easy.

SEE ALSO ENIAC—The First Digital Computer (1946), ARPANET (1969), Domain Name Service (DNS) (1984).

The World Wide Web continues to change society as we know it today.

http://www

1990

The Hubble Space Telescope

What if you want to build a telescope that is not affected by the Earth's atmosphere? The main reason for doing this is because the air that surrounds Earth blocks certain frequencies of light. Another reason is because changes in the atmosphere caused by wind and temperature differences distort the light that comes through, creating the twinkling effect we see in the night sky. It would also be nice to have a telescope that is not subject to light pollution and can see parts of the sky 24 hours a day.

The easy solution to these problems is to put a telescope in space. That sounds simple enough until you actually try to do it, however—then it becomes a massive engineering project.

The first problem is the mirror. Astronomers want it to be as big as possible, but there are limits based on size and weight restrictions in launch vehicles. Engineers chose a 94-inch (2.4 meter) mirror made of glass ground to a precision shape, slightly smaller than the mirror used in the **Hooker telescope**. A "Cassegrain" design reflects the light off the main mirror, to a smaller secondary mirror and back through a hole in the center of the main mirror to the Hubble's cameras. The different cameras can see infrared light, ultraviolet light, and visible light.

Engineers then had to package this telescope for space flight, like any other **satellite**. There are the solar panels and **batteries** of the electrical system, the antennas and radios of the communication system, the rate gyros and thrusters of the pointing system—especially important so the telescope can point precisely at targets for long periods of time.

In the final instrument, everything works amazingly well. For example, NASA's Hubble telescope was launched in 1990. It has precisely focused on one small section of space for more than a hundred hours to form the famous Hubble deep-field image. The image is mind boggling because it shows, for the first time, just how many galaxies are visible—thousands of them—in the tiniest patch of space.

SEE ALSO Hooker Telescope (1917), Space Satellite (1957), Lithium Ion Battery (1991).

An STS-125 crew member aboard the Space Shuttle Atlantis *captured this still image of the Hubble Space Telescope as the two spacecraft continue their relative separation on May 19, 2009, after having been linked together for over a week.*

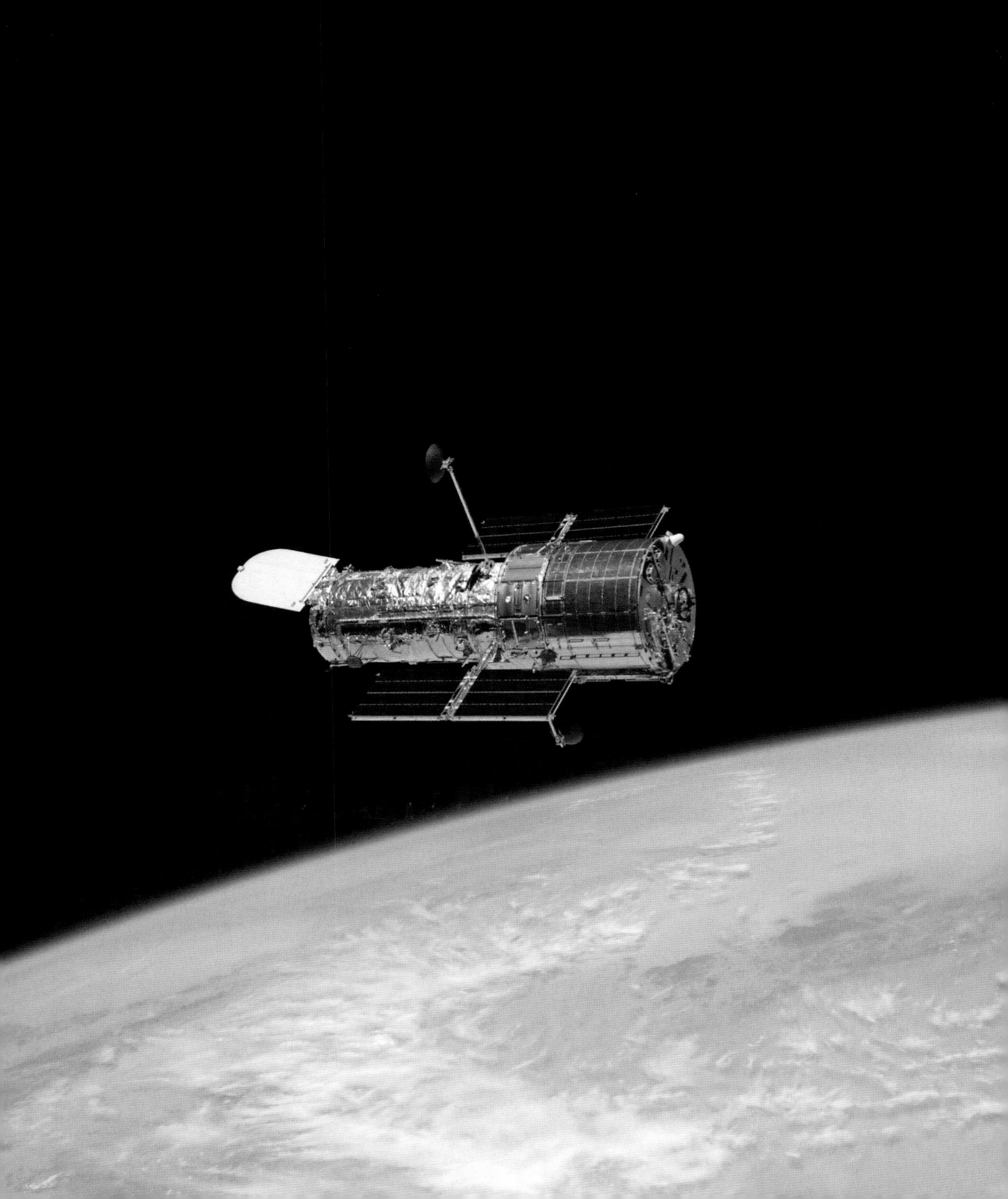

1991

Lithium Ion Battery

Everyone would like to have better batteries. Wouldn't it be great to have a **smart phone** that you recharge once a month, and the recharging time is less than a minute? Wouldn't it be great to have an **electric car** that can go 1,000 miles (1,600 km) on a charge and then takes only a few minutes to recharge? Wouldn't it be great to have a battery the size of a loaf of bread that could run an entire house for a day or two in case of a power failure? What about an electric jet airplane that can fly around the world on a single charge? It is easy to imagine batteries like this, but will we ever see them?

When we think of a battery, what we usually have in mind is a container full of chemicals. Chemical reactions inside the container produce electrons. In a disposable battery, the reaction happens once. In a rechargeable battery, the reaction is reversible.

Scientists and engineers have tried many different chemistries to create batteries: The first widely successful chemistry appeared in the crow's foot or gravity battery. It used copper and zinc electrodes and copper sulfate crystals in water. This battery powered the early **telegraph network** starting in the 1860s. Around the same time the rechargeable lead-acid battery first appeared. We still use this chemistry in just about every car battery today. Then came nickel-cadmium, nickel-metal hydride, and lithium ion chemistries. Lithium ion is the best battery technology available today in terms of energy density.

But the lithium ion battery of today, first commercially produced by Sony and Asahi Kasei in 1991, is only three times better than the lead acid battery in terms of its power-to-weight ratio and its power-to-size ratio. When you consider that scientists and engineers have been working for over a century on the problem, very little improvement has been seen compared to other technologies. If you compare a **747** to the **Wright Brothers' airplane**, or **ENIAC** to the computer inside a smart phone, you see that the rate of improvement in other engineering endeavors has been much more fruitful.

SEE ALSO Telegraph System (1837), The Wright Brothers' Airplane (1903), ENIAC—The First Digital Computer (1946), Boeing 747 Jumbo Jet (1968), Prius Hybrid Car (1997), Smart Phone (2007).

Batteries loaded in a row on an electric car.

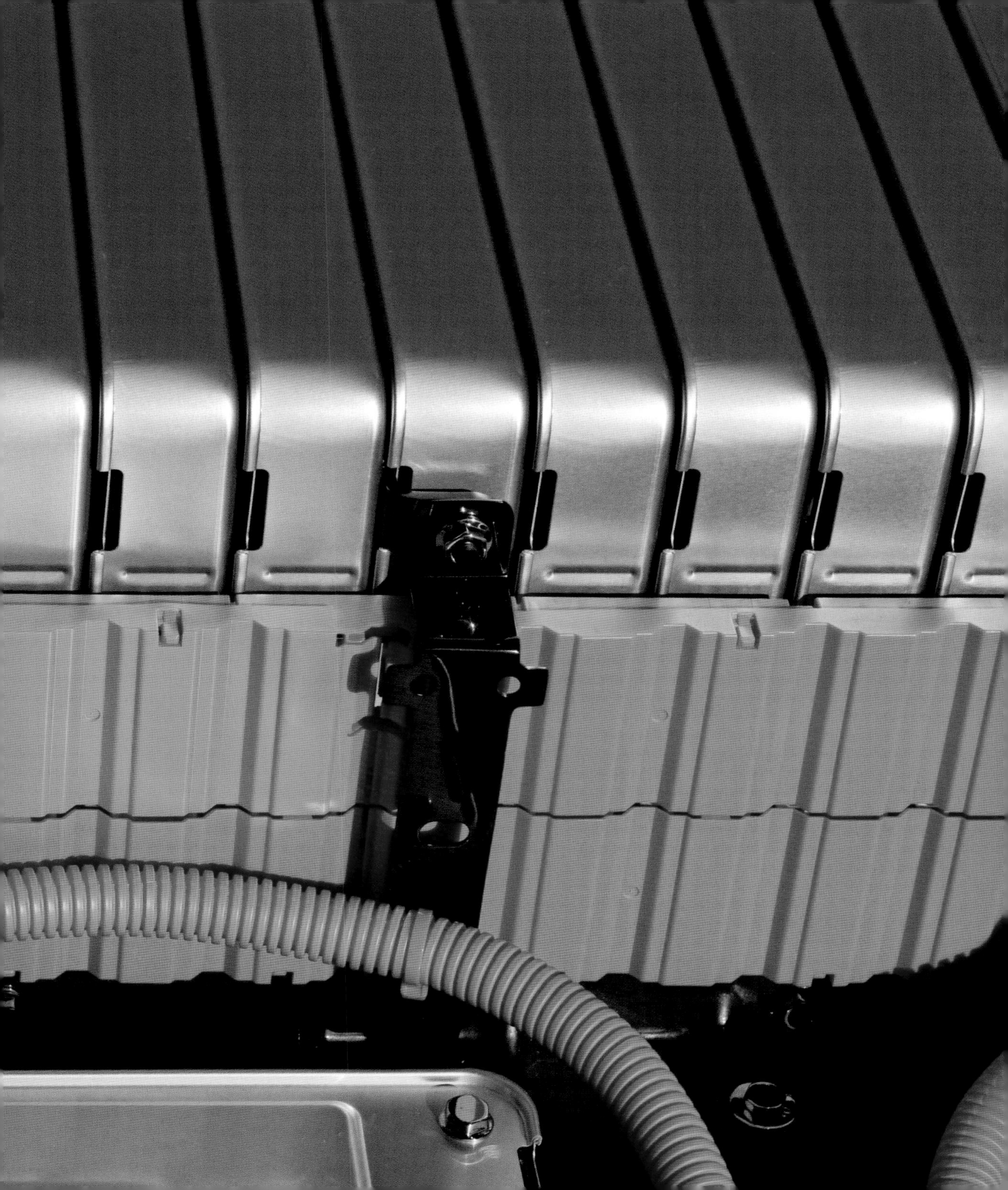

1991

Biosphere 2

John P. Allen (b. 1929)

Imagine what would happen if we decided to put a long-term, self-sustaining colony on the moon or **Mars**. Or if we wanted to create a spaceship, using current technology, that would fly to another star over the course of many years. To accomplish these goals, we would need to engineer an enclosed, self-sustaining biosphere—a sealed terrarium of sorts big enough to hold human beings. It would contain the food, water, and oxygen the people need, and also recycle all of the waste products. And the system would need to be able to run this cycle reliably for many years without significant problems—otherwise, the people inside would die.

Is a self-contained biosphere like this even possible? In one of the most fascinating experiments ever performed, engineer John P. Allen actually created a complete biosphere and ran it for two years. This large-scale science project, conducted by the University of Arizona and Space Biosphere Ventures, is called Biosphere 2, and the two-year run started in 1991.

First, a huge sealed container with a floorplan 3.15 acres in size and a glass enclosure high enough to hold a small rainforest was created. The structure is essentially a big ship, with the steel hull buried in the ground and the massive glass superstructure visible. Specialized systems were then designed to handle water and air.

Many things learned from the experiment have applicability in a real-life scenario. For example, CO_2 levels fluctuated wildly depending on time of day and season. In daylight the plants would suck up the CO_2 and then at night levels would rise. Levels got so high at one point that the crew grew and stored plant material to sequester carbon.

The crew of eight had trouble growing enough food to eat in year one. Everyone lost weight, but became healthier in the process.

Oxygen levels presented the biggest problem because O_2 was mysteriously disappearing from the air. This turned out to be caused by a reaction between concrete and CO_2, taking oxygen out of circulation. Eventually fresh oxygen supplies had to be injected for the crew's safety.

SEE ALSO Water Treatment (1854), Modern Sewer System (1859), Drip Irrigation (1964), International Space Station (1998), Mars Colony (c. 2030).

Biosphere 2 library and living quarters in Arizona, USA.

1992

Low-Flow Toilet

Think about the ordinary toilet that you find in a typical home. It is a low-cost device that uses gravity as its sole power source, yet it is highly reliable, virtually indestructible, easy and inexpensive to maintain, and it solves a huge sanitation problem.

Each time you use a typical home toilet you press the flush lever. This pulls on a chain that opens a small **rubber** valve at the bottom of the tank. This valve costs four dollars and needs replacing perhaps every five years. With the valve open, all of the water in the tank rushes into the bowl in a few seconds.

Because of the design of the bowl and the passageway leading out of it, the inrush of water causes a siphon to form in the passageway. The siphon efficiently sucks everything out of the bowl.

Meanwhile, with the tank empty, the flapper valve has closed over the opening in the tank. A flotation valve has realized that the tank is empty, starting the flow of water into the tank and the bowl so that both refill. When the tank is full, the flotation valve will shut off the water automatically and the toilet is ready for its next use.

Think about the alternatives. You could be using an outhouse. Or you could use a bucket indoors and then empty it into a human waste compost pile in the backyard. Which actually is not the worst idea environmentally, provided you have the land and the patience to handle it. But in a thriving city, both of those options are non-starters. The flush toilet is a major society-wide convenience brought to us by engineers.

In 1992, the United States went through an interesting transition. The old standard was 3.5 gallons per flush. To decrease water consumption, the US mandated a 1.6-gallon flush. Engineers had to create bowl designs and flow patterns that would empty the bowl, without clogging, on much less water. They rose to the challenge. This one small change now saves the US billions of gallons of water per day.

SEE ALSO Pompeii (79), Plastic (1856), Modern Sewer System (1859).

In 1992, toilets in the United States changed their standards from 3.5 gallons per flush to 1.6.

1992

Stormwater Management

Think about what happens as a city or suburb develops. Prior to the arrival of people, the land is likely forested and absorbs much of the rain that falls. Between the leaf litter, the friable soil under it, the dips in the terrain, etc., runoff is minimal. Creeks and rivers easily handle the runoff that does occur.

To build the city, developers cut down the trees and construct buildings. They also build parking lots, sidewalks, and roadways. All of these are impermeable surfaces like **asphalt** or **concrete**. When rainfall occurs on an impermeable surface, none of it gets absorbed or even slowed—it all runs off immediately.

In a dense urban area filled with impermeable surfaces, the effect can be devastating if there is a burst of rainfall. Imagine a city covering 100 square miles (260 square km). If an inch of rain falls and the entire area is impermeable, that's 401,448,960,000 cubic inches of water, or 1.7 billion gallons, or 6.6 billion liters. If engineers don't do anything about it, that water all immediately flows into the nearby river in an hour or two. If six inches of rain falls, as in a hurricane, 10 billion gallons (38 billion liters) ends up in the river, and there is no way the river can handle the pulse.

Therefore, in most urban and suburban areas, the idea of stormwater management has become a major aspect of engineering during development. For example, every subdivision, shopping center, and mall in a suburban area will typically now have a pond area that takes the runoff from roofs and parking lots, and stores it for slow release into local streams and rivers, or absorption into the ground. This approach mimics what used to happen when the land was forested.

In cities the same kind of thing happens, but often in massive underground vaults that catch and hold the stormwater. One vault in Saitama, Japan, built in 1992, measures 225 feet (78 meters) wide by 580 feet (177 meters) long by 83 feet (25 meters) tall.

Without stormwater management, flooding becomes a huge problem. With it, engineers have a great deal more control over flooding.

SEE ALSO Concrete (1440 AAA), Asphalt (625 AAA), Pompeii (79), Bath County Pumped Storage (1985), Venice Flood System (2016).

The underground stormwater system in Japan is awe-inspiringly huge.

1993

Keck Telescope

Jerry Nelson (b. 1944)

The 100-inch (2.5 meter) **Hooker telescope** reigned as the world's largest telescope for three decades, until the 200-inch (5 meter) Hale telescope on Mt. Palomar finally surpassed it. The Hale telescope was the world's largest for nearly three decades more until surpassed by the Keck telescope, which began to make observations in 1993. And to do it, engineers working on the Keck telescope had to completely reconceptualize the mirror at the core of the machine. The problem with the Hale telescope is that it uses a single piece of glass to form the mirror. Even though the back of this mirror is honeycombed to save some weight, the disk is incredibly large, heavy, and problematic. Making an even bigger mirror from a single piece of glass will probably never happen.

So how do engineers make bigger, precisely curved mirrors? The Keck telescope solves this problem by being made of 36 smaller hexagonal mirrors, designed by astrophysicist Jerry Nelson, that fit together to form one immense 10-meter (400 inch) disk. In addition, the curve of each hexagon can change at a nanometer level multiple times per second in order to accomplish something called active optics. The purpose of active optics is to give a mirror the correct shape despite the effects of factors like gravity, temperature changes, and mirror movement. Mechanisms on the back of the Keck's mirror segments change their shape.

This active system forms a perfectly shaped mirror and combines with an adaptive system to handle changes in the atmosphere over the telescope. The adaptive system looks at the twinkling of a bright star and uses it to adjust a mirror and remove the twinkling. If a bright star is not visible in the field of view, it is also possible to use a laser to create an artificial star in the upper atmosphere to serve the same purpose.

In 1996, a second Keck telescope was built, nearly identical to the first. The two can work independently. Precision engineering allows the light from the twin mirrors to combine to create an even bigger telescope. The combination gives the Keck telescope surprising power, and it has acted as a template for even larger telescope designs.

SEE ALSO Hooker Telescope (1917), The Hubble Space Telescope (1990).

Keck telescopes on the summit of Mauna Kea, Hawaii.

1993

Doom Engine

John D. Carmack (b. 1970)

In 1972, the game *Pong* appeared and for the first time people could see and play a video game in public. *Space Invaders* came in 1978, and then *PacMan* in 1980. All of these games are incredibly simple by today's standards: 2D sprite graphics with a few colors.

But then in 1993, with the release of a game called *Doom*, things changed dramatically. *Doom* was the first FPS (First Person Shooter) game to be played by millions of people, and it made immersive, realistic-feeling **3D** environments a desirable feature. John D. Carmack's "*Doom* Engine"—the first 3D game engine—made a huge impact. Since then, video games have improved in every way—more realistic, more detailed, faster, with extremely large worlds for players to explore.

From an engineering perspective, the advancements that made this all possible were faster CPUs, and then in 1999 the development of the first Graphics Processing Units (GPUs) to accelerate the 3D rendering. Since then, the power of the GPU has exploded.

The key concept in a GPU—the shader: small programs that manipulate vertexes or pixels. The GPU's power is measured in the number of cores it has for simultaneously executing shaders, along with the clock rate, memory available to them, and memory bandwidth. GPUs in desktop machines started with 20 million **transistors** in 1999 supporting one vertex shading core. Today there are billions of transistors supporting thousands of cores and gigabytes of memory.

Several engineering disciplines have worked together to make this possible. The increase in the number of transistors has been made possible by the engineers working in chip fabrication. The hardware engineers working for GPU companies have radically advanced their chips to take advantage of the transistors. Software engineers have created standards like OpenGL to make GPU access easier. And then the game developers create the games using the standards.

All of this GPU power has been transformative. **Automotive racing** games are approaching a level where they are indistinguishable from reality. FPS games have realistic worlds so large that they seem infinite. Engineers are creating artificial reality.

SEE ALSO Formula One Car (1938), Transistor (1947), 3D Glasses (1952), *Toy Story* Animated Movie (1995).

Two young people playing Doom *computer game on PlayStation video game system by Sony.*

WINS:00
55
WINS:00
CYRAX
SONY

1994

Channel Tunnel

Before the Channel Tunnel opened in 1994, the only way to get from Great Britain to France was by boat. **Airplanes** have been a possibility during the last century, but they don't really provide any advantages over boats. Engineers could build a **bridge**, but a bridge of that length at that location is problematic.

And so engineers chose a tunnel—the longest undersea tunnel ever at 31 miles (50 km) long. And it is not just one tunnel—it is actually three. There are two train tunnels to allow bi-directional travel and a service tunnel between them.

The original proposal for the construction was put forward by French mining engineer Albert Mathieu in 1899, but a detailed geological survey wasn't carried out until 1964. Construction commenced that year. As a result of this government-funded effort, engineers picked a chalk layer that offered strength but relatively easy digging. They used tunnel-boring machines, and started with the service tunnel because it is smaller: 16 feet (4.8 meters) in diameter. The two train tunnels are 25 feet (7.6 meters).

To speed up the drilling process, **tunnel-boring machines** started at both ends and headed toward the middle. This raises two obvious questions: How did engineers make sure they would exactly align, and what happened when the boring machines ran into each other? The first problem is particularly complicated. The tunnel changes angles both horizontally and vertically along its run because of the geology. The primary solution was old-fashioned surveying. The boring machines also used **laser**-controlled guidance systems. A large laser was bolted to the ground in the existing tunnel and aimed at a target at the back of the boring machine. In this way, the machine could stay aligned with its exact path. After a period of progress, the laser would move forward in the tunnel.

To keep the boring machines from colliding, the machines that started on the English side were pointed down into the earth as they approached the midpoint. That got them out of the way and entombed them. The French machines advanced to finish the job, and then they were dismantled.

Brilliant engineering solved a huge transportation problem. Twenty million people and twenty million tons of freight use it every year.

SEE ALSO Tunnel Boring Machine (1845), The Wright Brothers' Airplane (1903), Laser (1917), Kinsol Trestle Bridge (1920).

Boring machine, used to carve out rock to construct the Channel Tunnel.

1994

Digital Camera

If you think about a traditional film camera, it can be a pretty simple device. A pinhole camera is a sealed box with a piece of film inside and a tiny pinhole in the box to let in the light that forms the image.

A digital camera, on the other hand, is a combination of electrical engineering, computer engineering, and software engineering.

The world's first consumer digital camera is arguably the QuickTake, introduced by Apple Computers in 1994. And this was very probably the first year that engineers could have produced it at anything like a reasonable price. It was incredibly simple by today's standards, but had all of the essential elements. A CCD (charge-coupled device) sensor recorded a color image. An internal **microprocessor** could read the image off the sensor and store it in internal **flash memory**, which was just appearing in the marketplace. The camera could take 640 x 480-pixel color images, and the flash memory could hold eight of them. The camera's computer also operated a small liquid crystal display (**LCD**) and managed a serial interface to read out the images.

The idea of memory cards for cameras started later in 1994 with the CompactFlash standard, and really took off in 1999 with the introduction of much smaller SD cards.

The key element is the image sensor, either a CCD or CMOS (complementary metal-oxide semiconductor) sensor. Each pixel on the sensor can measure light intensity. Engineers use one of two systems to capture color information. The less expensive and therefore more common option is a bayer filter mosaic over the pixels on a single sensor. Every other pixel has a tiny green filter over it. The remaining pixels get red or blue filters. The internal computer interpolates to arrive at an RGB value for each pixel in the image. The other way is to split the incoming light from the lens into three beams. One beam goes through a green filter, the second through a red filter, and the third through a blue filter. These three beams hit three separate sensors, and the computer creates RGB values from them.

Since this first camera, engineers have radically improved every part of the digital camera. Image sensors now have tens of millions of pixels. Memory cards are measured in gigabytes. The internal computers and software can do amazing things.

SEE ALSO LCD Screen (1970), Microprocessor (1971), Flash Memory (1980), Smart Phone (2007).

A woman at a computer viewing a photograph of the Golden Gate Bridge taken with Apple's QuickTake 100 camera.

1994

Global Positioning System (GPS)

Ivan A. Getting (1912–2003), **Roger L. Easton** (1912–2014), **Bradford Parkinson** (b. 1935)

If you were an engineer working on the Global Positioning System (GPS), you were doing something incredible. You were proposing the creation of a new sense that could become available to every human being on the planet. Humans come equipped with the normal senses: vision, hearing, smell, taste, and touch. But humans definitely do not come equipped with a sense of direction, especially at night, especially on the open oceans, especially in bad weather where clouds, fog, and rain obscure every landmark.

The GPS engineers, a team including Ivan Getting, Roger Easton, and Bradford Parkinson, who were working for the United States Department of Defense, proposed to change all of that. By 1994, they had created a ubiquitous, instantaneous, and precise system by which any human could locate his or her exact position on the planet with roughly 30-foot (10 meter) accuracy, anytime, anywhere.

One of the most audacious parts of the proposed system would be the cost—approximately $12 billion for a constellation of 24 **satellites** funded by the US military that went into orbit between 1989 and 1994. Another audacious part was the technology. This new GPS system demanded small, accurate **atomic clocks** that could operate unattended for years in orbit—two of them per satellite. These clocks are not simple devices. And then there was the technique the engineers devised to determine location. A GPS receiver would need to be able to see at least four satellites overhead, know exactly where each is in orbit, and then determine exactly how far away each one is. Using the distance and location of the four satellites, the receiver could triangulate its exact position and altitude on earth. It could also derive the exact time with atomic clock accuracy without needing to have its own **atomic clock**.

Here on Earth, the arrival of the cheap consumer GPS receiver combined with the arrival of the cheap, pocket **cell phone** to make it seem like we were living in the future. It is an amazing pair of capabilities for any human to have.

SEE ALSO Atomic Clock (1949), Space Satellite (1957), Mobile Phone (1983), Self-Driving Car (2011).

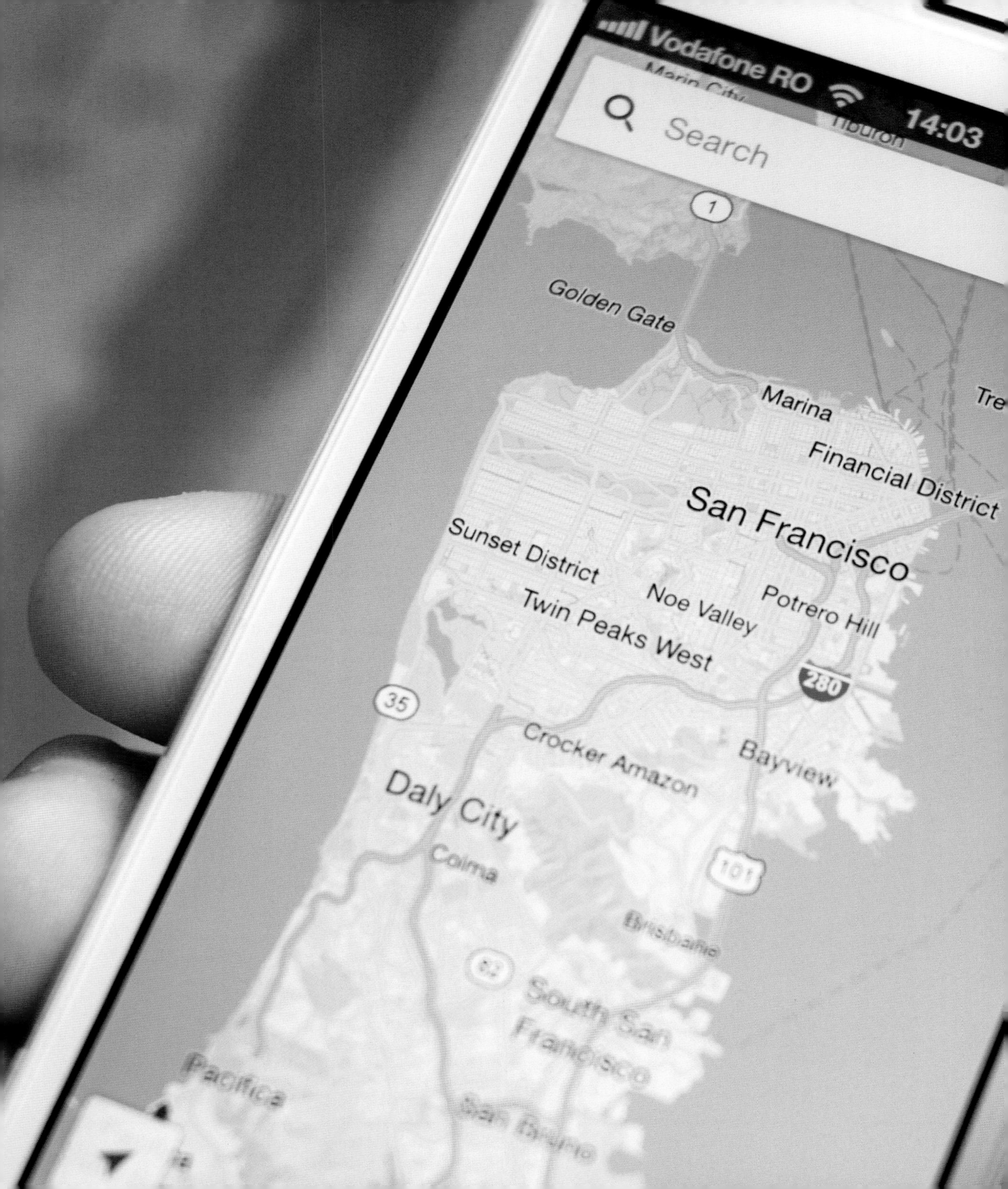

Vodafone RO
14:03
Search
Golden Gate
Marina
Financial District
San Francisco
Sunset District
Noe Valley
Potrero Hill
Twin Peaks West
Crocker Amazon
Bayview
Daly City
Colma
280
35
101
1

1994

Kansai International Airport

Imagine that a densely packed city in Japan needs a new airport. The problem is that the dense packing leaves no space for an airport. One solution might be the Dulles Airport approach used near Washington, DC: build the airport many miles out in the countryside. The problem with that is the inconvenience.

So the engineers in Japan chose a radically different path. With no land available, they decided to make their own by building a massive **artificial island** in Osaka Bay.

If you stand on the island, it seems impossible that it is artificial. It's huge, measuring 2.5 miles (4 km) by 1.6 miles (2.5 km) in total extent. Look at Central Park in Manhattan, which is itself massive. Now multiply by three. That's the size of this island. And then they built it over two miles (3 km) offshore to cut down on noise.

How do engineers build an island like this, in an area on the Ring of Fire well-known for earthquakes? Construction started with a sea wall around the perimeter. Then piles were driven into the clay under the island to stabilize it. Then they started dismantling nearby mountains to fill in the island. The water is about 100 feet (30 meters) deep, meaning they needed 27 million cubic yards (23 million cubic meters) of fill. Bringing in all the fill by barge, it took years to displace the sea.

Engineers didn't stop there. They also built the world's longest airport concourse, a train system, and an impressive bridge supporting the train track and highway to connect the island to the land. But one of the most interesting problems they had to solve was the settling. How do you build an incredibly long building on a new island that will definitely settle, and keep the building level? Engineers put a hydraulic cylinder under each column so they could raise each one separately by exactly the right amount.

Kansai International Airport, which opened September 4, 1994, is a great example of audacious engineering and the ability to foresee, mitigate, and solve the myriad problems that the audacity caused. It's one of the things that makes modern engineering so fascinating.

SEE ALSO Leaning Tower of Pisa (1372), Palm Islands (2006), Earthquake-Safe Buildings (2009).

High-angle view of Kansai International Airport.

1995

Toy Story Animated Movie

The year 1995 was impressive for engineers working in the movie industry. The film *Jumanji* featured a herd of animals running through town, and all of those animals were computer-generated. The movie *Casper* had the world's first computer-generated main character. The movie *Babe* used seamless and realistic computer-generated effects around each animal's mouth to make it look like the animals could talk.

And then there was the film *Toy Story*—the world's first completely computer-generated feature-length film. Every single image seen in the movie—all of the characters, all of the sets, all of the props, all of the scenery—came from **3D** artists and animators working with computer software.

This was Pixar's first feature-length film, and the production involved the collaboration of hundreds of people. Software engineers, network engineers, and hardware engineers played key roles. The software engineers wrote the code that made it possible for the animators to create the movie. Hardware engineers and network engineers built the server farm and storage systems that would render the movie frame by frame and put all of those frames together.

The amount of computing power necessary to make a film like this is staggering, and it keeps growing. Something like *Monsters Inc.*, in which a main character is covered in fur and the software has to compute the motion of the fur frame by frame, is a great example. *Toy Story*, even with its simplicity, took hours to compute each 300-megabyte frame of the movie.

How do people create movies like *Toy Story*? One part of the process involves the characters created using PhotoRealistic RenderMan, which is a program produced by Pixar to render all of their 3D animated movies. A character like Woody or Buzz Lightyear has an internal skeleton. Each piece of the skeleton has sliders that control its exact movement. The face has dozens of sliders to control lips, eyes, and eyebrows. The animator manipulates the sliders for each frame to make the character move in exactly the right way. Then the character appears in a digital scene with other characters and objects. **Virtual** lights and a virtual camera give the director complete control.

Engineers create an imaginary universe populated with imaginary characters, and with them filmmakers bring amazing stories to life.

SEE ALSO 3D Glasses (1952), Virtual Reality (1985), *Doom* Engine (1993).

Toy Story *represented a number of major software engineering innovations.*

LIGHTYEAR

1996

Ariel Atom

Niki Smart (b. 1973)

Jeremy Clarkson of the television show *Top Gear* describes the Ariel Atom this way: "I think this is one of the most beautiful cars in the world, partly because it is so elegant, and partly because it is such a wonderful piece of engineering." Why is it a wonderful piece of engineering? What does that phrase mean? It can mean that the pieces of metal and plastic have been put together in such a way that they beautifully accomplish the task. It can mean that the whole is greater than the sum of the parts. It can mean a minimalism that achieves the goal as efficiently as possible. It can mean an assembly that surprises people with its creativity and sophistication. In the case of the Ariel Atom, it is all of these things.

The Ariel Atom is produced by TMI Autotech, and its original design came from a British design student named Niki Smart. The Atom contains the essential essence of a car but not one thing more. It uses its exoskeleton frame not only as the core to which everything attaches, and as a truss to bridge front wheels to back, but also as a cage to protect the driver. That kind of multiuse stimulates a part of the engineer that can cause euphoria. When you look at a **Formula 1 car** and realize that the engine block is both an engine component and a structural, load-bearing member of the frame connecting the rear wheels to the rest of the car, we experience the same kind of euphoria. Or when looking at a Leatherman tool and realizing that the frame holding the blades also unfolds to become the handle for the pliers.

The Ariel Atom flaunts its sensuous exoskeleton frame, and to it attaches everything that combines to create a car: **engine**, transmission, differential, suspension, wheels, steering, **brakes**, fuel tank, seats, and driver controls. Thus the vehicle weighs just 1,100 pounds (500 kg). Yet it performs at a level that beats nearly all supercars on the road because the Atom does its one, single mission so well: it goes fast and it flawlessly obeys the driver's commands.

SEE ALSO Internal Combustion Engine (1908), Formula One Car (1938), Anti-Lock Brakes (1971), Bugatti Veyron (2005).

The Ariel Atom is an elegant piece of machinery—both beautiful to look at and efficiently designed.

03

1996

HDTV

In 1964, the NHK, or Japan Broadcasting Corporation, began to research a better method of **television** to engage with the "five human senses" after the Tokyo Olympics. This was the start of a long process leading to a transition from analog to digital transmission. In the United States, prior to HDTV, television stations broadcast analog signals, using the NTSC standard to define what the signal looked like. As long as everyone (TV stations and manufacturers) adhered to that same standard, stations knew that every television set would be able to decode their signal and consumers knew that they would be able to receive every TV station. The FCC (Federal Communications Commission) allocated standard frequencies for each TV station and everyone was happy.

Cable TV came along and transmitted the same NTSC signals on standard FCC frequencies along a cable. Everyone could hook their TVs to any cable system and it would work. **Video recorders** could record NTSC signals, or play them back, usually on channels 3 or 4 of the television. Video game consoles also played on channel 3 or 4. A gigantic economic ecosystem was built on these engineering standards, with billions of TVs and devices, thousands of broadcasters and cable systems. Everything worked universally with everything else.

To implement high-definition pictures, everything needed to change. The NTSC analog standard would be abandoned in favor of compressed digital signals with ten times better resolution. TV stations and cable systems began to send out these much better signals starting in 1996. Everyone needed to buy new HDTVs and hook them to new HD devices with new HDMI cables. All new equipment (cable boxes, Blu-ray players, video game consoles, etc.) used HDMI.

The world had never really seen anything like this change. There was a specific day (June 12, 2009) when all analog broadcasting ended in the United States. And yet engineers pulled this transition off and people willingly accepted it to get ten times better images on their TVs. It shows how much engineers can accomplish when everyone can get behind a new standard.

SEE ALSO Color Television (1939), Cable TV (1948), VHS Video Tape (1976).

Modern HDMI TV audio/video input connection panel.

HDMI IN
2
1
(DVI)
PC / DVI
AUDIO IN
PC IN
ANT IN

DIGITAL
AUDIO OUT
(OPTICAL)

LAN

R – AUDIO – L
COMPONENT IN

1997

Prius Hybrid Car

There was a call in the 1990s to reduce **automotive emissions**. One option: replace **gasoline engines** with a pure electric car. The technology wasn't quite there yet. The **batteries** were too expensive, didn't have enough range, and took too long to recharge. Engineers looked for some way to improve the efficiency to get much better mileage without compromises. Enter the Prius, the world's first mainstream hybrid car, made by Toyota. It is a normal-size sedan, introduced in 1997, that gets great gas mileage because of its hybrid drive train. Millions have been sold.

A hybrid car has more than one source of motive power. In the case of the Prius, the car combines a normal gasoline engine with an electric motor and a battery pack. There are three key engineering insights that improve the mileage of the Prius. First, most cars have a much bigger engine than they need for most situations. When the car is cruising on the freeway, it needs only a fraction of the engine's power. The reason why the engine is so big is for periods of acceleration, which are rare. If an electric motor can help with acceleration, then the gasoline engine can be much smaller and lighter. Second, idling and stop-and-go driving at slow speed wastes gasoline. If the electric motor handles these situations, the engine may not have to run at all. Electricity can also handle accessories like the air conditioner. Third, when the car brakes, the electric motor can become a generator and recharge the batteries. Putting these three things together gives the Prius significantly better gas mileage, especially in city driving.

Once the Prius proved to be so effective and popular, engineers came up with other hybrid flavors. At one end of the spectrum, electric motors and a battery handle accessories like **air conditioning** and instantly start the gasoline engine. Now the engine never needs to idle. At the other end, plug-in hybrids allow the car to recharge its battery overnight in the garage, and then the car can operate for 20 or 40 miles as a pure electric vehicle.

The hybrid vehicle marketplace is a great example of engineering creativity harnessed to create efficiency.

SEE ALSO Air Conditioning (1902), Internal Combustion Engine (1908), Lithium Ion Battery (1991), Self-Driving Car (2011).

Detail on a Toyota Prius plug-in hybrid on display during the 2012 Brussels Motor Show.

HYBRID
SYNERGY
DRIVE

1998

International Space Station

Imagine that you want to engineer a human outpost in the most abstract, inhospitable environment imaginable: space. Where do engineers start? Designing the International Space Station, or ISS, was a step-by-step process. They might begin with some kind of sealed container created by mechanical and structural engineers. The container can hold the air pressure that human inhabitants require. Next inhabitants need an atmosphere inside the container. Nitrogen and oxygen can arrive in high-pressure tanks, or oxygen can come from electrolysis or oxygen candles.

Next is temperature control. The space station becomes a big Thermos as it floats in the vacuum. A special radiator system is required to dump heat.

For humans to get into the container, engineers need to create a door—either a tightly sealed tunnel to an arriving spacecraft, or an airlock to space.

Once humans enter the station, several things become important. There needs to be a way to remove the carbon dioxide and humidity that humans exhale, and an oxygen replenishment system. There needs to be a restroom to handle liquid and solid waste—no simple task in a microgravity environment. There needs to be light, and the electricity to create it. Solar panels are the easiest way to create electricity for space stations in low Earth orbit, with batteries for the times when the panels are in the dark.

The solar cells need to orient toward the sun, and the station needs to avoid tumbling in space. An attitude control system solves these problems. A three-axis reaction wheel system, also known as a control moment gyro, is the standard solution because it does not require fuel like thrusters do.

One important design feature that engineers incorporated into the ISS has been its ability to grow incrementally, with components from several nations. Starting in 1998, it has taken more than three dozen launches to bring all of the various components into low Earth orbit. The final structure is about the size of a football field and provides inhabitants with roughly 35,000 cubic feet (1,000 cubic meters) of space. Imagine a 2,000-square-foot ranch-style house. It has eight-foot ceilings for a total of 16,000 cubic feet. The ISS is about twice that size. It lays the groundwork for other hopes of **colonizing space**.

SEE ALSO Lunar Landing (1969), Space Shuttle Orbiter (1981), Biosphere 2 (1991), Mars Colony (c. 2030).

International Space Station on August 20, 2001.

1998

Large Hadron Collider

What if scientists want to collide two streams of protons together traveling at nearly the speed of light? Why: To discover new subatomic particles. Built by the European Organization for Nuclear Research (CERN) between 1998 and 2008, the Large Hadron Collider, or LHC, is the largest proton accelerator ever and it exceeds all expectations in terms of size, complexity, cost, and engineering achievement. It actually accomplished its design goal—it proved the existence of a particle called the Higgs Boson. This is as big as Big Science gets.

Take, for example, the collider's tunnel, which they began digging in 1998. The LHC is housed in a huge tunnel bored at least 165 feet (50 meters) underground. The tunnel is shaped like a ring 17 miles (27.3 km) in diameter. The tunnel's interior is a **concrete** pipe 12 feet (3.6 meters) in diameter. This tunnel alone is an amazing accomplishment.

Inside the tunnel there is a steel pipe, and inside the pipe there are two smaller tubes that act as channels for the speeding protons. These tubes contain a near total vacuum. Two counter-rotating proton streams accelerate to nearly the speed of light in the vacuum. To curve the streams through the ring, more 1,200 helium cooled superconducting magnets help bend the streams around the ring. Each **magnet** is massive, weighing approximately 60,000 pounds (27,200 kg).

Then there are the detectors, where the actual science occurs. The proton beams cross and collide head on inside these detectors. The largest, called Atlas, is roughly the size of half of a football field and 82 feet (25 meters) tall. Atlas weighs 15 million pounds (6.8 million kg). It is like a gigantic **3D** movie camera for subatomic particles. Its job is to track the particles that get created when protons collide. The camera produces approximately 1,000 million million bytes (1 petabyte) of data per second. Atlas alone would be a major engineering achievement, and Atlas is just one of several detectors.

All of this engineering wonderment comes together in the LHC. It truly is one of the greatest things engineers have ever built.

SEE ALSO Concrete (1400 BCE), Tunnel Boring Machine (1845), 3D Glasses (1952), Neodymium Magnet (1982).

View of the Large Hadron Collider tunnel sector 3-4.

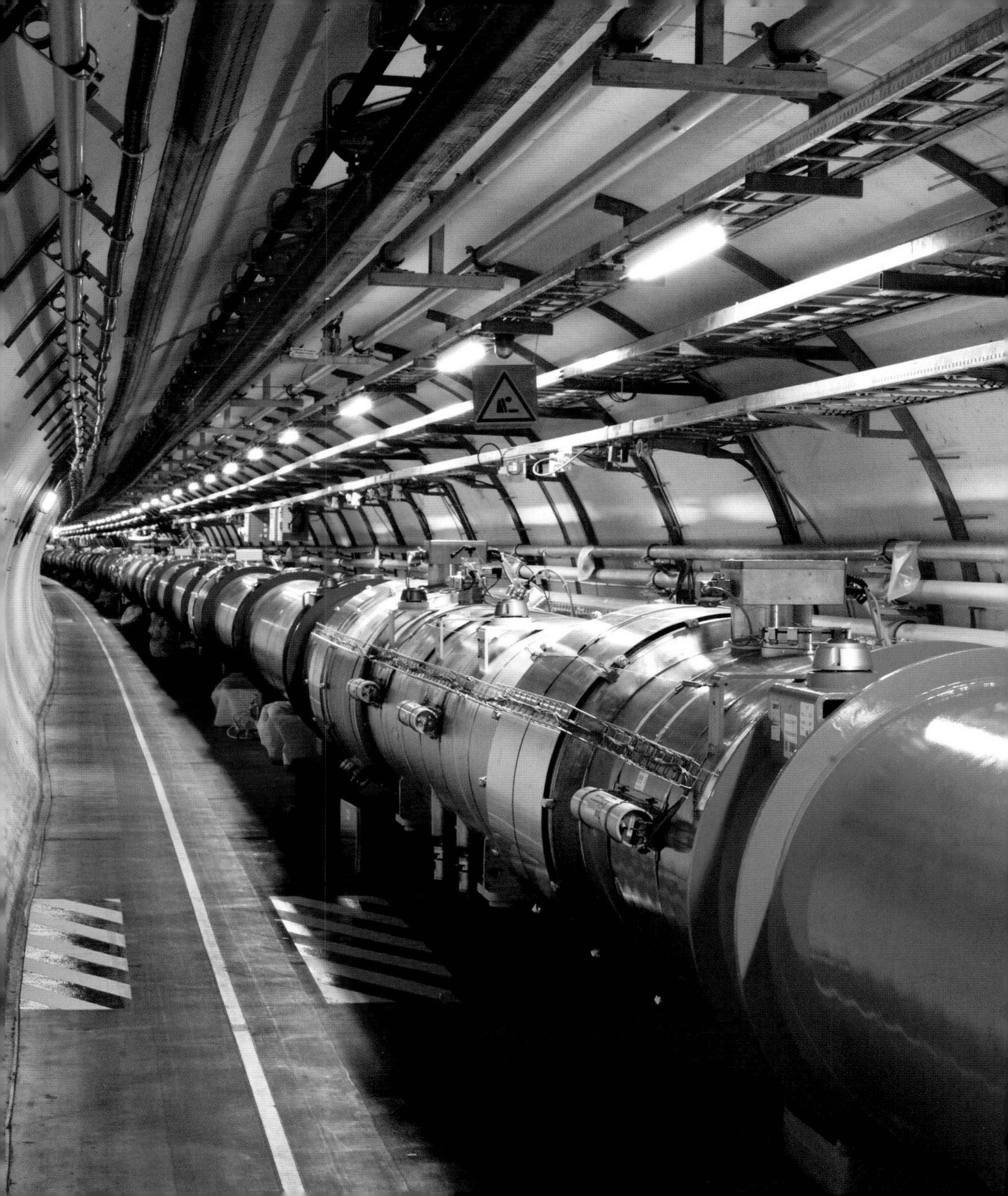

1998

Smart Grid

Massoud Amin (b. 1961)

If you want to, you could engineer and operate your own power grid. Buy a gasoline generator and use it to provide energy for your home, and run a line to your neighbor's house to provide her with power, too. If your generator is big enough, you could provide power to your whole neighborhood. But because of transmission distances, you might need to buy a transformer, convert the power from your generator to a higher voltage, run a single line to the next neighborhood, use another transformer there to step the voltage back down, and then run lines to the houses.

This is, in essence, how power grids got their start. But there is no data flowing in old-fashioned power grids. For 128 years grids were, for the most part, designed to move power only in one direction from large generators to customers. The "smart grid" is a term coined in 1998 by Massoud Amin who was working at the time for the Electric Power Research Institute (EPRI) in Palo Alto, California. Amin created and led a large research and development program, which funded 108 other professors and 240 researchers in 28 US universities along with 52 utilities, to promote research and development related to modernization and resilience of the grid. The smart grid's objective is also to allow for many sources of power, variable sources of power, power storage, and a great deal of information flowing to help optimize power delivery.

For example, people put **solar panels** on their houses and sell the power back to the grid. Old-fashioned **power grids** were not engineered for this. A neighborhood with a lot of solar panels could produce more power than the neighborhood needs. Then a cloud can pass by and instantly cut solar output in half.

What if the power company would like to charge different prices based on load, in order to encourage people to use less power during peak loads? What if major appliances could turn themselves off during peak loads? What if part of the grid fails in a storm, and the grid would like to send a signal shutting off every non-essential appliance? All of these kinds of situations can be handled with a self-healing grid.

SEE ALSO Power Grid (1878), Light Water Reactor (1946), Itaipu Dam (1984).

Smart meters monitor energy quality and provide real-time energy consumption data in a smart grid.

Silver Spring Networks
0013500100F6B4
FCC ID: OWS-NIC51
IC: 5975A-NIC51
01
00216
kWh
240
Type FOCUS AXR-SD
FORM 2S CL200 240V 3W 60Hz TA=30 Kh 7.2
NXL100801577711011
100801577
AR 60 min.
DS
PG&E
SmartMeter
107 681 57
-415
1010
Landis+Gyr
PATENT PENDING

1998

Iridium Satellite System

Barry Bertiger (Dates Unavailable), **Dr. Ray Leopold** (Dates Unavailable), **Ken Peterson** (Dates Unavailable)

The architecture of the current **cell phone** system is an example of engineering genius because of the way it optimizes things. By placing cell phone towers in a grid every few miles, it is possible for the phones to contain low-power transmitters that never need to send signals more than a mile or two. Low-power transmitters mean small size and low **battery** drain, meaning that a cell phone can fit in your pocket.

The towers-in-grid approach does have one drawback—it becomes economically unfeasible in places with low population density. Towers are expensive, and they can't be deployed unless there are enough people using them to recover the cost.

The Iridium system, owned by Iridium Communities Inc. in Virginia, asked the question: How do we provide cell phone coverage to all of the places that lack towers? The answer that Motorola engineers Barry Bertiger, Dr. Ray Leopod, and Ken Peterson came up with was amazing in scope—they would use dozens of **satellites** to provide planetwide coverage. Their system, conceived in 1987, became commercially available in 1998.

A satellite in geosynchronous orbit requires a stationary dish antenna because it is 24,000 miles (38,000 km) away. Iridium engineers wanted something far more portable and mobile. So they conceived of a constellation of satellites in low-Earth orbit, approximately 500 miles (800 km) overhead. At that altitude, there would need to be a lot of satellites so a phone would be guaranteed to always have a satellite in range. In fact, the number of satellites needed is 66, plus some spares. The satellites talk to each other so that conversations can route through a satellite-based mesh network to a ground station.

While the engineering is spectacular, this architecture has problems. Transmitting to a satellite 500 miles away requires a lot more power, so the phones are bigger, with bigger antennas and a high price tag. It also requires line-of-sight, meaning the phones have difficulty inside buildings. The system's price tag of $5 billion led to high-priced calls. Very few people signed up, and Iridium went bankrupt in 1999, though the system was purchased and still operates today.

SEE ALSO Telephone (1876), Space Satellite (1957), ARPANET (1969), Mobile Phone (1983), Lithium Ion Battery (1991), Smart Phone (2007).

An Inuit hunter using an Iridium satellite phone on the Melville Peninsula, Nunavut, Canada.

1999

Wi-Fi

Vic Hayes (b. 1941)

What if you had to use a network cable to hook your **tablet computer** up to your home network? Or imagine going to a coffee shop or airport and having to find an actual network cable to connect your laptop to the **Internet**. This painful image is eliminated by the widespread use of Wi-Fi, also known by its IEEE specification number: 802.11. In the same way that **Ethernet** is a standard that lets people create wired local area networks (LANs), Wi-Fi lets people create wireless LANs. To install a Wi-Fi LAN, you plug in a box with antennas that creates a Wi-Fi hotspot. This hotspot might spread out 100 feet (30 meters) in all directions from the box. Many call Dutch engineer Vic Hayes the "father of Wi-Fi" because he created the IEEE 802.11 Standards Working Group.

To make Wi-Fi work so well, engineers took a number of different factors into account. First they allow multiple machines (laptops, phones, tablets, etc.) to simultaneously connect to a single hotspot. Then, in a place like an airport, they allow multiple hotspots to simultaneously operate. So in an airport, there can be several companies providing Wi-Fi access, and then several people's **smart phones** can create their own hotspots. All of these hotspots can overlap while working flawlessly because of **spread spectrum** technology.

To solve the signal interception problem, Wi-Fi implements several encryption algorithms. WEP came first but was not secure enough, so WPA improved the situation.

Because so many people now use Wi-Fi on so many devices, bandwidth has improved over time. Today Wi-Fi is available on just about every portable device. Many homes have Wi-Fi hotspots, as do many businesses, airports, stores, arenas, restaurants, and apartment complexes. Even entire college campuses and cities have been blanketed with Wi-Fi signals.

Because of the engineering work done behind the scenes, Wi-Fi makes it much easier for people to connect to the Internet, and it lowers the cost of access by eliminating the wires.

SEE ALSO Spread Spectrum (1942), Ethernet (1983), Smart Phone (2007), Tablet Computer (2010).

A diagram depicting Wi-Fi range.

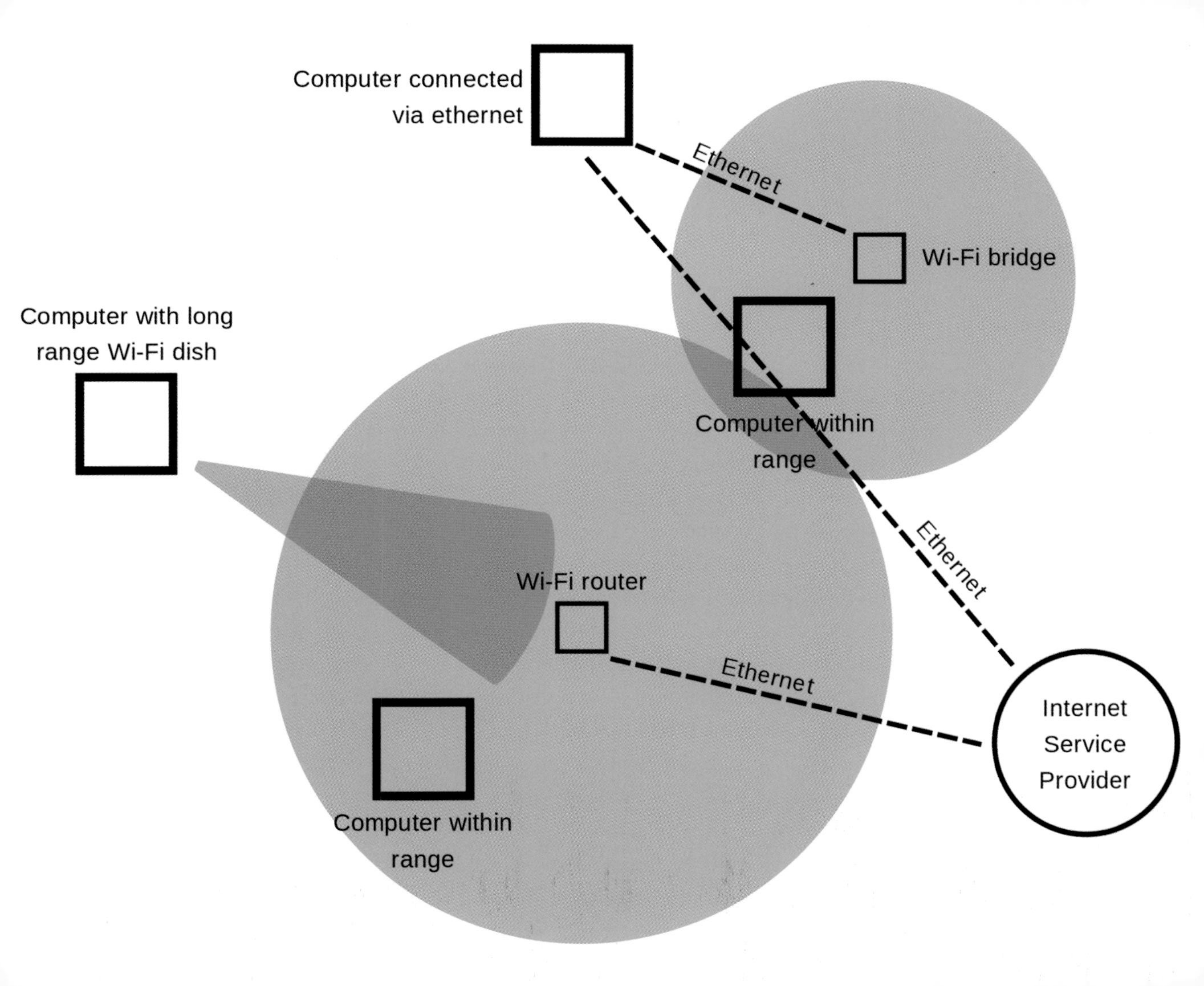
Computer connected
via ethernet
Ethernet
Wi-Fi bridge
Computer with long
range Wi-Fi dish
Computer within
range
Ethernet
Wi-Fi router
Ethernet
Internet
Service
Provider
Computer within
range

2001

Segway

Dean Kamen (b. 1951), **Doug Field** (Dates Unavailable)

Every now and again, engineers create something that absolutely amazes people when they first see it. Examples include the first Macintosh **computer**, the **microwave** oven, and, of course, the **nuclear bomb**.

And then there is the case of the Segway, invented by Dean Kamen and engineered by a team of people lead by chief engineer Doug Field. It generated a gigantic amount of speculation and hype *before* it was released, and before anyone knew what it was. The Segway generated so much hype that, unless it had been something truly spectacular like antigravity boots or a for-real phaser weapon, it could not possibly live up to its pre-release hype.

It is unclear how the hype started and got so much traction. But knowledgeable people (Steve Jobs for example) had secretly seen the Segway (code-named Ginger before release) and made statements about it that were both exciting and vague. "This will be bigger than the Internet," "More important than the PC," "It will completely revolutionize cities," and so on were the kind of things people heard.

Then, when the Segway made its actual debut in 2001, it was a huge letdown. It appeared to be a small electric scooter at an astronomical price. The hype and the reality were incompatible with one another.

But the Segway was and is a remarkably engineered device. It was the first big example of self-balancing technology that most people had seen. Accelerometers and gyros tell a computer when it is getting off-balance and the computer uses the electric motors to rebalance instantly. It has a remarkably simple user interface, where the driver's lean controls the speed. It also has small but powerful **batteries** that keep it compact. And its form factor means it can go just about anywhere a person on foot could go—airports, malls, sidewalks, **elevators**. But even that caused consternation because people were afraid of getting run over.

If nothing else, it shows the importance of publicity when announcing a revolutionary new piece of engineering. There really can be too much of a good thing.

SEE ALSO Elevator (1861), Trinity Nuclear Bomb (1945), ENIAC—The First Digital Computer (1946), Microwave Oven (1946), Lithium Ion Battery (1991).

Tourists riding Segways on July 2, 2014 in Copenhagen, Denmark.

Nyhavns Færgekro
www.stromma.dk
89

2003

A Century of Innovation

In the book, *A Century of Innovation: Twenty Engineering Achievements that Changed Our Lives*, written by George Constable and Bob Somerville and published in 2003, the National Academy of Engineering picked the twenty greatest engineering achievements of the twentieth century. It shows the massive positive impact that engineers have had on modern society. Here is the NAE's list:

Electrification	Highways
Automobile	Spacecraft
Airplane	Internet
Water Supply and Distribution	Imaging
Electronics	Household Appliances
Radio and **Television**	Health Technologies
Agricultural Mechanization	Petroleum and Petrochemical Technologies
Computers	**Laser** and **Fiber Optics**
Telephone	Nuclear Technologies
Air Conditioning and **Refrigeration**	High-performance Materials

When you look at the list, you realize how important the twentieth century was in terms of the advancement of modern society. Advancements like airplanes, automobiles, highways, spacecraft, oil refining, radio, TV, computers, electronics . . . they simply did not exist in any real form at the start of the twentieth century, and by the end of the century they are essential to modern life. It is safe to say that no other century in human history saw anything like this kind of advancement. And humanity is significantly better off for the work of the countless engineers who help bring these technologies forward.

SEE ALSO Air Conditioning (1902), Laser (1917), Radio Station (1920), Electrical Refrigeration (1927), Color Television (1939), ENIAC—The First Digital Computer (1946), Fiber Optic Communication (1970).

The computer was hailed as one of the greatest achievements of the twentieth century.

2004

Millau Viaduct

Michel Virlogeux (b. 1946)

Think about the simplest possible bridge. You can chop down a tree so it falls over a river and becomes a bridge. But it's not a very good bridge, it does not make efficient use of materials, and if the river is very wide this approach won't work. As the bridge gets longer, the loads become greater and costs become a factor, an engineer will get involved. An engineer will look at possibilities like I beams, **trusses**, arches, and suspension bridges. All have their place.

What if you need to span a deep and wide chasm at low cost in 2004? This is a situation that French structural engineer Michel Virlogeux faced when he set out to design a bridge near Millau, France, that ultimately became the Millau Viaduct. The chasm is 8,070 feet (2460 meters) wide, and incredibly deep at one point, where a tower needs to reach a record-setting 1,130 feet (345 meters) tall. Virlogeux's design uses seven towers and a cable stay design on each of them. Besides lowering the cost, this design minimizes the visual impact of the bridge on the landscape and made the bridge less expensive to build.

A cable stay bridge is fundamentally different from a suspension bridge in the way it handles load. In a suspension bridge, the main cables need massive anchor points at both ends. The two towers tend to be gigantic. The deck hangs below the cables on suspenders.

At Millau, no anchor points are necessary. But the deck must handle horizontal compression loads between each cable and its tower. A single set of cables radiating from a tower attaches to a huge steel box beam that runs down the center of the highway. The beam is 13 x 14 feet (4 x 4.2 meters). The roadway is cantilevered off the beam on both sides with triangular trusses. It is an aesthetically beautiful and structurally elegant design.

Wind is a big factor. The cross-section of the deck looks like an upside-down wing. The idea is that the wind adds tension to the cables, rather than lifting the deck, to avoid flutter.

SEE ALSO Truss Bridge (1823), Golden Gate Bridge (1937).

The Millau Viaduct: a stunning engineering achievement as well as a work of art.

2005

Bugatti Veyron

Hartmut Warkuss (b. 1941), **Wolfgang Schreiber** (b. 1959), **Jozef Kabaň** (b. 1973)

Imagine that you are one of the engineers working to create the new Bugatti Veyron. A major goal of the project is speed. The team, consisting of designers Jozef Kabaň and Hartmut Warkuss and engineered by a team headed by chief engineer Wolfgang Schreiber, is attempting to create the fastest street-legal production car in the world. In 2005, they produced a vehicle capable of a sustained velocity as high as 212 mph.

The key to this kind of speed is horsepower. The reason is because of the engineering reality of aerodynamic drag. When the speed of a car doubles, the power needed to move the car through the air goes up by a factor of eight. If a car needs 20 horsepower to go 50 mph, it will need 160 hp to go 100 mph.

To make it to 212 mph, the Veyron needs an amazing 1,001 hp. The Veyron's engineers created something unique to solve the problem: an unprecedented W-16 **engine** with four banks of four cylinders. An engine like this needs to breathe, so there are four **turbochargers**. The turbochargers pressurize extra air into the cylinders, meaning that more gasoline can burn on each power stroke.

All that horsepower then needs to get to the tires and onto the street, so there is a seven-speed electronic transmission and all-wheel drive. Since the Veyron can travel as fast as a **Formula 1** car, aerodynamic lift is a concern. So at 140 mph, the hydraulic suspension automatically lowers the car by four inches and a wing deploys to provide 700 pounds of down force.

At full speed, this engine sucks in a huge amount of gasoline. It gets about three miles per gallon. At 212 miles per hour it only takes 17 seconds to go a mile. So the car burns a gallon of gas in just 51 seconds. Creating a problem: What to do with all the heat from the burning gas? The Veyron has three water radiators and another three oil radiators to keep things cool. You can't drive for very long at full speed, but it's fun while it lasts.

SEE ALSO Supercharger and Turbocharger (1885), Internal Combustion Engine (1908), Formula One Car (1938).

The Bugatti Veyron Super Sport, the world's fastest car, driven on September 19, 2010, on the mountain roads around Jerez, Spain.

2005

Georgia Aquarium

Jeff Swanagan (1957–2009)

Atlanta wants to build an aquarium tank large enough to comfortably house four whale sharks, along with thousands of other fish and rays. Assume that the average whale shark is 32 feet (10 meters) long and weighs as much as several African elephants. How do you make an aquarium that big? These were issues faced by the engineers working on this structure, including the founding executive director, Jeff Swanagan, who was responsible for much of its creation and design.

The typical Olympic-size swimming pool (50 meters long, 10 lanes wide) holds about 660,000 gallons (2,500,000 liters). The Georgia Aquarium Ocean Voyager exhibit, at 6.3 million gallons (24 million liters), is roughly the size of 10 Olympic swimming pools. This immense tank, which opened in 2005, is 263 feet long (80 meters), 125 feet wide (38 meters), and 33 feet deep (10 meters). Its walls, some up to three feet (1 meter) thick, are made of Agilia **concrete**, a special mixture that flows almost like water yet does not allow the aggregate to settle out while it is curing. This means that when workers removed the formwork, there were no bubbles or voids anywhere in the concrete.

The main viewing window measures 23 feet (7 meters) high and 61 feet (18.6 meters) wide. The water behind that window has a combined weight of approximately 50,000,000 pounds (23,000,000 kg). How are engineers going to keep that water from bursting through the window? They made it out of acrylic measuring two feet (0.6 meters) thick. The advantage of acrylic, compared to glass, is that it is 17 times stronger while having half the weight.

The next challenge is filtration. At the Georgia aquarium, the total exhibit space is 10 million gallons (38 million liters). To filter this much water requires an immense pumping capacity. The total flow rate of the filtration system is 300,000 gallons (1.1 million liters) per minute, powered by over 500 pumps totaling 5,500 hp. Sand filters take out anything larger than 20 microns. There are also protein skimmers that use fine air bubbles to eliminate unwanted organics from excess food and fish waste.

SEE ALSO Concrete (1400 AAA), Water Treatment (1854).

The biggest aquarium in the world, Atlanta, Georgia, USA.

2006

Palm Islands

Mark Twain famously quipped, "Buy land, they're not making it anymore." But since he said this, engineers have proven him wrong. Large portions of San Francisco Bay wetlands have been filled to create new land for a growing city. Battery Park City in New York is 92 acres of new land formed when they dug the foundations for the **World Trade Center** towers. A third of Holland is made of land reclaimed from the sea.

The Palm Islands in Dubai represent one the most ambitious engineering projects of this kind. Here engineers built new islands for thousands of residents and tourists in the Arabian Sea—the first opened in 2006.

It would be one thing if the Palm Islands were simply big round mounds of rocks and sand. They are nothing like that. In order to maximize waterfront real estate, they are instead shaped like palm fronds—long, thin leaves branch off of a central stalk. Then the entire island is surrounded by a separate breakwater island to provide protection from storms. In the process, engineers created hundreds of kilometers of shoreline property that sells for maximal prices.

The process starts by laying seven meters of sand on the ocean floor, a process started in 2001 by Belgian civil engineering firm Jan De Nul and a Dutch company, Van Oord. On top of this is a second layer of rubble about the same height. Protecting the rubble is a layer of boulders. Then sand was sprayed over the top. This created long, narrow strips of new beachfront property.

Engineers also had to solve a number of other concerns. How to avoid water stagnating inside the breakwater? Deep channels make sure that water can circulate without diminishing the protection. How to make sure automobile traffic flows smoothly? Tunnels, bridges, and a broad highway up the spine provide plenty of capacity. How to keep wave action from eroding the islands into oblivion? Continual beach nourishment programs will likely be required, as on many other beaches around the world.

As the world population rises, is more new land an engineered option? It is definitely a possibility, as Holland proves. Floating cities that can move out of harm's way when storms are approaching are another possibility.

SEE ALSO World Trade Center (1973), Kansai International Airport (1994).

Jumeirah Palm Island development in Dubai.

2006

AMOLED Screen

Think about the kinds of screens that have been available in mobile devices over the years. In the 1980s, Sony engineers created a small, portable CRT screen with the electron guns below rather than behind the viewing area. The original GRiD laptop used a red plasma screen. Subsequent laptops used black and white and then color **LCD screens**, as did gaming devices like the Gameboy and Nintendo DS.

With the advent of **smart phones**, backlit LCD screens dominated initially. But engineers have made advances with OLED (Organic Light Emitting Diode) technology and then AMOLED (Active Matrix OLED) technology. The basic idea behind OLED screens is to create an array of tiny LEDs as a single sheet. Where LCD pixels act as shutters that open or close to let a backlight through, OLED screens turn their tiny LEDs on and off individually to provide light directly.

Two problems that OLED screens have faced are long-term reliability and size. Engineers have been working on these problems for years. Initially OLED screens were the size of postage stamps. Adding active matrix **transistors** to the display allowed the size to grow. Several manufacturing breakthroughs have worked together to make the pixels smaller and more reliable. So today, AMOLED displays are quite common on smart phones.

AMOLED screens, which first appeared commercially in 2006, demonstrate how engineers improve technologies over time to replace older technologies with newer ones. AMOLED screens are replacing LCD screens because AMOLED screens are thinner, lighter, and can even be printed on **plastic** so they can be flexible. In addition, engineers have been able to integrate the touch overlay into the OLED sandwich, making things thinner still.

SEE ALSO Plastic (1856), Transistor (1947), LCD Screen (1970), Smart Phone (2007).

Close-up of an AMOLED screen.

2007

Smart Phone

Steve Jobs (1955–2011)

The smart phone is something that we can pin down to an exact day of birth. Because, although there had been predecessors that had many of the attributes of smart phones, the smart phone came into widespread public consciousness on the day Steve Jobs announced the iPhone: January 9, 2007. This phone was the true beginning of the smart phone era in the same way that the Model T marked the start of the automotive era.

The iPhone is an engineering masterpiece, developed by many people at Apple. Think of the many things crammed into its tiny case—all of them the product of decades of engineering advancement. There is the low-power, tiny **CPU** with **RAM**—something that a decade earlier would have filled a breadbox and a decade before that would have filled a room. There is the **flash memory system** able to hold thousands of songs. Such a thing simply did not exist two decades earlier and this system is tiny compared to the hard disk it replaced. There is the **cell phone** radio system, digitized and shrunk down to tiny proportions, along with an antenna that uses the case itself to improve reception. The capacitive touch screen is light, thin, bright, and responsive to touch in ways never seen before. It even had a two-megapixel **camera**, along with an accelerometer and a proximity detector to keep your face from tapping the screen. And then there is the thin, light **battery** that provides enough electricity for a day of operation.

Thousands of engineers worked over the course of several generations to make the iPhone possible. Apple brought all of that hardware and technology together and their packaging and manufacturing engineers compressed it into an amazing, compact, lightweight product.

And then Apple added one more thing—software engineers wrote code that made this whole complex package so simple that just about anyone could use it intuitively.

The marketplace exploded with delight—millions of people had been waiting for someone to crack the code, and Apple engineers had done it.

SEE ALSO ENIAC—The First Digital Computer (1946), Dynamic RAM (1966), Flash Memory (1980), Mobile Phone (1983), Lithium Ion Battery (1991), Digital Camera (1994).

Apple CEO Steve Jobs holds up an iPhone at the MacWorld Conference in San Francisco, January 9, 2007.

2008

Quadrotor

If you look in Google Trends, no one searched for the word "quadrotor" prior to 2007. In 2007 there is a little blip. And then people start searching for it in earnest in 2008. This documents the birth of the quadrotor—a four-rotor helicopter—in public consciousness.

The modern quadrotor is made possible by a set of engineering advancements in the model airplane industry. Small, powerful **neodymium magnet** motors are essential to quadrotors. So are lightweight, inexpensive lithium polymer **batteries**, and precise motor controllers that feed battery power to the motors. And then there is the computer control system plus the new sensors that automatically stabilize the four-propeller platform.

Those elements all came together around 2008. Then, in 2010, the Parrot Company in France engineered a model priced at $300 that allowed for mass consumption of quadrotors when it appeared on the cover of Brookstone's catalog.

The quadrotor represents a major engineering reconceptualization of the **helicopter**, which had remained nearly unchanged in any major aspect since its invention in the 1930s. Instead of a single engine with rotor and swash plate plus a tail boom, the quadrotor uses four motors and four static propellers. The relative speed of the four motors controls the platform's motion.

Quadrotor control systems rely on inexpensive solid-state gyros, three-axis accelerometers, and clever software engineering. With these sensors, the vehicle can tell when it starts to spin, tip, and wobble—then it automatically corrects things to keep the platform stable. A human tells the stable platform where to go. Or, a computer-controlled quadrotor can have its own **GPS** unit and fly to a specific position, hover over a known point on earth, or fly from waypoint to waypoint.

Six- or eight-rotor vehicles are possible, allowing for large payloads like studio-quality cameras. An octorotor with camera can replace a helicopter with cameraman, dramatically lowering the cost of aerial shots. These vehicles are also useful for aerial surveillance, emergency response, etc.

Quadrotors are a perfect example of how advancement in several engineering disciplines can come together to create a reconceptualization. The next step will be full human-size vehicles with sixteen or more rotors.

SEE ALSO Helicopter (1944), Neodymium Magnet (1982), Lithium Ion Battery (1991), Global Positioning System (GPS) (1994).

Photo of a Dji Phantom with GoPro in flight taking aerial photos February 09, 2014 in Miami, Florida, USA.

HERO3+

2008

Carbon Sequestration

There is one funny aspect to engineering that can be seen either as its greatest attribute or its Achilles heal. It has to do with the fact that many engineering solutions create their own sets of problems, which then require engineered solutions themselves. If nothing else, it is a great source of job security for members of the engineering profession.

This aspect is evident in carbon emissions. Currently our society is highly dependent on fossil fuels for energy. Cars, **trains**, ships, and **airplanes** all use liquid fossil fuels. Electricity production is highly dependent on coal and natural gas. Engineers have made the production of fossil fuels dependable and inexpensive on a massive scale, to the point where humans produce and consume something like 85 million barrels of oil per day.

The problem this creates is the carbon dioxide it releases into the environment—gigatons of CO_2 every year. And one engineered solution is to capture and entomb as much of the carbon dioxide as possible.

There are lots of ideas for carbon sequestration that come in three main categories: 1) Compress carbon dioxide into a liquid and store it underground or deep in the ocean, 2) Encourage plant growth, for example algae in the oceans, and 3) Create chemical reactions, for example by adding crushed limestone to the ocean.

Carbon capture is most easily done at a large, stationary carbon producer like a big coal-fired power plant. One idea engineers have developed is to burn the coal in pure oxygen and then capture the nearly pure CO_2 byproduct that is available after moisture removal and **scrubbing**. The carbon dioxide is compressed to a liquid and injected into deep underground formations. The Schwarze Pumpe in Germany became the world's first coal plant to use this technique when it opened in 2008.

The other trick available to engineers is to create new technologies that cost less than fossil fuels. **Wind power** has recently fallen below fossil fuels in cost. **Solar power** is headed in that direction. Fusion power is another option further out on the horizon.

SEE ALSO Tom Thumb Steam Locomotive (1830), The Wright Brothers' Airplane (1903), Power Plant Scrubber (1971), Alta Wind Energy Center (2010), Ivanpah Solar Electric Generating System (2014).

First brown coal power plant in the world with carbon dioxide separation via CCS technology in Brandenburg, Germany.

2008

Engineering Grand Challenges

In 2008, the National Academy of Engineering publicized the Engineering Grand Challenges. It is a list of important problems society is currently facing from an engineering perspective. By focusing attention on these fourteen big problems, the NAE felt that engineers could make the world a better place.

The list is fascinating because it highlights the challenges we face today as well as big untapped ideas and areas of investigation that need creative engineering solutions. Here are some (see the website mentioned in "Further Reading" for a full list):

Economical **solar energy**—If its costs can be driven well below the cost of fossil fuels, solar energy use would explode.

Energy from fusion—Humankind has been dreaming about cheap fusion power for decades. **ITER** and **NIF** (National Ignition Facility) are two big efforts to make the dream a reality.

Urban infrastructure restoration—Roads, water systems, **sewers**, and **bridges** are aging. Providing funds to keep them modern is important.

Carbon sequestration—Cheap and easy sequestration would help lessen the impact of CO_2 in the environment.

Clean water—Universal clean water supplies would radically improve the lives of billions of people.

Cyberspace security—How do we keep cyberterrorists from crashing the **Internet** or infecting places like banks and **power grids**?

Health informatics—Can engineers create a system making everyone's medical information digitally and universally available to doctors and hospitals?

Nitrogen cycle—Humans fix a lot of nitrogen. Like CO_2, nitrogen has a negative impact on the environment.

Nuclear terror prevention—Nuclear materials and **bombs** in the hands of terrorists could create catastrophic events. Preventing that possibility is a major goal.

Tools of scientific discovery—Engineers build the tools for big science. The more tools we have, the more discoveries scientists can make.

Virtual reality—**Virtual reality** equipment is too expensive and there are major holes in the technology.

SEE ALSO Ivanpah Solar Electric Generating System (2014).

Many of the answers to engineering's greatest challenges have yet to be found.

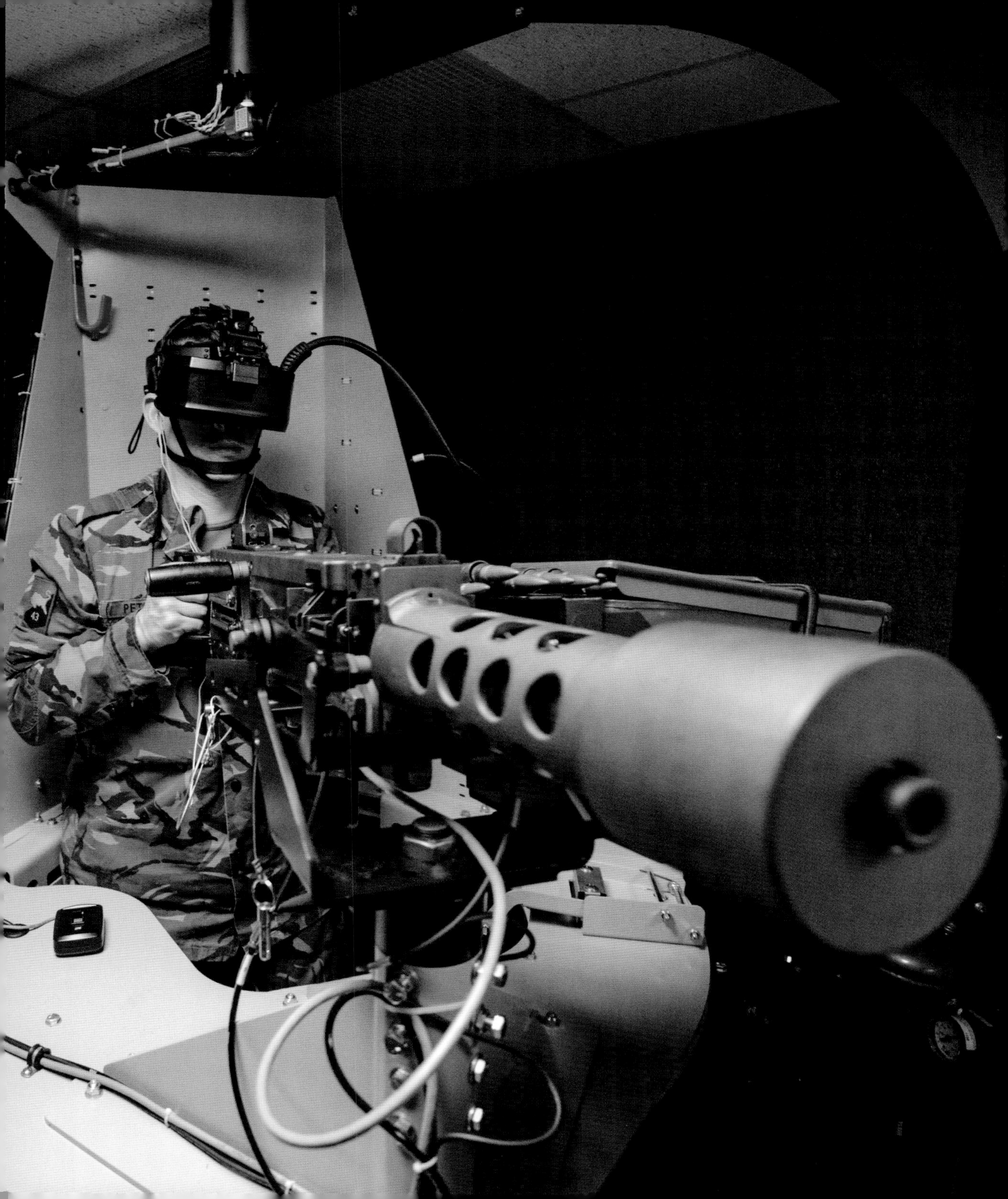

2008

Martin Jet Pack

Glenn Martin (Dates Unavailable)

It's one of those dreams that has been around since Buck Rogers made it popular—a rocket backpack of some sort that would let anyone fly anywhere they want to go. The concept is easy, but the engineering has been challenging.

The first successful backpack system that let people fly appeared in the 1960s, and the basic design principles have been copied and reengineered several times since. The basic idea goes like this: Take highly concentrated hydrogen peroxide. Not the drug store variety—it is only five percent H_2O_2. Ninety percent H_2O_2 or better is needed for this device. Put the hydrogen peroxide in tanks on your back, and provide a pressurization system to push it out of the tanks toward two nozzles mounted near the pilot's shoulders. Hydrogen peroxide has this useful property: when it comes into contact with a catalyst like silver, it immediately decomposes into water (in the form of high-pressure steam) and oxygen and can produce thrust.

This works. The only problem is that the fuel takes up a lot of space. The longest possible flight is just a minute or so.

The Martin Jet Pack is a much newer system that debuted in 2008 from the Martin Company in the United States. It had been in development for thirty years, when inventor Glenn Martin designed it in his garage. Despite the Jet Pack's name, it is not jet-powered, but instead uses two ducted fan propellers connected to a 200-hp (150 kW) reciprocating engine. And at 250 pounds, it is not really a "pack" either. It is more like a lightweight flying platform. The advantage of this design is that it can hold enough fuel to fly for about 30 minutes and it has good performance.

Fitting a 200-hp engine plus the ducted fans, a chassis, and controls into 250 pounds is an impressive accomplishment requiring the extensive use of lightweight materials like **carbon fiber** and **Kevlar**. Despite a lot of high-publicity demonstrations, the Martin Jet Pack is not yet shipping at the time of publication.

SEE ALSO Carbon Fiber (1879), Kevlar (1971).

Pictured: The Martin Jet Pack in action.

MARTIN
MARTIN

2008

Three Gorges Dam

Why do engineers build dams? Dams tend to be huge and expensive, so there must be a good reason. One of the best reasons is flood control. A wild river can be a powerful destructive force for the people who live nearby. A wet season can turn a peaceful river into a raging, flooding torrent. Or, if neighboring land is flat, floodwater can spread out over many square miles. The lake behind a dam can instead store the flood and release it at a reasonable rate, or save it for things like irrigation later in the year, as in India.

Another reason is drinking water. A big dam with a big lake behind it can provide a reliable supply of drinking water for millions of people. This is one of the major benefits of the **Hoover Dam**.

Another reason is hydroelectric power. A dam across a large river can create large amounts of electricity, like the **Itaipu Dam**.

Then there is navigation. A big part of the **Panama Canal** is not a canal at all, but is instead a route across a giant manmade lake.

What if engineers are able to achieve all of these goals with one dam? That is what the Three Gorges Dam in China, initially surveyed by American civil engineer John L. Savage in 1944, has been able to accomplish. Original plans for the dam began in 1932 under Chiang Kai-Shek's Nationalist government, but construction did not start until 1994.

The Three Gorges Dam is immense—7,660 feet (2,335 meters) long and 594 feet (181 meters) tall. It houses thirty-two 700-megawatt generators, giving it by far the largest electrical generating capacity in the world. By generating an expected 100 billion kilowatt-hours of electricity per year, the dam should be able to pay for itself in just 10 years.

The lake behind the dam is over 400 miles (600 km) long and holds 40 cubic kilometers (10 cubic miles) of water. To put that in perspective, if one billion people used 10 gallons of water from the lake every day for a year, it would only use one third of the water behind this dam.

SEE ALSO Hoover Dam (1936), Green Revolution (1961), Itaipu Dam (1984).

The Three Gorges Dam in China is the world's largest power station.

2009

National Ignition Facility

There is a class of scientific research project that goes by the name "Big Science." These are huge projects, costing in excess of one billion dollars, that explore important scientific questions. Examples include the **International Thermonuclear Experimental Reactor**, the **Large Hadron Collider**, the **International Space Station**, the **Curiosity rover**, the **Hubble** telescope, etc. One project that definitely falls into this category is the Lawrence Livermore National Laboratory's National Ignition Facility, which has the goal of creating energy-positive fusion reactions.

The NIF's facility is so large, so cool, and so sexy that people just want to look at it. Everything about the NIF, which first lit up in 2009, is impressive.

The basic idea goes like this: Take a tiny metal capsule filled with deuterium and tritium. Engineers suspend the capsule inside an immense spherical chamber that can blast the capsule with **lasers** from 192 different angles. The goal is for the lasers to help create a high enough temperature, and a high enough density in the fuel, for a long enough period of time, that the conditions at the heart of a star are replicated. If this happens, then fusion will occur. If it happens optimally, then a significant number of fusions occur and more energy comes out of the process than goes into powering the lasers.

How can beams of laser light compress the fuel? The tiny capsule turns to a plasma so fast that the reaction creates an implosion toward the fuel, compressing it. Therefore the process is called inertial confinement fusion.

The NIF is housed in an immense building. The bulk of the building is filled with the equipment needed to create, amplify, split, and combine laser beams. One end of the building holds the spherical chamber where all of the laser beams converge simultaneously on the target capsule.

With this facility, scientists hope to learn about the fusion process, and with that knowledge, find their way to creating viable fusion power plants.

SEE ALSO Laser (1917), International Thermonuclear Experimental Reactor (ITER) (1985), The Hubble Space Telescope (1990), International Space Station (1998), Large Hadron Collider (1998), Curiosity Rover (2012).

Vaughn Draggoo inspects a huge target chamber at the National Ignition Facility in California, a future test site for light-induced nuclear fusion.

2009

Earthquake-Safe Buildings

It is easy to erect a simple building. Stack a bunch of cinder blocks into a wall. Add three more walls and you have a room. Put **engineered lumber** over the top and you have a building. This takes no engineering at all. It could stand for many years. But the day an earthquake rumbles, this building becomes a pile.

During the Haiti earthquake in 2010, buildings collapsed on a massive scale. Tens of thousands of people died. In one city, only 10 percent of the buildings remained standing. The **concrete** they used contained too little cement. The walls were too thin. There was no reinforcing steel. The soil underneath was not stable. Engineers know how to build earthquake-proof buildings, and their knowledge is embodied in building codes. But when no one follows the building codes, disaster is inevitable.

Engineers have designed extremely large earthquake-safe buildings. One technique is called a base isolation system. The idea is to disconnect the base of the building from the ground so that, when the ground shakes, the building does not. But how is it possible to disconnect a building from the ground? A great example can be seen in Istanbul's airport—the largest earthquake-safe building in the world completed in 2009. In simple terms, every column in the building rests on plate riding atop a giant ball bearing. Now the ground can move back and forth while the building stays largely stationary by riding on top of its ball bearings.

Another system uses pads. The pads sit on the ground and the building sits atop the pads. The pads might allow the ground to move a foot (0.3 meters) or more side to side under the building. So when the ground shakes, the building does not shake nearly as much.

Engineers also strengthen the structure itself. Even a wooden frame house can gain dramatically improved strength with inexpensive additions that tie all the framing members together. Simple metal straps and tees significantly strengthen the frame at low cost. Engineers: Keeping the roof over your head even during earthquakes.

SEE ALSO Concrete (1400 AAA), Engineered Lumber (1905), Tuned Mass Damper (1977).

Earthquake-damaged building, following the Kobe earthquake, Japan, 1995.

盲人用信号

2009

US President's Limousine

The first car built for United States presidential use was a V12 Lincoln convertible designed for Franklin D. Roosevelt in 1939. The attack on Pearl Harbor had generated interest in adding protective features, and subsequent events like the assassination of President Kennedy in 1963 spurred engineers associated with the design and production of presidential vehicles to imagine myriad threats and find ways to protect against them. The culmination of these adaptations is the 2009 Cadillac limousine used by President Barack Obama.

What if someone tries to shoot out the tires? That's going to be difficult, since the tires are **Kevlar** reinforced and shred resistant, with steel tires underneath. Even if the rubber completely disintegrates, the car can roll at full speed.

What if someone fires a rocket-propelled grenade? The doors and body panels are 8 inches (24 cm) thick and designed much like **tank** armor. If you ever see the limo with a door open, the door looks more like the door of a passenger jet in terms of its thickness and latching mechanisms. The windows are 5 inches (12.5 cm) thick with bulletproof glass that can reject armor piercing rounds.

What if the car runs over a mine? Five inches (12.5 cm) of steel under the car can deflect the blast. What about a chemical or biological threat? The car can seal itself off and use its own internal oxygen supply. What if a huge crowd gathers? There are tear gas cannons on board.

Another factor in the president's safety is the convoy that the limousine travels with. There are approximately 45 vehicles in a typical convoy—everything from a special ambulance with surgeon onboard to communications vans, threat assessment vehicles, plenty of bodyguards in their own vehicles, even a spare limo just in case.

Then there are other precautions. When the motorcade goes by, the road, including all bridges over the road, have been cleared. This process can cause some serious traffic problems in big cities. Taken together, between the convoy, the procedures, and the limo itself, the secret service has kept the president safe for many years. It is an engineered system that can handle just about any contingency.

SEE ALSO AK-47 (1947), Kevlar (1971), M1 Tank (1980).

President Barack Obama and Vice President Joe Biden ride in the motorcade in Washington, DC, July 21, 2010, to sign the Dodd-Frank Wall Street Reform and Consumer Protection Act.

SEAL OF THE PRESIDENT OF THE UNITED STATES

2010

Solar Impulse Airplane

Bertram Piccard (b. 1958)

What if you want to engineer an airplane that is powered strictly by the sun. And this airplane has to meet two requirements: it must be able to fly for more than 24 hours— a full day/night cycle, and it must be able to carry two human beings and all of their gear. Those are the specifications for the Solar Impulse airplane, which flew for 26 hours in 2010. Developed by a multidisciplinary team of over fifty specialists from sixty countries, the construction was initiated and overseen by Swiss aeronaut Bertram Piccard, who worked in conjunction with the Swiss Federal Institute of Technology in Lausanne.

The engineering problem with this aircraft was that, while the sun does provide some energy, it is not constant and it is diffuse rather than concentrated. Collecting that energy requires equipment, which means weight, and weight is the enemy of **airplanes**.

The sun provides energy to the planet at a rate of roughly one kilowatt per square meter. But it only does this for six hours a day at an angle the airplane can use. Thin, light solar cells make this endeavor conceivable. The Solar Impulse uses cells only 135 microns thick with 22 percent efficiency. They cover the entire wing with approximately 200 square meters of collecting surface.

That power goes into four 10-horsepower (7.5 kilowatt) motors. What do these motors do at night? **Lithium ion batteries** handle part of the storage capacity. But these batteries are too heavy to store 18 hours of power.

To store the rest of the power, engineers use potential energy instead. When the sun is shining, the airplane rises as high as 30,000 feet (9150 meters). Then at night the airplane acts like a glider, sinking down as low as 3,000 feet (915 meters). Storing energy in this way adds no weight to the plane.

The next step is version two, with the plan to circumnavigate the globe strictly on solar power. With a cruise speed of less than 50 mph, it will be a long trip.

SEE ALSO The Wright Brothers' Airplane (1903), Human-Powered Airplane (1977), Lithium Ion Battery (1991).

The Solar Impulse experimental aircraft flying for demonstration in Le Bourget, France.

HB-SIA
SOLVAY
OMEGA
OMEGA
SOLVAY
SOLARIMPULSE
OMEGA
SOLVAY

2010

BP Blowout Preventer Failure

Engineers often see emergencies arise over and over again. Because of the frequency of recurrence, they develop safety systems. One example: sprinkler systems in public buildings. Fires are not uncommon, so sprinkler systems extinguish them automatically.

In the world of **oil drilling**, a common emergency is the blowout. Once a borehole has been drilled from the surface to the oil reservoir, the reservoir will frequently send oil and gas up the hole under pressure. If the pressure is moderate and consistent, it is manageable. But if the well kicks, creating a high-pressure surge, events can spiral out of control. The blowout preventer (BOP) is a piece of equipment engineers put in place to protect against surges.

The idea behind a ram-type blowout preventer: powerful hydraulic rams slide in from the side to close off the well. In some, shear plates cut through anything in the way as well. An annular-type BOP uses a rubber donut to seal around drill pipe. A well will have several blowout preventers in a stack attached to the wellhead. They may activate automatically, manually, or remotely.

What if the safety equipment fails? In one of the most famous cases, the well blew out at the site of the Deepwater Horizon rig in the Gulf of Mexico in 2010, ultimately releasing five million barrels of oil into the Gulf.

The blowout preventer was engineered to automatically seal the well in a case just like this. It should have activated and clamped the well shut. And it did try. One annular and three rams activated. But the engineers miscalculated. The blowout preventer completely failed at its assigned task.

What happened? A postmortem analysis is how engineers learn from mistakes to prevent repeats in the future. It appears that high-pressure flow from the well washed out rubber and metal pieces of the BOP. In addition, the sheer ram failed to sheer the drill pipe and seal. Other elements never activated. Using the results from the postmortem, engineers improve their designs, add redundancy, and eliminate points of failure.

SEE ALSO Oil Well (1859), Wamsutta Oil Refinery (1861), *Seawise Giant* Supertanker (1979).

vetco

2010

Burj Khalifa

Adrian Smith (b. 1944), **Bill Baker** (b. 1953)

The Burj Khalifa opened in Dubai in 2010, becoming the tallest building in the world by far—over half a mile (0.8 km) tall. There were a number of challenges in this project, faced by architect Adrian Smith at the firm Skidmore, Owings and Merrill (which also designed **One World Trade Center** in New York City), and structural engineer Bill Baker. An obvious one involves the stiffness of such a tall structure—how to keep it from swaying like a reed in the wind.

The foundation of any skyscraper is key, and with a building this tall it needs to be robust. So engineers created what is called a piled raft. They started by sinking nearly 200 steel piles 155 feet (47 meters) into the ground. They are 5 feet (1.5 meters) in diameter and filled with concrete. On top of the piles engineers designed a 12-foot-thick (3.7 meters) concrete raft. The building rises on top of this raft.

The core of the building is a hexagonal tube filled with elevators, stairs, and utilities. Around this tube are three buttresses that support the core and contain the actual floor space that tenants use. The buttresses taper until, at the top of the building, nothing but the central core remains.

Structurally, the entire building depends on concrete and steel. Each buttress is in essence a gigantic vertical I beam with 2-foot-thick (0.6 meter) steel-reinforced walls forming the flange and web of the beam. The steel is straightforward but the **concrete** required finesse. Construction using concrete is a challenge in such a hot climate. Engineers used ice in the mix and pumped concrete at night to prevent overheating and cracking.

The outer surface is a traditional curtain wall of glass, aluminum, and stainless steel. What is shocking is the amount of glass—it is said that the 26,000 panes of glass could cover 25 American football fields. The result is a building that is artistically stunning. It is an engineering marvel that is surprisingly beautiful.

SEE ALSO Great Pyramid (2550 BCE), Concrete (1400 BCE), Woolworth Building (1913), World Trade Center (1973).

Skyline view of downtown Dubai, UAE, showing the Burj Khalifa and Dubai Fountain in 2010.

2010

Harry Potter Forbidden Journey Ride

Imagine that you want to engineer the ultimate amusement park ride. You want it to have thrilling motion effects like a **roller coaster**. You want it to have theatrical effects and sets like a typical indoor ride such as Disney's Haunted Mansion. But then you want to blend those indoor sets into a **3D** video experience, for example to give people the thrill of virtual flying. And then to add to the 3D effect, you want to project the video image onto the inside of a dome so that it stretches around and fills the peripheral vision.

It sounds like a tall order, but this kind of fully integrated ride technology does exist thanks to some incredible engineering. Perhaps the best example of it is the Harry Potter ride called Forbidden Journey at the Universal Theme Park, which opened in 2010 in Orlando, FL.

As riders board, they are seated in four-person cars, with all four in a row facing the same direction. If riders could see beneath the floor, they would realize that the seating is highly unusual. It is mounted on a huge **robot** arm with a maximum reach of 25 feet (7.62 meters) or so. That whole robot arm, one for each car, rolls down a track.

As each robot arm with passengers goes down the track, one of two things happens. Either the arm is passing through a theatrical set, or it is passing in front of a dome screen. The domes are immense and are mounted on a carousel holding six of them.

As the car centers itself inside a dome, the car and dome can track together to provide a video experience lasting for many seconds. While the video is playing in the dome, the robot arm can tilt, dive, and rumble to simulate the experience of motion that correlates with the moving image. When the robot arm is in a theatrical section, the arm can move the car in any direction to focus the riders' attention.

This is a great example of engineers bringing incredible technology to millions of people to create an exciting, immersive experience.

SEE ALSO Under Friction Roller Coaster (1919), Robot (1921), 3D Glasses (1952).

Expo visitors ride a "robocoaster" in Hanover, Germany.

16
ORGANIZED BY
MATRADE

2010

Alta Wind Energy Center

The Alta Wind Energy Center, which opened in 2010 in the Tehachapi-Mojave region of California, is the world's largest wind farm. It has a planned capacity of 3 gigawatts on 9,000 acres. Hundreds of wind turbines will work together to generate the power.

The aeronautical engineering that goes into wind turbines is impressive. Currently, the largest wind turbine generates 8 megawatts. It does this using three massive blades that are each 262 feet (80 meters) long. To get an idea of the size, the full wingspan (both wings) on a **747** is only 225 feet (68 meters). Each wind turbine's tower is 460 feet (140 meters) tall.

The three hollow blades are generally made of fiberglass—E-glass fibers in a polyester resin. They attach to a hub, which spins a shaft that drives a generator. There is a sensor package that detects wind direction and then a motor that keeps the blades pointed directly into the wind.

The biggest problem engineers face with an operational wind turbine is self-destruction in high winds. If the blades spin too fast, centrifugal forces can rip the turbine apart. The key to survival is blade angle. A computer looks at wind speed and furls the blades as wind speed increases. There is also a heavy-duty braking system attached to the main shaft that allows the turbine to completely stop. The brake is also helpful during maintenance.

Strong, consistent winds are what engineers look for when siting a wind farm. Offshore locations are popular especially in Europe because ocean winds can be uniform. The plains areas of the United States, for example in Texas, Iowa, and the Dakotas, have uniform winds as well.

As turbines get larger they become more cost-effective, and wind power now costs less than power generated by coal or gas. There is but one fly left in the engineering ointment—the fact that the wind sometimes stops blowing. For decades, engineers have been working on systems to store power. Options include immense chemical batteries, flywheels, or compressed air and **pumped water systems**. Once engineers get the storage problem definitively solved in a cost-effective way, wind power may become the perfect power source.

SEE ALSO Power Grid (1878), Boeing 747 Jumbo Jet (1968), Bath County Pumped Storage (1985), Ivanpah Solar Electric Generating System (2014).

Wind energy is cleaner and less expensive than power generated by coal and gas.

2010

Tablet Computer

Prior to the iPad there had been hundreds of tablet computer designs devised by different manufacturers. Microsoft had even pushed a tablet computing initiative in 2001, based off of Windows XP, that had encouraged manufacturers to create tablet computers. But nothing really caught on.

Then the iPad tablet appeared in 2010, it took off like a rocket, and this tells us something about engineering. Sometimes the form that engineers create matters as much as the function.

There were several things that Apple engineers did with the iPad that were innovative or revolutionary. One was the impossible thinness and lightness of the package, while at the same time feeling remarkably solid. It was also balanced and easy to hold, and sized perfectly for holding. There was no fan, yet cooling was not an issue. Battery life was stunning at ten hours—for the first time users did not need to constantly think about recharging.

Then there was the fact that the iPad did many of the things a laptop could do, but did them so simply. You never had to think about an "operating system" or a "blue screen" or "rebooting" or "viruses." And when you wanted to install new software, you pushed one button and you had it in just a few seconds. Removal was just as easy. And the apps never corrupted each other or "crashed the machine."

Here, finally, was a simple device that anyone could use easily to browse the **web**, read e-mail, take notes, keep a calendar, etc. Unlike a laptop, you could curl up with an iPad to read in bed or on the couch. You didn't have to open it up and worry about a big power brick like a laptop. You took it out of your bag, or picked it up off the table, and pushed one button. It was instantly ready to go.

It was an engineering triumph. It contained unspeakable levels of thoughtful design and engineering touches. But the engineering was completely invisible to the user. Today many companies follow the path that Apple pioneered.

SEE ALSO Mobile Phone (1983), World Wide Web (1990), Smart Phone (2007).

Tablet computers, pictured here, are lightweight and portable.

5:17 PM
23%
FaceTime
Photos
Camera
Clock
Photo Booth
Friday
22
Calendar
Notes
Reminders
Newsstand
App Store
Game Center
Settings
Facebook
Twitter
Mail
Videos
Music
iPad
17:17
%31
Messages
FaceTime
Photos
Camera
Maps
Clock
Photo Booth
22
Calendar
Contacts
Notes
Reminders
Newsstand
iTunes
App Store
Game Center
Settings
Facebook
Google
Twitter
Instagram
Safari
Mail
Videos
Music

2010

The X2 and X3 Helicopters

Traditional **helicopters** have a single main rotor and a tail rotor. And they have a problem: top speed is much slower than a turboprop or jet airplane. In civilian situations that means longer flight times. In military situations it means that helicopters are easier to shoot down. Example: the **Apache helicopter** (AH-64) has a maximum speed of 182 mph (292 kph). If people are shooting at you, you would prefer to escape at a much higher speed.

This speed limitation is a well-known engineering problem. Consider the main rotor blades as they spin. The tip of each blade slices through the air at 500 mph (800 kph) in hover. Now the helicopter accelerates to 100 mph (160 kph). As a blade moves forward, in the direction of flight, the tip is actually traveling at 600 mph (960 kph). As a tip moves backwards (retreating), it is going 400 mph (640 mph). Two things happen as the forward speed increases: the advancing tip approaches supersonic speeds, which damage the rotor, and the retreating blade stops creating lift—it's called retreating blade stall.

The engineers of the X2 (Sikorsky) helicopter overcame this problem in 2008 in a creative way. First, they use two counter-rotating rotors. Lift remains balanced on both sides of the rotors as the speed increases. Second, they slow the rotational speed of the rotors as forward speed increases. Third, they add a large pusher prop to move the helicopter through the air faster.

The engineers of the X3 (Eurocopter) helicopter used a completely different approach in 2010: 1) the main rotor follows the tradition of a regular helicopter, 2) there is no tail rotor, 3) the helicopter has short wings, and 4) there are propellers on both wings. One of the wing propellers spins faster than the other to replace the action of the tail rotor. The main rotor has five blades and slows during high-speed flight, with the short wings picking up part of the load.

The X2 and X3 are able to go nearly 300 mph (480 kph) with these engineering innovations. Two engineering teams, two radically different solutions that both yield better performance.

SEE ALSO Helicopter (1944), Apache Helicopter (1986).

Pictured: An X2 helicopter in flight in Germany.

X3
EUROCOPTER
X3
333
EUROCOPTER
F-ZXXX

2011

Fukushima Disaster

What if engineers make assumptions that turn out to be untrue? Sometimes it doesn't matter—it is corrected before it becomes a big deal. Or the mistake occurs in a system that has enough play in it to compensate. Or there is a backup system that takes over. But every now and again, we see small engineering assumptions blow up into huge catastrophes. Such was the case with the Fukishima **nuclear reactor** disaster in 2011.

The Fukishima reactors were designed to handle earthquakes. When the magnitude 9.0 earthquake happened off of Japan's coast in 2011, the reactors shut down automatically by inserting their control rods. The reactor buildings were undamaged. Everything seemed fine. Even though the earthquake cut off the reactor facility from the **power grid**, there were multiple backup systems on site including batteries, diesel generators, and emergency cooling systems that required no power at all.

Engineers had even anticipated tsunamis. A sea wall surrounded the reactor facility to keep tsunamis out.

What the engineers had not anticipated was the size of the 50-foot (15 meter) tsunami that occurred, and the side effects it would have. Engineers assumed a maximum tsunami of 10 meters. The diesel generators, the batteries, the distribution gear, and the fuel tanks for them were all located in the basement of the reactor building. When a 15-meter tsunami came, it destroyed every source of backup power. The batteries shorted out. The diesel generators were inundated. The distribution gear was underwater so it was not possible to easily plug in new external sources of power. At Unit 1, the emergency cooling system that was supposed to operate without power failed because a closed valve could not be opened.

If engineers had considered the possibility of a 15-meter tsunami, things would have unfolded very differently at Fukushima. But because of an aberrant natural occurrence, all of the redundant systems in the world were useless.

SEE ALSO Light Water Reactor (1946), CANDU Reactor (1971), Chernobyl (1986).

Taken by an unmanned aircraft in 2011, this photo shows the Fukushima Daiichi nuclear plant in Okuma, Japan.

2011

Self-Driving Car

The idea of a self-driving car once seemed impossible. While self-driving cars using technology such as radio transmitters have been experimented on since 1925 (as in the case of the "Linriccan Wonder," which drove through a traffic jam in New York City), it wasn't until 2011 that Google engineers revealed that they had logged more than 100,000 miles with a self-driving car on normal public roadways.

How did engineers at Google make this happen? Several pieces of technology combine to let a self-driving car sense and understand the world. Lidar (distance sensing that uses laser light instead of microwaves) is key, letting the car form a 360-degree 3D view. Lidar detects stationary objects like parked cars, poles, curbs, etc., as well as moving things like other cars and pedestrians. There is also front- and rear-facing **radar**, which helps the car detect objects in its path at longer distances. An optical camera looks for things like changing traffic lights.

GPS gives the car a sense of its position in the world. An inertial guidance system gives a bit more accuracy, and sensors on the wheels let the car know exactly how far it has traveled.

One other technique engineers use is foreknowledge. Before a self-driving car gets on a road, another car has already driven the road to map out exactly where the lanes are located, where the car should halt for stop signs, where the traffic lights are located in the field of view, how the elevation changes, etc. This way, even if lane markings are obscured by rain, the car still knows where the lanes are located.

Putting all of these techniques and equipment together with powerful software creates a vehicle that can reliably drive itself. In fact, a self-driving car is a better driver than most human beings. That's because the sensors see everything all around the vehicle all the time and the computer never gets distracted.

SEE ALSO Radar (1940), Global Positioning System (GPS) (1994), Watson (2011).

A Lexus RX450h Google driverless car. The software powering Google's cars is called Google Chauffeur and is currently in the testing phase.

Google
self-driving car

2011

Instant Skyscraper

In the United States and many other countries, it can take quite a bit of time to build a skyscraper. For example, Freedom Tower—the new building at the site of the 9/11 disaster—took more than a decade for design and construction. Admittedly there was a lot of emotion involved in the design process and it was not the easiest construction site, but a decade is still a long time. Most skyscrapers take well over a year to build.

In 2011, a Chinese company named Broad showed a pathway to a completely new construction paradigm for skyscrapers: Nearly the entire building is engineered to be pre-fabricated in a factory, shipped to the site in modules, and then assembled on site in a few days. Using these techniques, the company erected a 30-story hotel in just 15 days.

Because it is pre-fabricated, much of the work can happen in a factory environment with standard parts, out of the weather. For example, floor modules are standardized steel sections reinforced with trusses. The concrete is poured, tile laid, ceiling and fixtures installed along with all utilities—plumbing, HVAC, electrical, etc.

At the construction site, there are only a few things left to do. The posts that support the building are all standardized and prefabricated. So posts go in, then floor sections, then posts, then floor sections, and so on. Standardized windows and exterior wall panels are attached. Then workers quickly erect all of the interior walls

Because the building is standardized, it is easy to erect many of them. And they can all contain the same engineered features like extreme **earthquake resistance**, energy efficiency, and air filtration. These features would be expensive in a custom building, but in a building like this, economies of scale and standardization bring prices down.

The company has plans to use the same engineering ideas to build the world's tallest building—taller than the **Burj Khalifa**—in just 90 days. It is easy to imagine engineers around the world applying these modular ideas to many different construction projects in the future.

SEE ALSO Empire State Building (1931), Earthquake-Safe Buildings (2009), Burj Khalifa (2010).

Chinese workers manufacture steel frameworks to be used to build the thirty-story Ark Hotel in the city of Changsha, which was constructed in fifteen days by Broad Group.

1-9
安全

2011

Watson

David Ferrucci (b. 1962)

Every now and again, engineers and scientists pop something into the world and it is completely unexpected. For example, we went from "no airplanes" to "airplanes" the day the **Wright Flyer** lifted off the ground. We went from "**self-driving cars** are way off in the future" to "self-driving cars are here today" the day Google announced 100,000 accident-free road miles for its self-driving technology.

And the same is true for Watson—IBM's Jeopardy-playing computer, developed by a team lead by principal investigator David Ferrucci. Out of the blue, Watson had trounced the best human players. It was quite surprising.

So how did IBM engineers pull this off? Part of it was a very large hardware problem, and part of it was innovative software. The key, and perhaps most surprising, decision the software engineers made was to work with untagged data. In other words, Watson would read in unmodified versions of things like Wikipedia, the Internet Movie Database, and the dictionary. Watson would employ things like natural language processing, machine learning, and semantic analysis to make sense of all of it. No one would have to structure or tag the data in any way. From that raw text information, the algorithms would sort through it all to form a knowledge base, and then combine it with a natural language processing front end to understand the Jeopardy questions. A synthetic voice then provides the answers.

And Watson is fast. The first system—the one that played on national TV in 2011—is a supercomputer filling a room. That's because it used nearly 100 servers, containing nearly 3,000 computing cores and 16 terabytes of memory, to run the software.

Watson's Jeopardy-playing ability is just the beginning. The same software techniques can grind through all kinds of text information—medical, legal, scientific, Internet—and perform the same kind of magic. For example, imagine a search engine that "understands" your question. Or a system where a doctor can ask any medical question and find all the related research results. Watson is a great example of engineering on the cutting edge.

SEE ALSO The Wright Brothers' Airplane (1903), Chess Computer (1950), Microprocessor (1971), Self-Driving Car (2011), Brain Replication (c. 2024).

Contestants Ken Jennings and Brad Rutter compete against Watson at a press conference on January 13, 2011 in Yorktown Heights, New York.

PIENSE
THINK
$0
WATSON
$0
BRAD

2012

Curiosity Rover

The request seems simple enough: "Let's put a rover on Mars!" But the instant the request is made, the engineering problems start piling up. The fact that multiple rovers and robots have made it to Mars and sent back impressive amounts of imagery and data is testimony to the engineering abilities we have available as a modern society.

Think about all of the problems faced by engineers employed by the Mars Science Laboratory. They had to design a rover for frigid temperatures of -100°F (-73°C) or less. The rover also needed power. So Curiosity is nuclear powered in the form of a radioisotope thermoelectric generator. Heat from 11 pounds (5 kg) of decaying plutonium turns into electricity, and the heat also keeps the rover warm.

The rover needed a communication system, even though Mars can be 250 million miles away. Also, both Earth and Mars are rotating. On Earth, there are huge antennas for the Deep Space Network positioned so that, wherever the Earth is in its rotation, one of the antennas can see Mars. If the rover is on a part of Mars facing Earth, it can talk directly to a DSN antenna. But it is easier to communicate with **satellites** we've positioned to orbit Mars when they pass overhead, and the satellite communicates with Earth.

There was also the problem of autonomy. It might take 10 minutes for a message to get to Mars and back because of the incredible distances. So the rover must do many things using its own onboard intelligence.

And just getting to Mars and landing was a problem. The rover had to get into Earth orbit, then spend months flying to Mars, then get into Mars orbit, then descend and land. Why not just open a big parachute for the landing? Because the Mars atmosphere is so thin. So there are retro-rockets that have to fire during descent, all autonomously, to place the rover on level ground.

Everything is difficult, down to the little lab that performs chemical analysis after scooping up soil. Yet the Curiosity rover successfully landed in 2012 and deployed in 2013 and is able to do science on a completely different planet.

SEE ALSO Space Satellite (1957), Lunar Rover (1971), Mars Colony (c. 2030).

This artist concept features NASA's Mars Science Laboratory Curiosity rover, a mobile robot for investigating Mars's past or present ability to sustain microbial life.

2012

Human-Powered Helicopter

Todd Reichert (Dates Unavailable), **Cameron Robertson** (Dates Unavailable)

Back in 1980, the excitement around human-powered aircraft was intense. An engineering marvel named the **Gossamer Condor** claimed the Kremer Prize (see "Human-Powered Airplane") in 1977 with a figure-eight flight. In 1979, the seemingly impossible happened—Bryan Allen peddled the Gossamer Albatross 22 miles (35 km) across the English Channel in 169 minutes.

So in 1980, when the American Helicopter Society announced the Sikorsky Prize, it seemed relatively straightforward. The winner would need to rise to a height of 3 meters with human power alone and hold the craft there for 60 seconds. The flight could even be indoors to eliminate the wind.

It would take engineers more than three decades to claim this seemingly simple prize. A big part of it is the fact that a **helicopter** that is hovering takes a lot more power than an **airplane** that is flying. The Gossamer Albatross took 0.4 hp (300 watts) to fly straight and level. It is said that the pilot of the human-powered helicopter trained so that he could output more than one hp (750 watts) for several minutes. He could peak at 1.5 hp (1,100 watts) to initially get off the ground. This would be enough power for takeoff, the 60-second hover, and landing.

The Atlas helicopter uses four rotors, with the pilot sitting in the center. Each rotor is 33 feet (10 meters) in diameter. The total of eight wings and four hubs is itself expensive in terms of weight. Then the rotors must link together. This requires a massive X-shaped truss structure made of thin carbon fiber tubes and Vectran (a polymer) bracing lines. Once assembled, the helicopter is a 154 feet (47 meters) across, but weighs only 122 pounds (55 kg). Instead of heavy chains to power the rotors, the hubs are wound with thin Vectran lines. When the pilot pedals, he is spooling in four lines from the four hubs. So flight duration is limited by the length of those lines.

On August 28, 2012, Todd Reichert and Cameron Robertson, working with a team from the University of Toronto, powered the Atlas Helicopter and won the Sikorsky prize—32 years later.

SEE ALSO Helicopter (1944), Human-Powered Airplane (1977), Quadrotor (2008).

Aerial view of the AeroVelo Atlas Human Powered Helicopter (HPH) developed for the AHS Sikorsky Prize. Taken from the "crow's nest" of the Ontario Soccer Center.

The City Above Toronto
UMBRO
PFAFF
VAUGHAN
GO FOR THE CUP!
Food Basics
ONTARIO CUP
metro
Food at its best.
1390 Major
Mackenzie Drive
FIELD
B
DANGER
DO NOT HANG
ON GOALS
NO
SPITTING
ON THE FIELD

2013

NCSU BookBot

If you think about a library from an engineering perspective—especially a large library like you would find at a university—you realize that it is not a very efficient way to use space. You have the shelves of books, along with human-size aisles between the shelves to allow patrons to access the books. All of those aisles waste a huge amount of space. In a big library, there can be 10 to 20 such floors filled with paper books and journals, but the majority of space in the library is devoted to aisles.

Engineered technology now affords a much more efficient way to store paper books and journals: the robotic book storage system.

The bookBot at North Carolina State University, which began operation in 2013, is one example of such a system. It consists of approximately 20,000 large drawers that can be accessed by four **robots**. Each drawer can hold approximately 100 books. A computer has a complete inventory of which books are located in which drawers. The bookBot gives NCSU an extremely dense way to store approximately 2 million books.

When a student or faculty member wants to borrow a book, he or she makes a request for the book through a web site. The bookBot dispatches a robot to retrieve the drawer containing the book. The drawer comes to a central place where a human being finds and retrieves the book from the drawer. Then the robot puts the drawer back. When the book comes back, the process is reversed.

Two million books could easily fill a 20-story building. By putting them in the bookBot, the total space consumed is 50 feet (15 meters) wide by 160 feet (50 meters) long by 50 feet (15 meters) high. It would take nine times more space to store the books in traditional shelving.

NCSU is using its bookBot in the following way: In its original library, it is taking floors that were once filled by bookshelves and emptying them. The books all move into the bookBot. Then the emptied floor is converted over to another use, such as study space, meeting rooms, etc. Engineers have essentially created a new building by compressing the books into the bookBot.

SEE ALSO Robot (1921).

North Carolina State University James B. Hunt Jr. Library automated book retrieval system ("bookBot").

2014

Ivanpah Solar Electric Generating System

Imagine that you want to collect sunlight. A *lot* of sunlight. So much sunlight, and so much heat, that you can power a small city with it. That is what this huge heliostat array near Las Vegas is doing starting in 2014. Owned by Google, NRG Energy, and BrightSource Energy, it is an engineered structure of immense proportions.

The basic idea is so simple: Take a flat mirror and use it to reflect sunlight toward a tower. Now scale up that flat mirror so it is 75 square feet (7 square meters) in size. Now put two of these mirrors on a moving axis—this assembly is called a heliostat. Now install 173,000 of these heliostats pointing at three different towers. That is the scale of operation at the Ivanpah Solar Electric Generating station.

The purpose of a heliostat is to track the sun. A heliostat can tilt and rotate the mirrors so they are always reflecting sunlight toward the tower. That would be a fairly easy task, except for the wind. Imagine that you stick your hand out of the window of a car traveling at 60 miles an hour. There is a lot of drag. Now imagine sticking a sheet of plywood out the window—the wind force is immense. The heliostat is holding two mirrors equaling roughly five sheets of plywood in size. Each heliostat has to be quite powerful to deal with that kind of wind loading and keep the mirror pointing properly on a windy day.

Inside the three towers there are custom boilers that turn the sun's heat into steam. The steam drives a **steam turbine** just as it would in a conventional **power plant**.

Consider that the sun delivers about 1 kilowatt of energy per square meter of the Earth's surface. Consider that this site has about 2.5 million square meters of mirror. That is 2.5 million kilowatts of heat being collected. About half of that heat is lost in the process of focusing it on the tower. It then converts to electricity at roughly 30 percent efficiency. So the site is able to produce about 400 megawatts when the sun is shining brightly—approximately 7 hours a day. If electricity sells for 10 cents per kilowatt-hour, that's about $100 million of electricity annually.

SEE ALSO Steam Turbine (1890), Light Water Reactor (1946), CANDU Reactor (1971).

Aerial view of the solar electric generating system in the Mojave Desert.

2016

Venice Flood System

Venice is a city where floods more than 5 feet (1.4 meters) high are not unusual. And it is going to get worse with rising sea levels. Eventually the city could be lost to the flooding. Many coastal cities may soon be experiencing similar problems. For example, in New York City, the chance of flooding is now more than 10 times greater than a century ago.

The entire city of Venice is fascinating from an engineering perspective. The city sits on a series of muddy islands in the center of a lagoon. To provide stability, millions of tree trunks were harvested and pounded into the mud hundreds of years ago.

The problem is that the level of the lagoon rises and falls freely with the tides. When storms occur, the lagoon can rise even higher. Even a low-pressure weather system reduces air pressure on the water and the lagoon rises.

To stop the flooding, Venice is installing a system called MOSE to close off the lagoon. A sea wall already separates the lagoon from the Adriatic Sea, with only three inlets to provide access for boats. A set of movable gates has been placed on the seafloor in the inlets. When higher than normal tides are predicted, the gates fill with compressed air and rise up to separate the lagoon from the sea. The gates can maintain up to a 10-foot (3 meter) level difference, keeping the lagoon below its flood stage.

Getting the gates positioned on the seafloor has been a special engineering challenge. The gates, their hinges, and compressed air supply rest in huge, modular **concrete** trays lowered to the seafloor. Engineers decided to cast them nearby in a purpose-made drydock. Then they float them into precise position and sink them into place.

There are 78 gates total, each 20 meters wide. Once the whole system is operational, expected in 2016, engineers should have complete control over floods and will have saved Venice.

SEE ALSO Concrete (1400 AAA), Leaning Tower of Pisa (1372), Zuiderzee Works (1891), Three Gorges Dam (2008).

Flood in Venice, Italy; acqua alta on Piazza San Marco.

c. 2020

Vactrains

There seems to be a speed limit on human transportation. Cars get up to about 200 mph (320 kph) and they seem to stop there. Trains too. Jet airplanes do better, but they cap out around 500 mph (800 kph). Why do engineers hit these barriers?

The answer is air and the drag it causes. To get a **Bugatti Veyron** to go 200 mph, it takes about 1,000 hp (750,000 watts) and 2 gallons of fuel per minute. Almost all of that power goes into pushing air out of the way. The same is true for high-speed trains. Jet airplanes can do better because they cruise at 30,000 feet (9,000 meters) where air density is one-quarter of its value at sea level.

What is the solution if people want to go faster? Engineers need to eliminate the air. That is where the idea of vactrains comes from—and research currently being conducted at China's Southwest Jiaotong University suggest that they could become a reality as soon as 2020. A vactrain is a **magnetically levitated** vehicle running inside a tube that has had much of the air removed. In theory a vactrain can run at 3,000+ mph (4,800 kph) if the vacuum is good enough. At that speed, a trip across the United States takes an hour. A trip from New York to Beijing takes four hours. Vactrains would make long distance travel much easier and also, potentially, far less expensive.

There are several engineering hurdles that arise, however. One is the turning radius at 3,000 mph. A vehicle going that fast needs very gentle curves in order to avoid excessive g-forces for the passengers. The turning radius is measured in hundreds of miles. That same rule applies for elevation changes. The track needs to be very smooth and level.

Another potential problem is the ocean. How can track cross large bodies of water? One innovative solution is to let the tube float several hundred feet below the surface, with anchor cables tying it to the bottom.

The vactrain is one of those ideas that could revolutionize our lives. And we already possess all of the technology to do it. All that is needed is the will and the investment.

SEE ALSO Magnetically Levitated Trains (1937), Bugatti Veyron (2005).

Released by Tesla Motors, this is a conceptual design rendering of the Hyperloop passenger transport capsule.

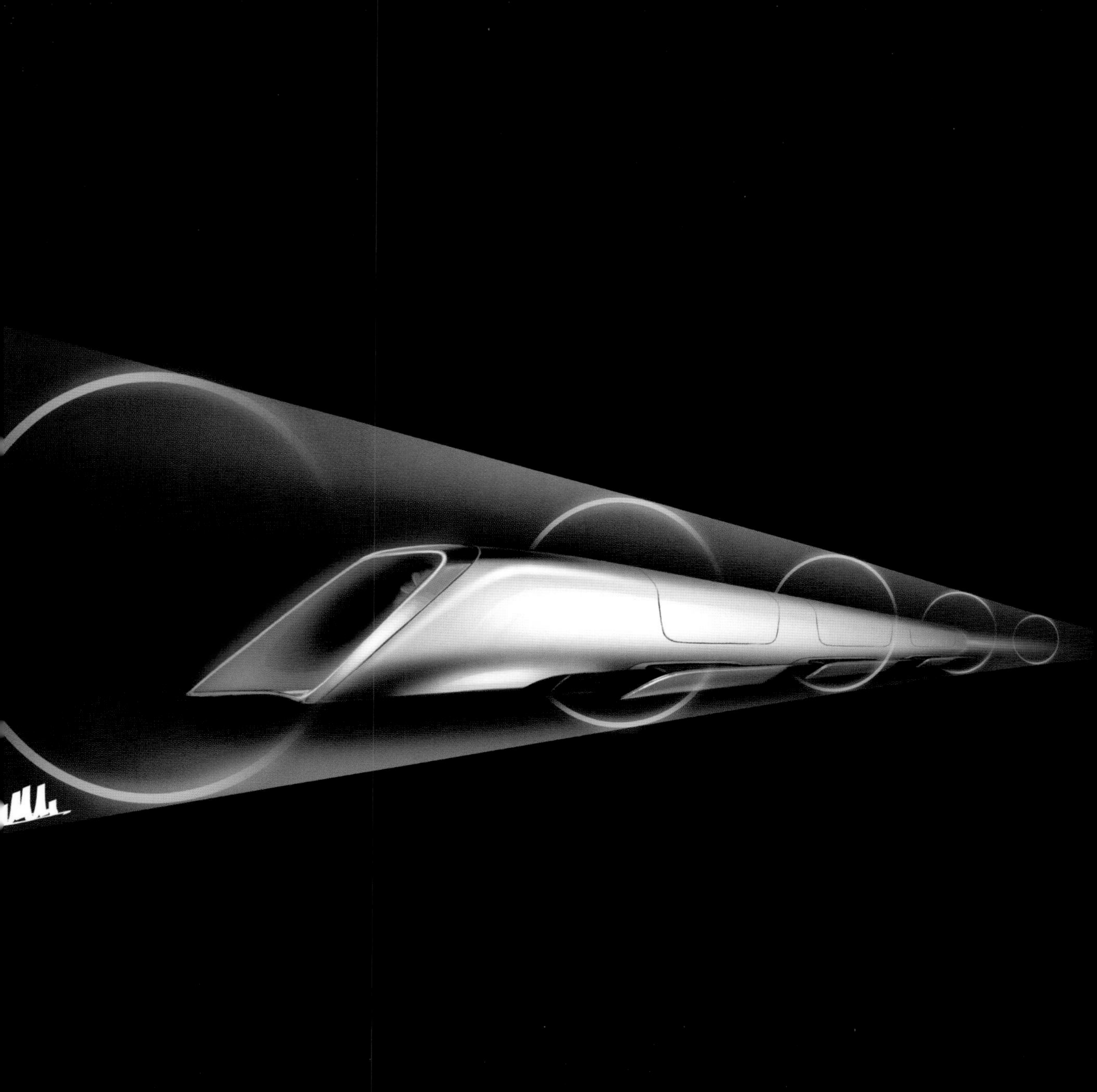

C. 2024

Brain Replication

Compared to the computers we all use today, the human brain is absolutely stunning in its power and efficiency. Since a silicon **computer** and a human brain are based on such different technologies (silicon transistors in one case, biological neurons and synapses in the other), it is difficult to do a perfect apples-to-apples comparison. But roughly speaking, scientists think the human brain might do the equivalent of 10 petaops (peta = one quadrillion) of computation per second and have a petabyte or more of storage capacity.

Yet the human brain only consumes about 20 watts and fits inside your head. Compare that to a typical laptop of today. The laptop uses roughly the same amount of power, but it has one-millionth the processing power and one-millionth the memory. And your laptop, at least today, cannot learn a new human language, look out at the world and recognize things, program itself, or do many of the other things human beings do quite trivially. Nor will your laptop say to itself, "I think, therefore I am."

So engineers look at the human brain and ask, "Is there a way to replicate this computing architecture?" Replicating the human brain would be beneficial in many different ways. And if the replication is close enough, in theory scientists and engineers will also replicate consciousness, human learning capabilities, and all the rest.

So how might they do it? One way is a complete software simulation running in a supercomputer. The human brain contains something like 86 billion neurons and 100 trillion synapses—a number that is daunting to simulate with current technology, but imaginable, and getting easier to imagine every year. Maybe we need a supercomputer that can perform a quintillion operations per second to do it?

Assembling enough hardware is one part of the equation. Understanding how to connect all the simulated neurons together is another. And then we need to get everything to actually work. Humans have yet to simulate even the simple neural networks found in insects.

Europe has the Human Brain Project. The US has the Brain Research Through Advancing Innovative Neurotechnologies project. The hope is that scientists and engineers can crack this nut in a decade or so.

SEE ALSO Robot (1921), ENIAC—The First Digital Computer (1946), Chess Computer (1950), ARPANET (1969), Microprocessor (1971), Watson (2011).

Brain replication is just one example of engineers looking to push the limits of traditional mortality.

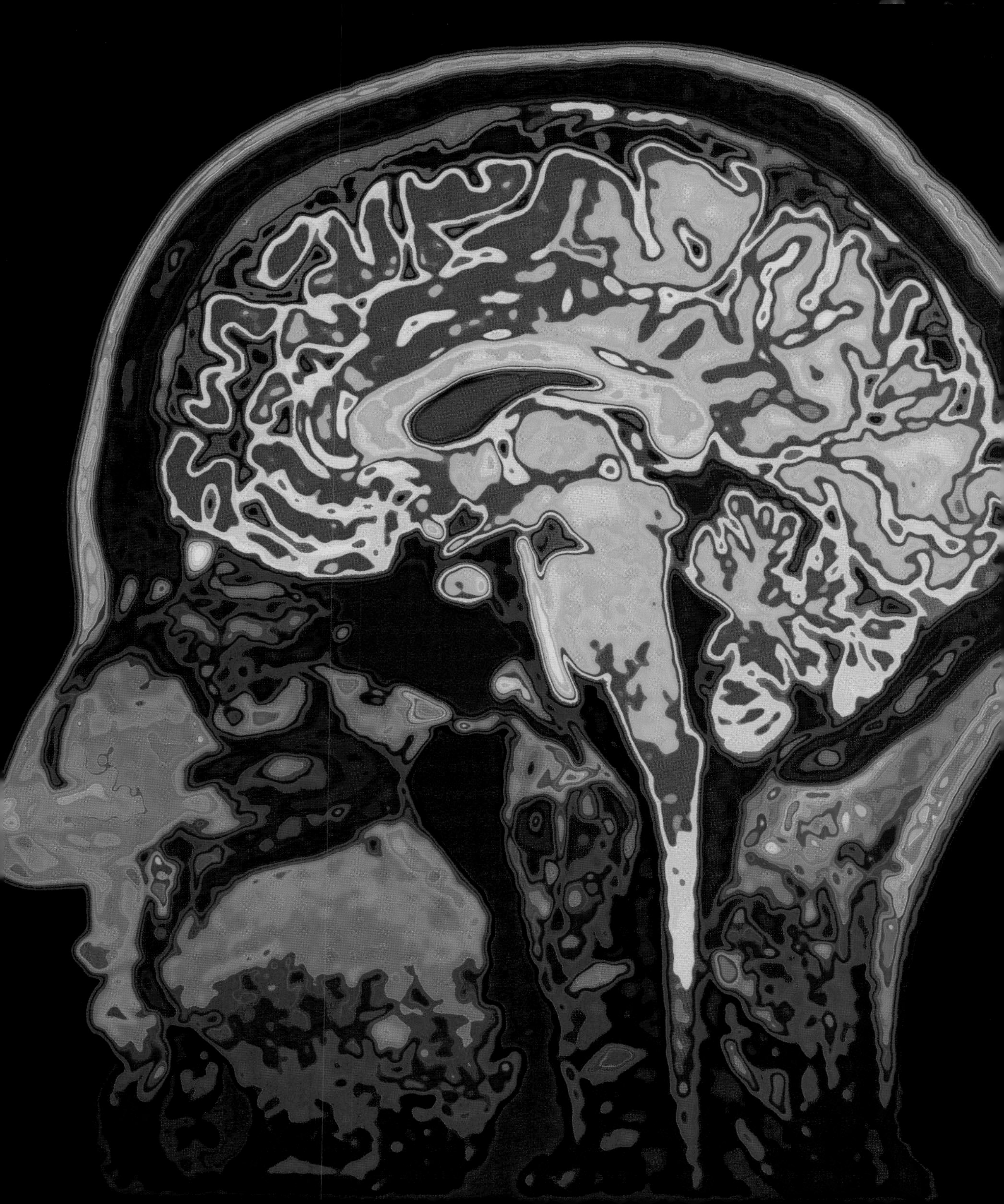

c. 2030

Mars Colony

No Mars colony exists yet. There may never be a Mars colony for a variety of reasons that include economics, inhospitability, and low gravity. But engineers have spent a great deal of time imagining a Mars colony. And the idea of setting up a self-sustaining human colony on Mars is fascinating to think about.

It's one thing to imagine setting up a human colony on, say, a deserted island. That's because we can take certain things for granted. There will be adequate air pressure, air temperatures, and oxygen. Rain will fall and provide fresh water. Animals roaming the island and fish in the sea provide an initial food source. Plants also provide food as well as fibers for clothing, ropes, and building materials. If a group of humans were dropped naked on a deserted island, they have a great chance of surviving.

On Mars, none of this is true. The temperature, air pressure, availability of oxygen, and water supply are all missing. There are no plants or animals of any kind. Even the gravity is wrong, being less than half Earth gravity, and there is no magnetosphere and insufficient atmosphere to protect the surface from UV radiation or other forms of radiation. The atmosphere contains so much CO_2 that it is toxic to humans and plants alike.

So a Mars colony would need to be completely engineered. It would be a sealed system, with its own artificial atmosphere, radiation protection, and heat. This probably means building underground. Subsurface ice could provide a water source, but colonists would need to bring their own initial food supply and then engineer a way to grow or manufacture additional food.

There are two things that might help. Engineers could create a robotic construction crew to arrive first and prepare everything. And terraforming might be possible, although there are a number of roadblocks and it takes centuries.

If humans ever do colonize Mars, one thing is sure—they won't do it without a lot of engineering talent brought to bear on the many problems that Mars poses.

SEE ALSO Lunar Landing (1969), Biosphere 2 (1991).

Artist impression of a Mars settlement with cutaway view.

NASA

Things We Have Yet to Engineer

This book has covered a wide range of amazing engineering advancements. Some of them started out as science fiction fantasies and became reality. The **cell phone**, for example, is a lot like the communicator seen in the *Star Trek* TV series in 1966. It only took 30 years for engineers to create a cheap, pocket-sized communication device.

Lots of other sci-fi ideas await implementation in the real world. Many are stalled right now because we don't have the fundamental scientific principles to support them, or the money. Here is a list of some of the things that engineers may figure out in the future:

Flying car—stalled by economics, stability, and weight.concerns

Time machine—stalled by fundamental science (SBS), if even possible

Immortality—SBS

Transporter room—may never happen, but **virtual reality** is the next best thing

Instant healing—long a staple of sci-fi and video games, SBS

Hoverboard—needs repulsorlifts

Easy underwater breathing—**SCUBA** and artificial gills get partway there

Starships, warp drive—stalled by engine technology, economics, and science

Suspended animation—SBS

Vertebrane—SBS

Space elevator—stalled by materials science and economics

Really good **batteries**—SBS

End of poverty—stalled by greed, inaction

Solution for global warming—stalled by inaction, economics

Food machine—SBS, but artificial meat is close

Easy interplanetary travel—stalled by engine technology, power sources, economics, gravity issues, radiation concerns, etc.

There are many innovations engineers have no way to start because the science doesn't exist. For example, the fictional *Star Wars* universe uses repulsorlifts on speeder bikes, land speeders, hovering robots, and spacecraft. If repulsorlift technology existed, engineers could exploit it in a thousand different ways. So we wait for scientists to deliver the goods. Then engineers will jump into action.

SEE ALSO Robot (1921), SCUBA (1944), Mobile Phone (1983), Virtual Reality (1985), Lithium Ion Battery (1991).

Conceptual illustration of a futuristic flying car.

Notes and Further Reading

General Reading

Blockley, D., *Engineering: A Very Short Introduction*, London: Oxford UP, 2012.

Constable, G. and Somerville, B., *A Century of Innovation*, Washington, DC: Joseph Baker, 2003.

Wikipedia Encyclopedia, *www.wikipedia.org*.

Books by Marshall Brain

Brain, M., *How God Works*, New York: Sterling, 2015.

Brain, M., *How Stuff Works*, New York: Chartwell, 2010.

Brain, M., *More How Stuff Works*, Hoboken, NJ: Wiley, 2002.

Brain, M., *How Much Does the Earth Weigh?*, Hoboken, NJ: Wiley, 2007.

Brain, M., *The Teenager's Guide to the Real World*, Cary, NC: BYG, 1997.

Brain, M., *Manna*, Cary, NC: BYG, 2012.

Brain, M., *The Meaning of Life*, Cary, NC: BYG, 2012.

Introduction

Merriam-Webster.com. http://tinyurl.com/q37olct.

30,000 BCE, Bow and Arrow

Dollinger, A. *http://tinyurl.com/3pbdqav.*

3300 BCE, Hunter/Gatherer Tools

Fowler, B., *Iceman*, Chicago, IL: Chicago UP, 2001.

2550 BCE, The Great Pyramid

Brier, B., *The Secret of the Great Pyramid*, NY: Harper, 2009.

2000 BCE, Inuit Technology

Living Dictionary. *http://tinyurl.com/nvzjzqj.*

1400 BCE, Concrete

Moorehead, C., *Lost and Found*, NY: Penguin, 1997.

625 BCE, Asphalt

Marozzi, J., *The Way of Herodotus*, NY: Da Capo, 2010.

438 BCE, Parthenon

Beard, M., *The Parthenon*, Cambridge, MA: Harvard UP, 2010.

312 BCE, Roman Aqueduct System

Rinne, K., *The Waters of Rome*, New Haven, CT: Yale UP, 2011.

100 BCE, Waterwheel

Wikander, Ö., *Handbook of Ancient Water Technology*, Leiden: Brill, 1992.

79, Pompeii

Beard, M., *Fires of Vesuvius*, Cambridge, MA: Belknap, 2010.

1040, Compass

Vardalas, J., *http://tinyurl.com/oz3rrm3.*

1144, Basilica of Saint Denis

Honour, H. and Fleming, J., *A World History of Art*, London: Laurence King, 2009.

1300, Catapult

Gurstelle, W., *The Art of the Catapult*, Chicago, IL: Chicago Review, 2004.

1372, Leaning Tower of Pisa

McLain, B., *Do Fish Drink Water?*, NY: Morrow, 1999.

1492, Square-Rigged Wooden Sailboats

Ship Wiki, *http://tinyurl.com/kflx47n.*

1600, The Great Wall of China

Bloomberg News, *http://tinyurl.com/pehjxd2.*

1620, Gunter's Chain

Linklater, A., *Measuring America*, NY: Penguin, 2003.

1670, Mechanical Pendulum Clock

Rawlings, A., *The Science of Clocks and Watches*, Upton, UK: British Horological Institute, 1993.

1750, Simple Machines at Yates Mill

Anderson, W., *Physics for Technical Students*, NY: McGraw-Hill, 1914.

1773, Building Implosions

Blanchard, B., http://tinyurl.com/nz9pjcg.

1784, Power Loom

Marsden, R., *http://tinyurl.com/mpq88kd.*

1790, Cotton Mill

Hunt, D., *http://tinyurl.com/morw2on.*

1794, Cotton Gin

Roe, J., *English and American Tool Builders*, New Haven, CT: Yale UP, 1916.

1800, High-Pressure Steam Engine

Kirby, R., *Engineering in History*, Mineola, NY: Dover, 1990.

1823, Truss Bridge

The Science Museum, *http://tinyurl.com/ntm2qnt.*

1824, Rensselaer Polytechnic Institute

Rensselaer Polytechnic Institute, *http://tinyurl.com/pwy2u9n.*

1825, Erie Canal

Finch, R., *http://tinyurl.com/mvll2fa.*

1830, Tom Thumb Steam Locomotive

Stover, J., *History of the Baltimore and Ohio Railroad*, West Lafayette, IN: Purdue UP, 1987.

1837, Telegraph System
Connected Earth, *http://tinyurl.com/lgntn64*.

1845, Mass Production
Hounshell, D., *From the American System to Mass Production, 1800–1932*, Baltimore, MD: John Hopkins UP, 1984.

1845, Tunnel Boring Machine
Bagust, H., *The Greater Genius?* Birmingham, UK: Ian Allan, 2006.

1846, Sewing Machine
ISMACS International, *http://tinyurl.com/qhj7t9f*.

1851, America's Cup
America's Cup, *http://tinyurl.com/qb9psfo*.

1854, Water Treatment
EPA.gov, *http://tinyurl.com/4ftm2y*.

1855, Bessemer Process
Ponting, C., *World History, A New Perspective*, New York: Pimlico, 2000.

1856, Plastic
Dreher, Carl, *http://tinyurl.com/k8hsxtk*.

1858, Big Ben
Hill, R., *God's Architect*, New Haven, CT: Yale UP, 2009.

1859, Oil Well
Tarbell, I. M., *The History of the Standard Oil Company*, Gloucester, MA: Peter Smith, 1963.

1859, Modern Sewer System
Halliday, S., *The Great Stink of London*, Gloucestershire, UK: History Press, 2001.

1860, Louisville Water Tower
Amies, N., *http://tinyurl.com/lewuhkk*.

1861, Wamsutta Oil Refinery
Schmidt, B., *http://tinyurl.com/lzt4c6x*.

1861, Elevator
Bellis, M., *http://tinyurl.com/ao5qkqo*.

1869, Transcontinental Railroad
Cooper, B., *Riding the Transcontinental Rails*, Philadelphia, PA: Polyglot, 2005.

1873, Cable Cars
Thompson, J., *http://tinyurl.com/lxtd7f5*.

1876, Telephone
John, R., *Network Nation*, Cambridge, MA: Harvard UP, 2010.

1878, Power Grid
Energy Graph, *http://tinyurl.com/mxayh62*.

1879, Carbon Fiber
Kopeliovich, D., *http://tinyurl.com/qbfn2tg*.

1885, Supercharger and Turbocharger
McNeil, I., *Encyclopedia of the History of Technology*, London: Routledge, 1990.

1885, Washington Monument
Savage, K., *Monument Wars*, Oakland, CA: California UP, 2011.

1886, Statue of Liberty
Khan, Y., *Enlightening the World*, Ithaca, NY: Cornell UP, 2010.

1889, Eiffel Tower
Harvie, D., *Eiffel*, Gloucestershire, UK: Sutton, 2006.

1889, Hall-Héroult Process
American Chemical Society, *http://tinyurl.com/kbmurtf*.

1890, Steam Turbine
Encyclopedia Britannica, *http://tinyurl.com/ncrj8q7*.

1891, Carnegie Hall
Carnegie Hall, *http://tinyurl.com/kelly3a*.

1891, Zuiderzee Works
Kimmelman, M., *http://tinyurl.com/qbhr4l6*.

1893, Two-Stroke Diesel Engine
Sloan, A., *Diesel Engine Design*, London: George Newnes, 1953.

1893, Ferris Wheel
World Digital Library, *http://tinyurl.com/myzq959*.

1897, Diesel Locomotive
Churella, A., *From Steam to Diesel*, Princeton, NJ: Princeton UP, 1998.

1899, Defibrillator
Nature magazine archives, *http://tinyurl.com/kap38f9*.

1902, Air Conditioning
Bergen Refrigeration, *http://tinyurl.com/lrer5sw*.

1903, EKG/ECG
Cooper, J., *http://tinyurl.com/jw3ub7k*.

1903, The Wright Brothers Airplane
National Parks Service, *http://tinyurl.com/mkkd4et*.

1905, Engineered Lumber
APA—The Engineered Wood Association, *http://tinyurl.com/kn9rqwh*.

1907, Professional Engineer Licensing
NSPE, *http://tinyurl.com/kz9yuyb*.

1908, Internal Combustion Engine
For a video of the engine in action, see *http://tinyurl.com/q9f9wla*.

1910, Laparoscopic Surgery
Martin Hatzinger, et al., *http://tinyurl.com/mpdpm96*.

1912, *Titanic*
Freer, A., and Griffiths, D., *http://tinyurl.com/o7tov7w*.

1913, Woolworth Building
Douglas, G., *Skyscrapers*, Jefferson, US: McFarland, 1996.

1914, The Panama Canal
Greene, J., *The Canal Builders*, New York: Penguin, 2009.

1917, Laser
Siegman, A., *Lasers*, Sausalito, CA: University Science, 1986.

1917, Hooker Telescope
Mount Wilson Observatory, *http://tinyurl.com/qeffczm*.

1919, Women's Engineering Society
Nicholson, V., *Singled Out*, Oxford, UK: Oxford UP, 2008.

1919, Under Friction Roller Coaster
Coker, R., *Roller Coasters*, New York: Metrobooks, 2002.

1920, Kinsol Trestle Bridge
Webb, W., *Railroad Construction*, Hoboken, NJ: Wiley, 1917.

1920, Radio Station
Nebeker, F., *Dawn of the Electronic Age*, Hoboken, NJ: Wiley, 2009.

1921, Robot
Roberts, A., *The History of Science Fiction*, New York: Palgrave Macmillan, 2006.

1926, Heart-Lung Machine
Kohn, L., *http://tinyurl.com/ntf8mpw*.

1927, Electric Refrigeration
Burstall, A., *A History of Mechanical Engineering*, Cambridge, MA: MIT Press, 1965.

1931, Empire State Building
Wagner, G., *Thirteen Months to Go*, San Diego, CA: Thunder Bay, 2003.

1935, Tape Recording
Onosko, T., *Wasn't the Future Wonderful?*, New York: Plume, 1979.

1936, Hoover Dam
Hiltzik, M., *Colossus*, New York: Free Press, 2010.

1937, Golden Gate Bridge
Starr, K., *Golden Gate*, London: Bloomsbury, 2012.

1937, *Hindenburg*
Lawson, D., *Engineering Disasters*, New York: ASME, 2005.

1937, Turbojet Engine
NASA, *http://tinyurl.com/orxbfxd*.

1937, Magnetically Levitated Trains
Post, R., *http://tinyurl.com/lhzjqd8*.

1938, Formula One Car
De Groote, S., *http://tinyurl.com/72fayvs*.

1939, Norden Bombsight
St. John, P., *Bombardier*, Nashville, TN: Turner, 1998.

1939, Color Television
A timeline of the introduction of color television in a variety of countries: *http://tinyurl.com/nfeobs*.

1940, Tacoma Narrows Bridge
Petroski, H., *To Engineer is Human*, New York: Vintage, 1992.

1940, Radar
Bowen, E. G., *Radar Days*, UK: Taylor & Francis, 1987.

1941, Doped Silicon
Brain, M., *http://tinyurl.com/kov5tve*.

1942, Spread Spectrum
Petersen, A., *http://tinyurl.com/lts47sn*.

1943, Dialysis Machine
Pendse, S., Singh, A., and Zawada, E., *Handbook of Dialysis*, 4th ed., NY: Macmillan, 2008.

1944, SCUBA
Cousteau, J., *The Silent World*, New York: Nat Geo, 2004.

1944, Helicopter
Chiles, J., *The God Machine*, New York: Bantam, 2007.

1945, Uranium Enrichment
US Nuclear Regulatory Commission, *http://tinyurl.com/opubbot*.

1945, Trinity Nuclear Bomb
Monk, R., *Robert Oppenheimer*, Toronto: Doubleday, 2012.

1946, ENIAC—The First Digital Computer
Rojas, R. and Hashagen, U., eds., *The First Computers*, Cambridge, MA: MIT, 2000.

1946, Top-Loading Washing Machine
Stanley, A., *Mothers and Daughters of Invention*, New Brunswick, NJ: Rutgers UP, 1995.

1946, Microwave Oven
Cowan, R., *More Work for Mother*, New York: Basic, 1985.

1946, Light Water Reactor
Bunker, M., *http://tinyurl.com/oflyu7o*.

1947, AK-47
Rottman, G., *The AK-47*, New York: Osprey, 2011.

1947, Transistor
Riordan, M., *http://tinyurl.com/yaufccm*.

1948, Cable TV
Eisenmann, T., *http://tinyurl.com/kckooou*.

1949, Tower Crane
Liebherr, *http://tinyurl.com/nlfjtzt*.

1949, Atomic Clock
Ost, L., *http://tinyurl.com/nllbz9j*.

1949, Integrated Circuit
A short film on Jack Kilby: *http://tinyurl.com/k74ed7p*.

1950, Chess Computer
Computer chess, a film by Andrew Bujalski, provides some interesting background information: *http://tinyurl.com/k4cql25*.

1951, Jet Engine Testing
For more on the chicken gun used to test jet engines: *http://tinyurl.com/3dehuo*.

1952, Center-Pivot Irrigation
Snyder, C., *http://tinyurl.com/nkyfbyj.*

1952, 3D Glasses
Zone, R., *Stereoscopic Cinema and the Origins of 3-D Film*, 1838–1952, Lexington, KY: Kentucky UP, 2007.

1952, Ivy Mike Hydrogen Bomb
Formerly classified, this short film on the bomb is now available online: *http://tinyurl.com/pqbxlrm.*

1953, Automobile Airbag
Bellis, M., *http://tinyurl.com/2vror8.*

1956, Hard Disk
Mueller, S., *Upgrading and Repairing PCs* (21st Ed.), Upper Saddle River, NJ: Que, 2013.

1956, TAT-1 Undersea Cable
Burns, B., *http://tinyurl.com/menaxz3.*

1957, Frozen Pizza
Hulin, B., *The Everything Pizza Cookbook*, Avon, MA: F+W, 2007.

1957, Space Satellite
Jorden, W., *http://tinyurl.com/lc6rm6s.*

1958, Cabin Pressurization
Larson, G., *http://tinyurl.com/md7oveb.*

1959, Desalination
Barlow, M. and Clarke, T., *http://tinyurl.com/m3bykcl.*

1960, Cleanroom
Leary, W., *http://tinyurl.com/lzsb534.*

1961, T1 Line
How Stuff Works, *http://tinyurl.com/jlppv.*

1961, Green Revolution
Jain, H., *The Green Revolution*, Houston, TX: Studium, 2010.

1962, SR-71
Video with SR-71 pilot Richard Graham: *http://tinyurl.com/lt7wftd.*

1962, Atomic Clock Radio Station
More information on the station: *http://tinyurl.com/bum65ma.*

1963, Retractable Stadium Roof
Video demonstrating the retractable roof at Canada's West Harbour stadium: *http://tinyurl.com/k9jshef.*

1963, Irradiated Food
FDA's food irradiation FAQ: *http://tinyurl.com/837e83n.*

1964, Top Fuel Dragster
A 10,000 horsepower top fuel dragster: *http://tinyurl.com/qaqesv5.*

1964, Drip Irrigation
Freedman, C., *http://tinyurl.com/qc6gctl.*

1964, Natural Gas Tanker
Barden, J., *http://tinyurl.com/ltsmzfl.*

1964, Bullet Train
Hosozawa, A. and Hiroshi, N., *http://tinyurl.com/lz467ek.*

1965, Gateway Arch in St. Louis
Campbell, T., *The Gateway Arch*, New Haven, CT: Yale UP, 2013.

1965, Cluster Munition
Clancy, T., *Fighter Wing*, London: Harpercollins, 1995.

1966, Compound Bow
Paterson, W., *The Encyclopaedia of Archery*, London: St. Martin's, 1985.

1966, Parafoil
Popular Science, *http://tinyurl.com/p7zy4ap.*

1966, Pebble Bed Nuclear Reactor
Bradsher, K., *http://tinyurl.com/orrumbd.*

1966, Dynamic RAM
Wang, D., *http://tinyurl.com/kjp5th7.*

1967, Automotive Emission Controls
The International Council on Clean Transportation has updated information on global emissions standards: *http://tinyurl.com/6yo97xz.*

1967, *Apollo 1*
The New York Times, *http://tinyurl.com/3c65on5.*

1967, Saturn V Rocket
Tate, K., *http://tinyurl.com/afo3foz.*

1968, C-5 Super Galaxy
A useful infographic illustrating the Galaxy's massive size: *http://tinyurl.com/pgupdqr.*

1968, Boeing 747 Jumbo Jet
Pealing, N. and Savage, M., *Jumbo Jetliners*, Osceola, WI: Motorbooks, 1999.

1969, Lunar Landing
A video of the first moon landing: *http://tinyurl.com/2znnoa.*

1969, ARPANET
Stewart, W., *http://tinyurl.com/dd4mzc.*

1969, Space Suit
For an excellent selection of space suit photos: *http://tinyurl.com/mex5r7f.*

1970, LCD Screen
Castellano, J., *Liquid Gold*, Hackensack, NJ: World Scientific, 2005.

1970, *Apollo 13*
Lattimer, D., *All We Did was Fly to the Moon*, Cedar Key, FL: Whispering Eagle, 1985.

1970, Fiber Optic Communication
Keiser, G., *Optical Fiber Communications* (4th Ed.), New York: McGraw-Hill, 2011.

1971, Anti-Lock Brakes
Lincoln Continental car brochure from the year anti-lock brakes became standard: *http://tinyurl.com/m7to74t.*

1971, Lunar Rover
Lunar Rover manual: *http://tinyurl.com/pmagdv5*.

1971, Microprocessor
Augarten, S., *State of the Art*, New York: Houghton-Mifflin, 1983.

1971, CANDU Reactor
Whitlock, J., *http://tinyurl.com/2kakzd*.

1971, CT Scan
CT scan medical animation: *http://tinyurl.com/p6oeylv*.

1971, Power Plant Scrubber
A useful video: *http://tinyurl.com/qh2554v*.

1971, Kevlar
Pearce, J., *http://tinyurl.com/nlhzb3l*.

1972, Genetic Engineering
Voosen, P., *http://tinyurl.com/l7a4edl*.

1973, World Trade Center
An archival video about the construction of the original World Trade Center: *http://tinyurl.com/lxgmzhs*.

1975, Router
Some recollections from other engineers who worked with Virginia "Ginny" Strazisar on the development of the router: *http://tinyurl.com/mbapf9u*.

1976, The Concorde
Conway, C., *High Speed Dreams*, Baltimore, MD: John Hopkins UP, 2005.

1976, CN Tower
Fulford, R., *Accidental City*, Ottawa, CN: Macfarlane Walter & Ross, 1995.

1976, VHS Video Tape
IEEE History Center, *http://tinyurl.com/k54eptr*.

1977, Human-Powered Airplane
Roper, C., *http://tinyurl.com/qffjswj*.

1977, Tuned Mass Damper
The Taipei 101 tuned mass damper moving during an earthquake: *http://tinyurl.com/n47rc5m*.

1977, Voyager Spacecraft
Clark, S., *http://tinyurl.com/kadbgmu*.

1977, Trans-Alaska Pipeline
McPhee, J., *Coming Into the Country*, New York: FSG, 1976.

1977, MRI
Mayo Clinic, *http://tinyurl.com/lrz3ayz*.

1978, Nitrous Oxide Engine
NOS Systems, *http://tinyurl.com/nxxjxjt*.

1978, Bagger 288
Murray, P., *http://tinyurl.com/7cg7ma8*.

1979, *Seawise Giant* Supertanker
Trex, E., *http://tinyurl.com/pob5l3y*.

1980, Flash Memory
Computer engineer David Woodhouse presentation about how flash storage works: *http://tinyurl.com/mq7ebjn*.

1980, M1 Tank
Orr, K., *King of the Killing Zone*, New York: WW Norton, 1989.

1980, Stadium TV Screen
Mercer, B., *ManVentions*, Blue Ash, OH: Adams, 2011.

1981, Bigfoot Monster Truck
Borelli, C., *http://tinyurl.com/m5vtx5u*.

1981, Space Shuttle Orbiter
NASA, *http://tinyurl.com/3xkpjbj*.

1981, V-22 Osprey
Whittle, R., *The Dream Machine*, New York: Simon & Schuster, 2010.

1982, Artificial Heart
Long, T., *http://tinyurl.com/kjqjd3m*.

1982, Neodymium Magnet
Swain, F., *http://tinyurl.com/7l4ww9o*.

1983, RFID Tag
Angell, I. and Kietzmann, J., *http://tinyurl.com/nczuwcr*.

1983, F-117 Stealth Fighter
Video of the F-117 Stealth Fighter crashing: *http://tinyurl.com/n2wrmvb*.

1983, Mobile Phone
Agar, J., *Constant Touch*, London: Totem, 2004.

1983, Gimli Glider
Hoffer, M. and Hoffer, W., *Freefall*, New York: Simon & Schuster, 1989.

1983, Ethernet
The original patent can be found online here: *http://tinyurl.com/mr846j4*.

1984, 3D Printer
Hart, B., *http://tinyurl.com/8xosxgm*.

1984, Domain Name Service (DNS)
Ball, J., *http://tinyurl.com/pvqz53x*.

1984, Surgical Robot
Pransky, J., *http://tinyurl.com/kmqmh3c*.

1984, Container Shipping
Cudahy, B., *Box Boats*, New York: Fordham UP, 2006.

1984, Itaipu Dam
BBC, *http://tinyurl.com/lkn32w*.

1985, Bath County Pumped Storage
Koronowski, R., *http://tinyurl.com/mnkd9le*.

1985, International Thermonuclear Experimental Reactor (ITER)
Khatchadourian, R., *http://tinyurl.com/lqpcbya*.

1985, Virtual Reality
Lanier, J., *You Are Not a Gadget*, New York: Vintage, 2011.

1986, Chernobyl
Mahaffey, J., *Atomic Accidents*, New York: Pegasus, 2014.

1986, Apache Helicopter
Macy, E., *Apache*, New York: Grove, 2010.

1990, World Wide Web
Hafner, K., *Where Wizards Stay Up Late*, New York: Simon & Schuster, 1998.

1990, Hubble Space Telescope
Hubblesite.org, *http://tinyurl.com/dmnpr5*.

1991, Lithium Ion Battery
List of lithium ion batteries at dmoz.org: *http://tinyurl.com/klvyssp*.

1991, Biosphere 2
Allen, J., *Me and the Biospheres*, Santa Fe, NM: Synergetic, 2009.

1992, Low-Flow Toilet
Nash, J., *http://tinyurl.com/d8tsosl*.

1992, Stormwater Management
Lee, E., *http://tinyurl.com/k4x9fhp*.

1993, Keck Telescope
Yarris, L., *http://tinyurl.com/n3dttn2*.

1993, *Doom* Engine
Kushner, D., *Masters of Doom*, New York: Random, 2004.

1994, Channel Tunnel
Flyvbjerg, B. and Rothengatter, N., *Megaprojects and Risk*, Cambridge: Cambridge UP, 2003.

1994, Digital Camera
Kaplan, J. and Segan, S., *http://tinyurl.com/5gutm4*.

1994, Global Positioning System (GPS)
Lagunilla, J., Samper, J., Perez, R., *GPS and Galileo*, New York: McGraw-Hill, 2008.

1994, Kansai International Airport
Watkins, T., *http://tinyurl.com/n6hly9d*.

1995, *Toy Story* Animated Movie
Price, D., *The Pixar Touch*, New York: Vintage, 2009.

1996, Ariel Atom
Nusca, A., http://tinyurl.com/lr8j79v.

1996, HDTV
Princeton University, *http://tinyurl.com/p3wv67z*.

1997, Prius Hybrid Car
Berman, B., *http://tinyurl.com/m8xgzm4*.

1998, International Space Station
To see a full list of the countries that have contributed to this space station, see *http://tinyurl.com/psz9wfr*.

1998, Large Hadron Collider
Kolbert, E., *http://tinyurl.com/nw38lt6*.

1998, Smart Grid
For the first articulation of this vision and the resultant funding please see *http://tinyurl.com/nobgsum* and for consortia and their foci *http://tinyurl.com/mwe75qg*.

1998, Iridium Satellite System
McIntyre, D., *http://tinyurl.com/m2k3smp*.

1999, Wi-Fi
Cox, J., *http://tinyurl.com/pbd3ufp*.

2001, Segway
This is one example of an invention that is often attributed to one or two people, but is actually the result of hard work by numerous people. For a list of the full team, including dynamics engineers, programmers, electrical engineers, mechanical engineers, and industrial designers, see *http://tinyurl.com/nfcopmy*.

2003, *A Century in Innovation*
The catalog page for this book contains a link to an informative podcast: *http://tinyurl.com/nccp693*.

2004, Millau Viaduct
This document on the construction of the bridge includes the names of the full team: *http://tinyurl.com/q6xd4ds*.

2005, Bugatti Veyron
A fascinating documentary on the process of designing the Bugatti Veyron can be found online at *http://tinyurl.com/ougbzad*.

2005, Georgia Aquarium
Rothstein, E., http://tinyurl.com/n69csdv.

2006, Palm Islands
Webb, M., *http://tinyurl.com/m3az4fc*.

2006, AMOLED Screen
Pohlmann, K., *http://tinyurl.com/ljz6zqe*.

2007, Smart Phone
PC Magazine, *http://tinyurl.com/kg6jpmg*.

2008, Quadrotor
Mellinger, D., *http://tinyurl.com/opg3t5k*.

2008, Carbon Sequestration
One of the Engineering Grand Challenges is a better method of carbon-capture: *http://tinyurl.com/pbxy5n9*.

2008, Engineering Grand Challenges
A full list of the Challenges is available online at *http://tinyurl.com/pbxy5n9*.

2008, Martin Jet Pack
The Martin Jet Pack website has a great gallery of videos: *http://tinyurl.com/lsl58of*.

2008, Three Gorges Dam
An animation showing the interior of the dam: *http://tinyurl.com/nol5zcj*.

2009, National Ignition Facility
Sample, I., *http://tinyurl.com/ly3lmrr*.

2009, Earthquake-Safe Buildings
Hart, M., *http://tinyurl.com/pdqst7q*.

2009, US President's Limousine
Strong, M., *http://tinyurl.com/l9bhfzy*.

2010, Solar Impulse Airplane
Boyle, A., *http://tinyurl.com/kl23okx*.

2010, BP Blowout Preventer Failure
Gold, R., *http://tinyurl.com/mylqamk*.

2010, Burj Khalifa
The construction timeline can be found online here: *http://tinyurl.com/n3u2pu7*.

2010, Harry Potter Forbidden Journey Ride
A full rundown on the RoboCoaster can be found on KUKA's website: *http://tinyurl.com/l8kqoym*.

2010, Alta Wind Energy Center
The ten largest wind farms in the world: *http://tinyurl.com/pkdj39y*.

2010, Tablet Computer
PC Magazine, *http://tinyurl.com/nk4wfp3*.

2010, The X2 and X3 Helicopters
Paur, J., *http://tinyurl.com/n293wlx*.

2011, Fukushima Disaster
Beech, H., *http://tinyurl.com/lt4wdw7*.

2011, Self-Driving Car
A video of one of the first driving tests is available here: *http://tinyurl.com/muunl4w*.

2011, Instant Skyscraper
Brennan, L., *http://tinyurl.com/kt8ufkx*.

2011, Watson
BBC, *http://tinyurl.com/mpgx69c*.

2012, Curiosity Rover
Stromberg, J., *http://tinyurl.com/n56z7n8*.

2012, Human-Powered Helicopter
An article about the winners with links to video: *http://tinyurl.com/oc2np3n*.

2013, NCSU BookBot
The bookBot website provides more information: *http://tinyurl.com/p3ny2em*.

2014, Ivanpah Solar Electric Generating System
Weiner-Bronner, D., *http://tinyurl.com/k4cdwzn*.

2016, Venice Flood System
ATPN, *http://tinyurl.com/q3gb2td*.

c. 2020, Vactrains
Stewart, J., *http://tinyurl.com/bwcwkel*.

c. 2024, Brain Replication
Horgan, J., *The Undiscovered Mind*, New York: Free Press, 2000.

c. 2030, Mars Colony
Maynard, J., *http://tinyurl.com/kn5tpd7*.

c. 3000, Things We Have Yet to Engineer
Kanani, R., *http://tinyurl.com/qb2zdb9*.

Index

Accidental discoveries of, 218, 326
A *Century of Innovation: Twenty Engineering Achievements that Changed Our Lives*, 446
Active Matrix OLED (AMOLED) screens, 240, 308
Adler, Dankmar, 124
Aerodynamics, 188, 300, 450
Air conditioning, 136
Airports, 134, 136
AK-47, 222
Alajuela Lake, 154
Alfa Romeo, 188
Allen, Bryan, 500
Allen, Holless Wilbur, 284
Allen, John P., 408
Alta Wind Energy Center, 484
Aluminum, 120, 198
Alyeska, 346
America's Cup, multi-disciplinary engineering and, 82
American Helicopter Society, 500
American Society of Civil Engineers, 390
American Society of Mechanical Engineers (1880), 160
Amin, Massoud, 438
AMOLED (Active Matrix OLED) technology, 456; OLED (Organic Light Emitting Diode) technology, 456; screens, types of, 456
Andrews, Thomas, 150
Antheil, George, 202
Anti-lock brakes, 314
Apache helicopter, 184, 400, 488
Apollo 1 tragedy, 294; initial design, 294; pure oxygen atmosphere, 294; review and redesign, 294
Apollo 4, 296
Apollo 11, 302
Apollo 11, 12, 14, 16, 17, successful missions, 310
Apollo 13, 150, 310
Apollo 15, 310, 316
LEM (Lunar Excursion Module), 310; on-the-fly problem solving, 310; oxygen tank explosion, 310
Appius Claudius Caecus, 30
Apple Computers, 420
Apple, 370, 458, 486
Ariel Atom, 428; Formula 1 car vs. 428; TMI Autotech, 428
ARPANET, 304; Digital Equipment Corporation (DEC), 304; IMPs (Interface Message Processor), 304; NCP (Network Control Program), 304; PDP-8, 304; routers, 304; TCP/IP, 304
Artificial heart, 138, 148, 170, 368; Abiocor heart, 368; Syncardia heart, 368
Asphalt, 16, 200; in Babylon, 26; U.S. recycling and, 26
Associate Member of the Institution of Mechanical Engineers, 160
Association of German Engineers (1856), 160
AT&T's Bell Labs, 224
Atomic clock, 230; accuracy of, 230; pendulum clock, 230
Atomic clock radio station, 266; Daylight Savings Time and, 266; NIST-certified (National Institute of Standards and Technology) atomic clock , 266; processors, 266; radio waves and, 266; WWVB, Colorado, 266
Auchinleck, Geof, 386
Aus, Gunvald, 152
Automotive emission controls, 108, 292, 432; catalytic converter, and unleaded gasoline, 292; crankcase ventilation, 292; exhaust gas recirculation, 292; Federal Air Quality Act of 1967, 292; gas tank and, 292; Prius hybrid car and, 432
Automotive racing games and, 416
Axles, 52
Babe, 426
Bagger 288 (bucket wheel excavator), 352
Baker, Bill, 480
Balcom, Homer G., 174
Bandel, Hannskarel, 280
Baran, Paul, 304
Bar codes/QR (Quick Response) codes and, 372
Bardeen, John, 224
Bartholdie, Frédéric Auguste, 116
Basilica of Saint Denis, 20; cathedrals, 38; flying buttress, 38; Gothic arch (pointed arch), 38; Gothic cathedrals, 38
Batelli, Frédéric, 134
Batteries: chemical, 293; energy density and, 406; Hubble space telescope and, 404; Lunar rover and, 316; storage, 392; various chemistries for, 406
Bath County pumped storage, 316, 392, 406
Bazalgette, Joseph, 94
Bell Helicopters, 366
Bell, Alexander Graham, 106
Bent pyramid, 20
Berg, Paul, 328
Berle, Korte, 152
Berners-Lee, Tim, 402
Bertiger, Barry, 440
Bessemer, Henry, 86
Bessemer process, 26, 44, 86, 146, 152
Big Ben, 90; mechanical pendulum clocks and, 90; Westminister Chimes, 90
Big Science, definition of term, 470
Bigfoot monster truck, 362; money/public interest and innovations, 362; pickup trucks and, 362; trusses, 362
Biosphere 2, 408; food production in, 408; oxygen levels and, 408
Blass, Simcha, 274
Blass, Yeshayashu, 274
Bluetooth source, 202
Boeing, 366
Boeing 707, 254
Boeing 747 Jumbo Jet, 300
Boeing 767 scenario, 378
Boggs, David, 380
Borlaug, Norman, 262
Bow and arrows, 16; archers, 16; longbows, 16; projectile weapons, 16; single arched bow, 16
Bowen, Edward George, 196
BP blowout preventers (BOP), 478
Brain replication, 510
Brain Research Through Advancing Innovative Neurotechnologies project, 510; computers vs., 510; Human Brain Project, 510
Brattain, Walter, 224
Brukhonenko, Sergei, 170
Bugatti Veyron, 450
Building implosions, 54; Holy Trinity Cathedral in Waterford, Ireland, 52; shaped charges, 54
Bullet train, 278
Burj Khalifa, Dubai, 42, 86, 174, 480
Bwana Devil, 240
C-5 Super galaxy, 298
Cabin pressurization, 254; B-7 bombing runs and, 254; Boeing 707 and, 254; Douglas DC-8 and, 254; evolution of new technologies and, 254; jet engines and, 254; Pan Am, 254
Cable cars, 104
Cable TV, 226, 430; antenna usage and, 226; HBO and, 226; HDTV and, 226; TV channels and, 226; WTBS and, 226
Cailliau, Robert, 402
Canadian Society for Civil Engineering (1887), 160
CANDU (CANada Deuterium Uranium) reactor, 320; heavy water, 320; light water reactors, 320; natural uranium burning, 320; Pickering Nuclear Generating Station (Ontario), 320
Capek, Karel, 168
Carbon fiber, 86, 110, 340; chassis, 188; strength, 110
Carbon sequestration, 462
Carmack, John D., 416
Carnegie Hall, 124
Carothers, Wallace, 88
Carrier, Willis, 136
Cartwright, Edmund, 56
Casper, 426
Cassegrain design, 404
Catapult, 40; Castel of Castelnaud, 38; mangonel, 40; Stirling Castle, 40; trebuchet, 40; War Wolf (trebuchet), 40
Cell phone: gps and, 422; improvements, 514; system, 10, 12, 440
Cell phone radio system, 458
Center-pivot irrigation, 238; lowered costs/improved efficiency and, 238; standard irrigation vs., 238
Chandler, Bob, 362
Channel tunnel, 418
Chernobyl explosion, 398; containment building, 398; control rods, 398; failure, reasons for, 398; steam turbine and, 398
Chess computer, 234; Deep Blue (IBM), 234; humans vs. 234
Chrysler Building, 118
Chrysler Defense, 358
Ciza Necropolis, 20
Clarkson, Jeremy, 428
Cleanrooms, 258
Cluster munition, 282; BLU-108 submunitions, 282; CBU-105, 282; Sensor Fuzed Weapon, 282

CN Tower, Toronto, 42, 336
Coal, 352
Colombo, Gioacchino, 188
Color Television, 192; CRT technology, 192; HDTV, 192; Tournament of Roses Parade and, 192
Columbus's ships, 44
Combine harvester, 72; agriculture and, 72; center-pivot irrigation and, 72; drip irrigation and, 72; human productivity, improvement in, 72; labor reduction and, 72
Communication cables, 312
Compass, 36
Compound bow, 284; energy storage and, 284; reconceptualization of, 284
Concorde, 334
Concrete, 24, 86, 480
Constable, 446
Container ships, 128
Container shipping, 388; intermodal containerized shipping, 388; international shipping process, 388; two-stroke diesel engine, 388
Cook, William Fothergill, 74
Cooper, Martin, 376
Corning, 312
Corps of Topographical Engineers, 102
Cotton Gin, 60; cotton manufacturing process, effect upon, 60; in Israel, 274
Cotton manufacturing, 56, 58, 60
Cotton mill, 58
Cousteau, Jacques-Yves, 206
CT (Computed Tomography) scan, 322, 348; engineering disciplines and, 322; MRI, 322; X-ray machine, 322
Cuban Missile Crisis (1962), 202
Curiosity Rover, 498
Daimler, Gottlieb, 112
Damadian, Raymond Vahan, 348
David Ferrucci, 496
Davies, Donald, 304
Day, Brian, 386
De Nul, Jan, 452
de Rochas, Alphonse Beau, 146
Deep Space Network, 498
Deepwater Horizon rig, Gulf of Mexico, 478
Defibrillator, 134, 138; AED (Automatic External Defibrillators), 134; cardiac arrest, 134; microprocessors, 134; portable, 134
Dennard, Robert, 290
Desalination, 256; distillation, 256; environmental issues and, 256; MSF (Multistage Flash) desalination, 256; reverse osmosis (RO), 256
Dialysis machine, 204
Diesel locomotive, 132
Diesel, Rudolf, 128, 132
Digital camera, 420; CCD (charge-coupled device) sensor, 420; CCD sensor, 420; CMOS (complementary metal-oxide semiconductor) sensor, 420; CompactFlash, 420; QuickTake, 420; traditional film cameras, 420
Dishwashers, 160
Domain Name Service (DNS), 384; early Internet connections, 384; IP (Internet Protocol) address, 384; superefficiency of, 384; web browsers, 384
Doom engine, 416; artificial reality and, 416; engineering disciplines and, 416; FPS (First Person Shooter) game, 416; Graphics Processing Units (GPUs), 416; OpenGL, 416
Doped silicon, 200
Dr. Strangelove, 170
Drake, Edwin, 92
Drip irrigation, 274; drip-irrigation emitter, 274; in Israel, 274; irrigation techniques, 274
Dynamic RAM (DRAM); CPU (Central Processing Unit), 290; RAM (Random Access Memory), 290
E-mail, 402
Earthquake-safe buildings, 52, 472, 494; Haiti earthquake, 472; Istanbul's airport, 472
Easton, Roger L., 422
Eaton, Amos, 66
Eckert, J. Presper Jr., 214
Edison, Thomas, 110
Egypt's Old Kingdom, 16
Eiffel tower, 114, 116
Eiffel, Gustave, 116
Einstein, Albert, 156
Einthoven, William, 138
EKG/ECG machines, 138
Electrical Association for Women, 160
Electric refrigeration, 172; Freon and, 172; General Electric and, 172; ozone destruction and, 172
Elevators, 100
Ellis, Charles Alton, 180
Empire State Building, 118, 174; assembly process, 174; parallelizing the work, 174
Engineered lumber, 142; oriented strand board (OSB), 142; plywood, 142; Portland Manufacturing Company, 142; standard lumber, 142; trusses, 142; wooden I beam, 142
Engineered objects, 12
Engineered systems (architectures), 12
Engineering; defined, 11; disciplines, 11, 322, 416; future of, 514; goals, 222; specialized areas, 12
Engineering achievements, greatest in the twentieth century, 446
Engineering grand challenges (2008), 464
Engineers, 18; cost reduction and, 76; defined, 40; material efficiency and, 64; role of, 10, 12
ENIAC, 214
Environmental issues, 256, 292, 306, 324
Environmental Protection Agency, sulfur dioxide laws and, 324
Eppelsheimer, William, 104
Erie Canal, 66, 68; canals, maintenance, 68; freight and, 68
Essen, Louis, 230
Ethernet, 380; bus topology, 380; LAN (Local Area Network), 380; multiple computer connection, 380; Xerox PARC, 380
European Organization for Nuclear Research (CERN), 436
F-117 Stealth Fighter, 374
Faggin, Federico, 318
Failures in engineering, 108, 150, 194; fail-safe designs and, 378; Fukushima (2011), 490; reasons for, 398; backup mechanisms scenarios, 316
Federal Communications Commission; radio frequencies and, 376; standard frequencies and, 430
Ferris, George Washington Gale Jr., 130
Ferris Wheel, 130
Fiber optic communication, 106, 156, 312
Field, Doug, 444
Flash memory, 356, 420, 458; EEPROM, or Electrically Eraseable Programmable Read-Only Memory, 356; EPROM, or Eraseable Programmable Read-Only, 356; PROM, or Programmable Read-Only Memory, 356; RAM (Random Access Memory) chips, volatility of, 356; Read-Only Memory ROM, 356
Flatiron Building, New York, 152
Food, 250, 408; frozen, 136, 172, 250, 270; world food production, 262
Formula One Cars, 188, 428; carbon fiber chassis, 188; speed of, 188; turbocharged engine, 188; weight, importance of, 188
Freight, 68, 70
Freon, 172
Frozen foods, 136, 172, 250
Frozen pizza, 250; assembly lines, 250; food scientists, 250; manufacturing engineer and, 250; mass production, 250; wastewater treatment plants and, 250
FTP (File Transfer Protocol), 402
Fukishima nuclear reactor (2011), 150, 490
Gagarin, Yuri, 306
Gagnan, Emile, 206
Garwin, Richard, 242
Gas emission, environmental problems and, 292
Gateway Arch (St. Louis), 280; adverse weather conditions and, 280; elevator, 280; world's tallest stainless steel monument, 280
Genetic engineering, 328; *E. coli bacteria*, 328; gene gun, 328; genomes, 328; selective breeding, 328
Georgia Aquarium, 452
Getting, Ivan A., 422
Gilbert, Cass, 152
Gimli Glider, 378; Boeing 767 scenario, 378; fail-safe designs and, 378; ram air turbine (RAT), 378
Global Positioning System (GPS), 422, 492
Goethals, George Washington, 154
Golden Gate Bridge, 12, 64, 180
Google Trends, 460
Grant, Ulyssses S., 62
Great pyramid of Giza, 20
Great Wall, of China, 46; Manchus, 46; Mongols, 46; purpose of, 46
Green Revolution, 262; fertilizers and, 262; genetic engineering and, 262; irrigation projects and, 262; new strains of plants, 262; world food production, 262
Guns, 16
Gunter, Edmund, 48
Gunter's chain, 48, 102; plane table, surveying compass, 48; surveying and, 48; theodolite, 48; topographer's rod, 48; total station, 48
Gunzberg, Milton, 240
Hale telescope, 414
Hale, George Ellery, 158
Hall-Héroult Process, aluminum and, 120

Hall, Benjamin, 90
Hall, Charles Martin, 120
Hallidie, Andrew Smith, 104
Hard disks, 246; cost-lowering/improvement process, 246; IBM and, 246
Harry Potter Forbidden Journey Ride, 482
Haslett, Caroline, 160
Hayes, Vic, 442
HD devices, 430
HDMI cables, 430
HDML, 430
HDTV, 192, 226, 318, 430; five human senses research and, 430; NHK, Japan Broadcasting Corporation, 430
Heart-lung machine, 170
Helicopters, 208, 366, 370, 488
Héroult, Paul, 120
Hertz, Heinrich, 196
Hetrick, John, 244
High-pressure steam engine, 62, 70, 100, 104, 122; boiler systems, 62; Centennial Exhibition in Philadelphia (1876) and, 62; Corliss steam engine, 62; Industrial age, transition to, 62; San Francisco cable car system, 62; Sultana explosion (1865), 62; *Titanic*, 62
Hindenburg, 182
Hoff, Marcian, 318
Hoffmann-LaRoche, 308
Holmes, Verena, 160
Hooke, Robert, 50
Hooker Telescope, 158, 404, 414
Hoover Dam, 178
Hounsfield, Godfrey, 322
Howe, Elias, 80
HTML, 402
HTTP, 402
Hubble space telescope, 404
Hubble Telescope, 158
Hull, Chuck, 382
Human brain, engineering mindset and, 18
Human-powered airplanes, 340; Kremer Prize, 340; MIT Daedalus HPA, 340
Human-powered helicopters, 500; Atlas helicopter, 500; Gossamer Albatross, 500; Gossamer Condor, 500; Sikorsky Prize, 500
Hunter/gatherers, 18
Hydrogen bomb, 214
Industrial Revolution in America; cotton, 56, 58, 60; high-pressure steam engine and, 62; mass production and, 76; spinning frame, 58; textile industry and, 56, 58; water and, 32
Institution for Civil Engineers, 160
Institution of Mechanical Engineers (London, 1847), 160
Integrated circuit technology, 232, 258
Internal Combustion Engine, 146; four-stroke engine, 146; Model T engine, 146; Otto cycle engine, 146
International Space Station (ISS), 434, 470
International Thermonuclear Experimental Reactor (ITER), 394, 470; conventional nuclear bomb, 384; fusion process, 394; thermonuclear bomb, 384
Internet search; ARPANET and, 304; Domain Name Service (DNS), 384; IP (Internet Protocol) address, 384; packet switching and, 332; T1 lines and, 260; tools, 402
Internet of things, 200
Inuit technology, 22; clothing, 22; crafting by, 22; igloo, 22; kayak, 22; snow goggles, 22; toggling harpoon head, 22
Iridium satellite system, 440; Iridium Communities Inc., Virginia, 440; Motorola engineers, 440
Iron age, 86
Irradiated food, 270; bacteria elimination, 270; canning and, 270; FDA approval, 270; fear of, 270; frozen pizza and, 270; gamma-ray, 270; NASA and, 270
ISO 1 cleanroom, 258
Itaipu Dam, 390; harnessing of electricity and, 390; seven wonders of the modern world and, 390
Ivanpah Solar Electric Generating System, 504; BrightSource Energy, 504; Google, 504; heliostat and, 504; NRG Energy, 504
Ivy Mike hydrogen bomb, 242; nuclear bomb mechanics, 242; Soviet Union's Tsar bomb, 242
Jacobaeus, Hans Christian, 148
Jacobi, Werner, 232
Jalbert, Domina, 286
Jarvik, Robert, 368
Jet engine testing, 236; bird strikes, 236; US Airways flight 1549, 236
Jobs, Steve, 458
Judah, Theodore, 66
Jumanji, 426
Kabaň, Jozef, 450
Kalashnikov, Mikhail, 222
Kamen, Dean, 444
Kansai International Airport, 424; artificial island, 424; world's longest airport concourse, 424
Kasei, Asahi, 406
Keck Telescope, 158
Keck, Donald, 312
Kemper, Hermann, 186
Kennedy, John F., 474
Kennis, Adrie, 18
Kennis, Alfons, 18
Koechlin, Maurice, 116, 118
Kevlar, 326, 88; accidental discovery, 326; as nylon replacement, 326
Kibbutz Hatzerim, 274
Kilby, Jack, 232
Kinsol Trestle Bridge, 164
Knibb, Joseph, 50
Kolff, Willem Johan, 204
Kremer Prize, 500
Kremer, Henry, 340
Krupp, 352
Kwolek, Stephanie, 326
Lake Mead 178
Lamarr, Hedy, 202
Lampson, Butler, 380
Lanier, Jaron, 396
Laparoscopic surgery, 148, 386; cytoscope, 148; thorascopic diagnosis, 148
Large Hadron Collider (LHC), 436, 470
LASER, Light Amplification by Stimulated Emission of Radiation, 156; fiber optics and, 156; Stimulated Emission process, 156
LCD screens, 308, 456; Active Matrix OLED and, 308; mass production, 308; nematic field effect, 308; Pulsar 1, 308
Leaning Tower of Pisa, 42; foundations, importance of, 42; soil extraction, 42
Lely, Cornelius, 126
LEM (Lunar Excursion Module), 302, 310
Leon Moisseiff, 180, 194
Leopold, Ray, 440
Levers, 52
Liebherr Group, 228
Liebherr, Hans, 228
Light water reactor, 220; moderator, 220; radioactivity, safety precautions and, 220
Linderer, Walter, 244
Lithium ion battery, 406, 476
Loeb, Sidney, 256
Louisville water tower, 96
Low-flow toilet, 410
Lowell, Francis, 56
Lunar landing, 302; command/service module (CSM), 302; LEM (Lunar Excursion Module), 302; mission architecture, 302; Saturn V, 302
Lunar rover, 316; batteries, 316; dish antenna, 316; failure scenarios, backup mechanisms for, 316
Madden Dam, 154
Madium pyramid, 18
Magnetically levitated train, 186, 508; friction, elimination of, 186; magnetic effects, 186; Shanghai Maglev Train, 186
Manhattan Bridge, 194
Manhattan Project (1942), 210, 212
Mars Colony, 22, 512; possibilities of, 512; robotic construction crew, 512; terraforming, 512
Mars Science Laboratory, 498
Martin jet pack, 466
Martin, Glenn, 466
Mass production, 76; guns and, 76; sewing machines and, 80
Masuoka, Fujio, 356
Mauchly, John, 214
Maurer, Robert, 312
Maus, Henri-Joseph, 78
Mazor, Stanley, 318
McCready, Paul B., 340
McEwan, James, 386
Mechanical pendulum clock, 50; accuracy of, 50; anchor escapement mechanism, 50
Metcalfe, Robert, 380
Microprocessor, 318, 420; computer circuits, 318; Intel 4004 chip, 318
Microsoft, 486
Microwave oven, 218, 318; accidental discovery of, 218; cost of, 218; radar and, 218; Radaranges (Raytheon), 218
Millau Viaduct, 12, 86, 118, 154, 448
Miller, David S., 268
Miller, John, 162
Mobile phones, 376; radiophone, 376; radio towers, cost of, 376
Mockapetris, Paul, 384
Moisseiff, Leon, 180
MOSE system, 506
M1 tank, 184, 358
Monsters Inc., 426
Moore, Hiram, 72
Motorola, 376, 440

MRI scanners, 322, 348
Mylar, usage, 340
National Academy of Engineering, 446, 464
National Ignition Facility, Lawrence Livermore National Laboratory 470
Natural gas tanker, 276; liquefied natural gas (LNG), advantages of, 276; *Methane Princess* (British Gas), 276; natural gas shipping, 276; Q-Max class, 276; super-cold temperature problems and, 276
NCK, 336
NCSU bookBot, 502
Nelson, Jerry, 414
Neodymium magnet, 370; magnets, manufacturing process, 370; Nd2 Fe14 B alloy, General Motors, 370; Sumitomo Special Metals, 370
Netherlands, 126
Nitrous oxide engine, 112, 350
Norden Bombsight, 190; advertising campaign, 190; guidance systems, 190; target sighting, 190
Norden, Carl, 190
North Carolina State University, 502
Nouguier, Émile, 118
Noyce, Robert, 232
NTSC standard, 430
Nuclear bombs, 200, 220
Nuclear fuel, 220
Nuclear power plants, 122, 200
Obelisks, 114
Ocean Voyager exhibit, 452
Oil wells, 92; economics, dependence on, 92; Seneca Oil company, 92; Titusville, PA, 92; Trans-Alaska Pipeline, 92
Oppenheimer, Robert, 212
Otis Elevator Company, 100
Otis, Elisha, 100
Otzi, 18
Oxford High School for Girls, 160
PacMan, 416
Palm Islands, Dubai, 454
Panama Canal, 154
Pantridge, Frank, 134
Parafoil, 286
Paris World's Fair, 108
Parkes, Alexander, 88
Parkinson, Bradford, 422
Parsons, Sir Charles, 122
Parthenon, 20, 28; Athena and, 28; built in "wrongness" of, 28
Pawtucket, Rhode Island, 58
Pebble bed nuclear reactor, 288; plans for, in China, 288
Pedro, Don, 62
Peter Carl Goldmark, 192
Peterson, Ken, 440
Pfleumer, Fritz, 176
Piccard, Bertram, 476
Pisano, Bonanno, 42
Plastic, 86, 88; Kevlar, 88; Parkesine, 88; polyethylene, 88; types of, 88
Pompeii, 24, 34; building materials, 34; heating systems (hypocaust), 34; Naples, Italy, 34; public baths, 34; roads, 34; sidewalks, 34; water and sewer systems, 34
Pong, 416
Postel, Jon, 384
Power grid, 108, 216, 392; alternative power sources, 108; failures in, 108; Zénobe Gramme dynamos, 108
Power loom, 56; industrialization, effect upon, 56; production of cloth and, 56; secrecy in Britain and, 56; in the US, 56
Power plant scrubber, 324; Environmental Protection Agency and, 324; sulfur dioxide removal process, 324
Prévost, Jean-Louis, 134
Prius hybrid car, 432; accessories, 432; automotive emissions and, 432; motive power, sources of, 432; Toyota, 432
Professional engineer licensing, jobs and, 144
Pulleys, 52
Pyramid of Djoser, 20
Quadrotor, 460; neodymium magnet motors, 460; Parrot Company, France, 460
Radar (Radio Detection And Ranging), 196, 218; Chain Home system, 196; electromagnetic waves and, 196; radar evasion, 374
Radio stations, 166; AM, 166; Golden Age of Radio, 166; Great Depression and, 166; KDKA, Pittsburgh, 166; radio frequencies, 376; radio waves, 266; RMS *Titanic* and, 166; towers, cost of, 376; World War I and, 166
RAM (Random Access Memory), 290, 356
Ramps, 52
Raskob, John J., 174
Reactors, 320
Red pyramid, 20
Refineries, 292
Reichert, Todd, 500
Rensselaer Polytechnic Institute, 66; ABET (Accreditation Board for Engineering and Technology), 66; engineering degree programs, 66; Troy, New York, 66
Retractable roofs, 124
Retractable stadium roof, 268
RFID (Radio-frequency identification) tags, 372
RMS *Titanic*, 62, 122, 150; radio stations and, 166; sinking of, 150
Roberts, Lawrence, 304
Robertson, Cameron, 500
Robots, 168, 482; first usage of, 168; general vision capability and, 168; robot arms and, 482; robotic construction crew, 512; role of, 10, 12; surgical, 386
Roller coasters, 163, 482; Hulk, Orlando, 162; Miller Under Friction Wheel, 162; wooden, trestles and, 164
Roman Empire, 34, 64, 94
Roman Empire aqueduct system, 30; Aqua Appia, 30; Arelate (present-day Arles), 32; modern sewer system and, 30
Roof trusses, 142
Roosevelt, Franklin D., 474
Rosling, Hans, 216
Routers, 304, 332
Saarinen, Eero, 280
San Francisco, 62, 180
Saturn V rocket, 296, 302; F-1 engine, 296; Instrument Unit, 296
Savage, John L., 178
Schreiber, Wolfgang, 450
Schultz, Peter, 312
Scientists vs. engineers, 13
Screws, 52
Scuba (Self Contained Underwater Breathing Apparatus), 206; air supply in a backpack (1943), 206; Aqualung, 206; rebreather (closed-circuit system), 206; regulation of pressure, 206
Seawise Giant supertanker, 354
Secret Communications System, 202
Segway, 444; disappointing debut, 444; prerelease buzz for, 444; self-balancing technology, 444
Self-driving cars, 196, 496; foreknowledge technique, 492; Google engineers and, 492; Lidar (distance sensing), 492; Linriccan Wonder, 492; in Pompeii
Sewer systems, 94, 216; Great Stink, in London, 94; modern, 94; in Pompeii, 34; Roman aqueduct systems, 30
Sewing machine, 80; lockstitch, 80; mass production and, 80; Singer sewing machines, 80
Shannon, Claude Elwood, 234
Shima, Hideo, 278
Shima, Masatoshi, 318
Shreve, Lamb, and Harmon, 174
Siemens, von Werner, 100
Sikorsky, 208; helicopters, 208, 488; prize, 500
Simple machines, 52; water-powered gristmill and, 52; mechanical systems diagram, 52; Yates Mill, 52
Skidmore, Owings and Merrill, 480
Skyscrapers, 152, 330, 494; Broad, China and, 494; Freedom Towers, 494; instant, 494
Slater Mill, 58
Smart grid, 438; Electric Power Research Institute (EPRI), Palo Alto, 438; research and development program, 438
Smart phones, 176, 234, 246, 458; costs, 106, 248; intuitive simplicity of, 458; iPhone, 458
Smart, Niki, 428
Smith, Adrian, 480
Snow, John, 84
Society of Women Engineers (SWE), in the United States, 160
Solar impulse airplanes, 476
Solar power, 252, 344, 462
Somerville, Bob, 446
Song Dynasty, in China, 36
Sourirajan, Srinivasa, 256
Space Biosphere Ventures, 408
Space Invaders, 416
Space satellite, 252, 258, 440; inaccessible, harsh environments and, 252; orientation and, 252; reliability testing, 252
Space shuttle orbiter, 364
Space suit, 306; PLSS or Primary Life Support System, 306; SK-1, 304
Space toilet, 364
Spencer, Percy, 218
Sperry Gyroscope Company, 200
Spread spectrum technology, 202
Square-rigged wooden sailboats, 44
SR-71, 264; inlet cone, 264; military surveillance and, 264; titanium alloy and, 264
St. Louis Arch, 42
Stadium, 136

Stadium TV screen, 124, 360; Diamond Vision Board, 360; Dodgers stadium and, 360; LEDs (light-emitting diodes), 360; RGB light source, 360
Star Trek TV, 514
Star Wars, 514
Statue of Liberty, 116
Steam engine, 196
Steam turbine, 122, 398; nuclear power plants and, 122; *Titanic* and, 122
Steel, 86, 152, 198
Stevens, John Frank, 154
Stormwater management, 412
Storm surge barriers, 126
Strauss, Joseph, 180
Strazisar, Virginia, 332
Sumitomo Heavy Industries, 354
Supercharger/turbocharger, 112, 188, 272; nitrous oxide engines, 112; two-stroke diesel engines, 112
Supertanker, 128, 346
Surgical robots, 386
Sutter, Joe, 300
Swanagan, Jeff, 452
Swiss Federal Institute of Technology, Lausanne, 476
T1 lines, 260, 304
T3 lines, 260
Tablet computer, 486; iPad, dominance of, 486; various manufactures, 486
Tacoma Narrows Bridge, 150, 194; aeroelastic flutter, 194; failure of, 194
Tape recordings, 176; AEG Magnetophon K1, 170; NTSC signals and, 430
Tarlton, Robert, 226
TAT-1 (Transatlantic 1) undersea cable, 248; durability, 248; phone calls, costs, 248
Taylor, Charles, 140
Technology; engineers and, 18; new developments, 18
Telegraph system, 74, 102; batteries and, 406; undersea telegraph cables, 74
Telephone, 106; fiber optic cables, 106; mechanical switches, 106; phone costs, 106; VoIP, 106
Television, 226, 338
Thacker, Chuck, 380
Thermostatic oven, 160
3D glasses, 240; Active Matrix OLED (AMOLED) screens, 240; Goggles, 240; LCD shutter glasses, 240; polarized lenses, 240; red-cyan glasses, 240
3D printer, 20; ABS plastic, 382; extrusion, 382; rapid prototyping, 382
Three Gorges Dam, China, 154, 468
Tiber River, 30
Tiryns, Greece, 24
Titanium, 26, 198; aluminum and steel vs., 198; SR-71 aircraft and, 198
Tokaido Shinkansen, 278
Tokyo National Railways, 278
Tom Thumb steam locomotive, 70; Big Boy engine, Union Pacific railroad and, 70; transportation systems, effect upon, 70
Top fuel dragster, 272; as fastest piston-engine vehicle, 272; direct-drive system, 272; engine power, 272; fuel, types of, 272; National Hot Rod Association (NHRA), 272; Nitromethane, 272; supercharger, 272; wrinkle wall technology, 272
Top Gear, 428
Top-loading washing machines, 216; current automatic, 216; mass marketing of, 216; Rural Electrification Act (1935, 1944), 216
Tour de France athlete, 340
Tower crane, 228; limitations, 228; sliding action of, 228
Toy Story Animated Movie, 426; PhotoRealistic RenderMan, 426; Pixar, 426; Virtual lights/camera, 426; world's first completely computer-generated feature-length film, 426
Trans-Alaska Pipeline, 346; Denali Fault, 346; oil transportation, 346
Transcontinental railroad, surveying and, 102
Transistors, 224, 232; in the marketplace, 224; TRADIC, 224
Trevithick, Richard, 62
Trinity nuclear bomb, 212
Truss bridge, 64; Golden Gate Bridge, 64; Hyde Hall Covered Bridge, New York (1823), 64; Kingpost truss, 64; Roman Empire and, 64; World Trade Center, 64
Trusses, 64, 142, 362
Tulip Trestle, Southern Indiana, 164
Tuned mass damper, 342; John Hancock building, 342; Taipei 101 tower, 342
Tunnel boring machine (TBM), 78, 418; Channel Tunnel, 78, 418; Mountain Slicer, 78
Turbojet engines, 184; efficiency of, 184; multistage compressor, 184; ramjet engines, 184
Turing, Alan, 234
Tuthill, William Burnett, 124
Twain, Mark, 454
Two-stroke diesel engine, 112, 128, 338, 354
University of Arizona, 408
Uranium enrichment, 210; gaseous diffusion process, 210; K-25 building, 210; Manhattan Project (1942), 210; separation processes, 210
US President's Limousine, 474; 2009 Cadillac limousine, 474; precautions for, 474; V12 Lincoln convertible, 474
Vactrains, 508; China's Southwest Jiaotong University, 508; magnetically levitated vehicle, 508
Van Oord, 454
van Rensselaer, Stephen, 66
Venice flood system, 506
VHS videotape, 338; Betamax by Sony, 338; helical scan approach, 338; sound recording process, 338; TV and, 338; VHS by JVC, 338; video format wars, 338
Virlogeux, Michel, 448
Virtual reality, 396; three-dimensional images, 396; VPL Research, 396
Voyager spacecraft, 344; communication system, 344; power sources and, 344; Radioisotope Thermoelectric Generator (RTG), 344; solar power, 344
V-22 Osprey, 366; engine failure, 366; turboprop airplanes, 366; wing requirements, 366
Wake County North Carolina, 52
Wallace, John Findlay, 154
Walson, John, 226
Walton, Charles, 372
Wamsutta Oil Refinery, 26, 98; butane, 98; gasoline, 98; hydrocarbon, 98; kerosene, 98; oil process, 98; paraffin, 98; primary goals of, 98
Warkuss, Hartmut, 450
Washington Monument, 42, 114
Washington, George, 114
Water supply, 30
Water towers, 96
Water treatment, 84, 250
Waterwheel, 32; flour mill (Barbegal), 32; Portogruaro, 32; Roman city of Arelate (present-day Arles), 32; sawmills, 32; treadwheels, 32; water-powered gristmill, 52
Watson-Watt, Robert, 196
Watson, 496; IBM's Jeopardy-playing computer, 496; untagged data and, 496
Watt, James, 196
Wedges, 52
Weinberg, Alvin, 220
Wheatstone, Charles, 74
Wheels, 52
Whitfield, Willis, 258
Whitney, Eli, 60
Whittle, Frank, 184
Wi-Fi, 202, 442; IEEE 802.11 Standards Working Group, 442; IEEE specification number: 802.11, 442; signal interception problem, 442
Wigner, Eugene, 220
Wilkins, Arnold Frederic, 196
William Justin Kroll, 198
Wind power, 462, 484
turbines, 108, 484
wind farm, 484
Woman Engineer, 160
Women's Engineering Society, 160
Woodyard, John Robert, 200
Woolworth Building, 24, 152; steel and, 152; skyscrapers, 152
World Trade Center, 64, 330
World Wide Web, 402; core ideas for the, 402; hypertext project, 402
Wright brothers, 58
Wright brothers' airplane, 140, 246; engine, 140; wind tunnel, 140; Wright Flyer, 496
Wright, Benjamin, 68
Wright, Orville, 140
Wright, Wilbur, 140
X-ray machine, 322, 348
X2 (Sikorsky) and X3 (Eurocopter) helicopters, 488
Yablochov arc lamps, 108
Yablochov, Pavel, 108
Yamasaki, Minoru, 330
Yuster, Samuel, 256
Zepplin company, 182
Zuiderzee Works, 126
Zybach, Frank, 238

Photo Credits

age footstock: © Jim West: 505

Alamy: © Martin Shields: 19; © Paul Williams/funkyfood London: 39; © Bernhard Classen: 463; © Roy Garner: 473; © Marcus Brandt/european pressphoto agency b v: 483; © ZUMA Press: 491

© Boeing: 301

Corbis: © Bettmann: 133, 249; © Brian Cahn/ZUMA Press: 191; © Ira Wyman/Sygma: 369; © Hans Blossey/imageBROKER: 393; © HO/Reuters: 479; © Peng hongwei/Imaginechina: 495

Dreamstime: © Babar760: 41; © Alacrityp: 121; © Kristiana007: 199; © Actionsports: 273; © Photozirka: 275; © Damirfajic: 291; © Ensuper: 315; © Redwood8: 351; © Falun1: 363; © Annedave: 415; © Chrisstanley: 429; © Jibmeyer: 449; © Novikat: 453; © Nfx702: 489

Dutch National Archives: Hans van Dijk/Anefo/gahetNA: 335

DVIDS: US Air Force: 299; US Marines: 359; S1C Andrew Kosterman/US Army: 401; Matthias Fruth/Training Support Center Grafenwo/US Army: 465

European Southern Observatory: Y. Beletsky: 157

Courtesy Everett Collection: © Buena Vista Pictures: 427

Courtesy Flickr: naoK: 413

Getty Images: © Popperfoto: 197; © Yvonne Hemsey: 235; © The LIFE Picture Collection: 241; © Gerard Julien/AFP: 395; © Michael L. Abramson/The LIFE Images Collection: 417; © National Railway Museum/SSPL: 419; © John Harding/The LIFE Images Collection: 421; © David Paul Morris: 459; © Joe McNally: 471; © Ben Hider: 497; © Victor Habbick/Visuals Unlimited: 515

Courtesy HEIST: 107

Courtesy IET Archives: 161

Image Works: © TopFoto: 195

Courtesy Intel Free Press: 319

iStock: © ivanmateev: 29; © malerapaso:37; © Charles Mann: 75; © Lisa-Blue: 105; © Roel Smart: 135; © Nils Kahle - 4FR Photography: 137; © monkeybusinessimages: 139; © typhoonski: 187; © Rob Byron: 217; © Archive Photos: 219; 257, 267, 327; © Mayumi Terao: 277; © Bukharov Denis & Yana: 321; © baranozdemir: 323; © Mark Winfrey: 325; © Imnature: 337; © David Pedre: 355; © Tom Prout: 357; © kreci: 381; © seraficus: 383; © nicolasdecorte: 391; © joel-t: 407; © Lo Chun Kit: 425; © glo: 431; © Sjoerd van der Wal: 433; © AAA-pictures: 447; © hocus-focus: 487; © Jason Doiy Photography: 493; © MachineHeadz: 511

Kobal Collection: © 20th Century Fox: 377

Library of Congress: 23, 43, 57, 59, 63, 71, 77, 81, 87, 93, 97, 115, 141, 147, 153, 165, 167, 283; Carol M. Highsmith: 117, 163, 175, 281, 331; Jet Lowe: 181;

© Martin Jetpack/www.martinjetpack.com: 467

Mary Evans Picture Library: © Institution of Mechanical Engineers 2013

NASA: 253, 265, 295, 297, 303, 307, 311, 317, 341, 345, 365, 405, 435, 513; James Blair/JSC: 397; Caltech/JPL: 499

National Archives: 61, 101, 179, 243; Franklin D. Roosevelt: 209

Courtesy National Institute of Standards and Technology Digital Collections, Gaithersburg, MD: Atomic Clock Photographic Collection: 231

Private Collection: © Chef Boyardee: 251

Science Source: © Claus Lunau/Science Photo Library: 53; © Cordelia Molloy: 271; © Peter Menzel: 387; © B. & C. Alexander/Photo Researchers: 441

Shutterstock: © Nagib: 17; © Waj: 21: © Lev Kropotov: 25; © InavanHateren: 33; © Andrea Muscatello: 35; © Hung Chung Chih: 47; © Dario Sabljak: 49; © Danny E Hooks: 51; © mihalec: 73; © VILevi: 79; © spirit of america: 83; © Shcherbakov Ilya: 85; © hxdbzxy: 89; © Mapics: 91; © Kekyalyaynen: 95; © ArtisticPhoto: 109; © Radu Razvan: 111; © Dmitry Brizhatyuk: 119; © herjua: 149; © Chris Jenner: 155; © Everett Collection: 173; © HomeArt: 177; © ID1974: 185; © MrSegui: 189; © sspopov: 201; © Picsfive: 205; © Sergiy Zavgorodny: 207; © Atelier_A: 227; © WDG Photo: 229; © pedrosala: 233; © B Brown: 239; © Vitaly Korovin: 247; © karinkamon: 263; © Tim Roberts Photography: 269; © Sean Pavone: 279; © Dostoevsky: 313; © fotoslaz: 333; © SeDmi: 339; © Roger Asbury: 347; © Kondor83: 349; © Eric Broder Van Dyke: 361; © Dan Hanscom: 371; © Albert Lozano: 373; © Faraways: 389; © Oliver Sved: 399; © Jakub Krechowicz: 403; © Visions of America: 409; © B Calkins: 411; © Jeff Wilber: 439; © Evikka: 445; © Max Earey: 451; © Haider Y. Abdulla: 455; © FCG: 457; © Felix Mizioznikov: 461; © Thomas Barrat: 469; © Frederic Legrand: 477; © Sophie James: 481; © javarman: 507

Superstock: © Science MuseumSSPL: 159; 193; © Corbis: 261

Courtesy Tesla Motors: 509

Thinkstock: © jocic: 27; © Medioimages/Photodisc: 31; © lagereek: 99; © uatp2: 245; © Matt-Gush: 293; © RobertCorse: 423

US Air Force: A1C Michael Shoemaker: 237; 375

US Marine Corps: 367

US Navy: Gus Pasquarella: 183

US Patent Office: 203, 287

Courtesy Vestas Wind Systems: 485

© Wärtsilä Corporation: 129

West Point Collection: 223

The White House: Pete Souza: 475

Courtesy Wikimedia Foundation: 45, 67, 255, 289, 385, 443, 501; Karsten Ratzke: 55; California Historical Society: 65; Christopher Ziemnowicz: 113; Siemens AG Pressebilder/Presspictures: 123; Jorge Royan: 125; Joop van Houdt: 127; From Official Views Of The World's Columbian Exposition by C. D. Arnold and H. D. Higinbotham: 131; Cephas: 143; F.G.O. Stuart: 151; DARPA: 169; Jerry Hecht/National Institutes of Health: 171; U.S. DOE: 211; Fastfission: 213; Tosaka: 221; Marcin Wichary: 225; University College London Faculty of Mathematical & Physical Sciences: 259; Pierre-Yves Beaudouin: 285; FastLizard4: 305, 309; www.glofish.com: 329; Armand du Plessis: 343; Raimond Spekking: 353; Andrew Thomas from Shrewsbury, UK: 379; Maximilien Brice/CERN: 437; Rhododendrites: 503

Courtesy Yale University: Beinecke Rare Book and Manuscript Library: 69, 103